Innovative Renewable Energy

Series Editor

Ali Sayigh, Flat 3, Building 5/6,
World Renewable Energy Congress Flat 3,
Building 5/6, Brighton, UK

The primary objective of the Innovative Renewable Energy book series is to highlight the best-implemented worldwide policies, projects, and research dealing with renewable energy and the environment. The books are developed and published in partnership with the World Renewable Energy Network (WREN). WREN is one of the most influential organizations in supporting and enhancing the utilization and implementation of renewable energy sources that are both environmentally safe and economically sustainable. Contributors to books in this series come from a worldwide network of agencies, laboratories, institutions, companies, and individuals, all working together towards an international diffusion of renewable energy technologies and applications. With contributions from most countries in the world, books in this series promote the communication and technical education of scientists, engineers, technicians, and managers in this field and address the energy needs of both developing and developed countries.

Each book in the series contains contributions from WREN members and covers the most up-to-date research developments, government policies, business models, best practices, and innovations from countries all over the globe. Additionally, the series publishes a collection of best papers presented during the annual and bi-annual World Renewable Energy Congress and Forum.

Ali Sayigh
Editor

Transition Towards a Carbon Free Future

Selected Papers from the World Renewable Energy Congress (WREC) 2023

Springer

Editor
Ali Sayigh
WREN, UK
World Renewable Energy Congress
Brighton, UK

ISSN 2522-8927 ISSN 2522-8935 (electronic)
Innovative Renewable Energy
ISBN 978-3-031-61659-4 ISBN 978-3-031-61660-0 (eBook)
https://doi.org/10.1007/978-3-031-61660-0

This Springer imprint is published by the registered company Springer Nature Switzerland AG
The registered company address is: Gewerbestrasse 11, 6330 Cham, Switzerland

If disposing of this product, please recycle the paper.

Introduction

This Proceeding is a departure from our usual WREN format. It presents a selection of papers delivered by keynote and invited speakers and some participants at the recent successful WREN and University of Kebangsaan Malaysia, World Renewable Energy Congress-XXII, July 16–20, 2023, held in Kuala Lumpur Convention Centre (KLCC). The selected papers cover a wide spectrum of renewable and sustainable energy fields and were submitted by participants from 19 countries, namely: Nigeria, Armenia, Australia, United States of America, United Kingdom, Morocco, Greece, Canada, Russia, Oman, India, South Africa, Indonesia, Philippine, Germany, China, Malaysia, Saudi Arabia, and Spain.

Contents

Chapter 1
Gender Pay Gap in the Nigerian Renewable Energy Sector

Amina Batagarawa and Dominica Una

1.1 Introduction

Access to clean and sustainable energy is a critical global challenge that holds significant implications for social, economic, and environmental development. At present, between 600 and 733 million people, or 43% of the total global population, lack access to electricity, most of them in sub-Saharan Africa (IEA, 2022a, b; IRENA, 2022; News24, 2022) (Fig. 1.1).

In Africa, as the population increases, the share without electricity access but especially clean cooking is also increasing (Fig. 1.2).

In recent years, renewable energy sources have emerged as promising alternatives to traditional fossil fuels, offering the potential to mitigate climate change and improve energy security. As of May 2022, countries representing more than 70% of global CO2 emissions have committed to reach net zero emissions by around mid-century. This includes 12 African countries that represent over 40% of the continent's total CO_2 emissions (IEA, 2022a, b). In the SAS, around four million additional energy-related jobs are needed across the continent by 2030, largely to reach universal energy access in sub-Saharan Africa (IEA, 2021a, b).

However, the intersectionality of gender and renewable energy access remains a pressing concern, particularly in regions such as Nigeria and other western parts of Africa, where gender disparities persist (UN Women & UNIDO, 2018). In Nigeria and across Africa, women are disproportionately affected by energy poverty (Paola et al., 2023) and limited access to modern energy services. According to the International Energy Agency (IEA, 2022a, b), there is a substantial gender gap in energy access.

A. Batagarawa (✉)
Department of Architecture, Baze University, Abuja, Nigeria

D. Una
Women in Renewable Energy Association, Abuja, Nigeria

A. Sayigh (ed.), *Transition Towards a Carbon Free Future*, Innovative Renewable Energy, https://doi.org/10.1007/978-3-031-61660-0_1

Fig. 1.1 Universal energy access. (International Renewable Energy Agency, 2022)

Women's limited participation in the renewable energy sector further perpetuates gender disparities. Women face various barriers, including limited technical and entrepreneurial skills, inadequate access to finance, and cultural biases that restrict their involvement in decision-making processes and entrepreneurship opportunities within the sector. This marginalization of women hampers the development of inclusive and sustainable energy solutions, limiting the potential for transformative change in the renewable energy landscape.

Addressing the challenges of gender inequality in renewable energy access requires a multifaceted approach. It involves promoting gender-responsive policies and regulatory frameworks, enhancing access to financing and capacity-building opportunities for women, and fostering an enabling environment that encourages women's participation and leadership in the renewable energy sector (IRENA,

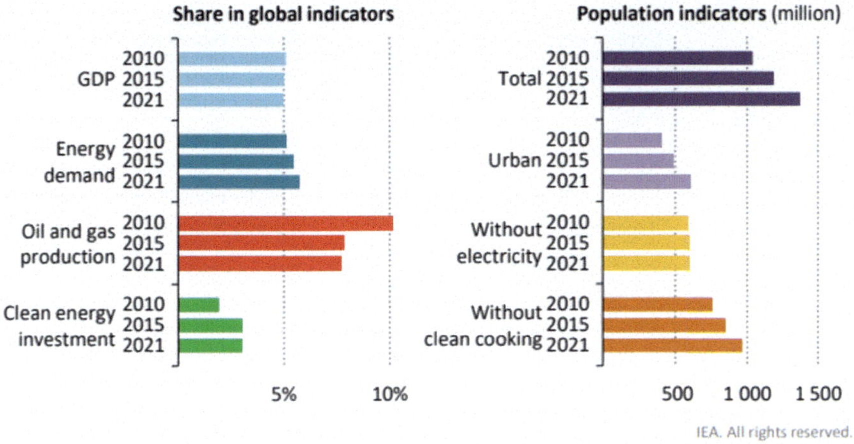

Fig. 1.2 Historic global energy indicators in Africa. (International Energy Agency, 2023)

2019). Moreover, addressing social norms and cultural stereotypes that hinder women's engagement in energy-related activities is crucial for achieving gender equality and advancing sustainable development goals in Nigeria, Africa, and beyond.

Gender statistics and disaggregated data are needed to create a more comprehensive assessment of energy value chains, which not only clarify differences in energy access but also reveal the often invisible framework of needs and impacts that compound inequities in access for underrepresented groups (UN Women and UNIDO, 2023).

The objective of the survey is to gather quantitative and qualitative insights on the current salary status and working conditions of women in the renewable energy sector, exploring existing challenges and potential solutions to improve gender diversity. By examining the existing literature, policy frameworks, and initiatives, it seeks to identify the key challenges faced by women and propose recommendations for promoting gender equality and women's empowerment in the renewable energy sector. Through an in-depth analysis, this study intends to contribute to the ongoing dialogue on sustainable development, gender equality, and clean energy transitions in Nigeria and Africa.

1.2 Methodology

The objective of the survey is to gather quantitative and qualitative insights on the current salary status and working conditions of women in the renewable energy sector, exploring existing challenges and potential solutions to improve gender diversity. The method of data collection is via a survey using Google Forms. The

Table 1.1 Inquiry topics

S. no.	General questions	Gender-related questions
1.	Location	Salary disparity
2.	Gender	Annual salary
3.	Marital status	Gender diversity
4.	Age	Gender policy or regulations
5.	Educational qualification	Gender mainstreaming strategies
6.	Experience	Gender mainstreaming barriers
7.	Size of organisation	
8.	Area of specialisation	
9.	Employment status/cadre	

survey was shared in the public Women in Renewable Energy Association Social media groups and other sister groups like Renewable Energy Association of Nigeria and Council for Renewable Energy Association of Nigeria social media handles. It was also emailed to a comprehensive list of Renewable Energy Companies operating in Nigeria and personally administered by the author.

The inquiry is classified into general and gender-related topics as shown in Table 1.1.

Simple statistics are used to calculate indices such as average salary, age groups, gender, experience, etc., while descriptive analysis is employed to interpret gender policy or regulations, mainstreaming strategies and barriers.

1.3 Results

Out of the 41 participants of the study, 15 work in the Federal Capital Territory (FCT), followed by 6 in Lagos, 4 in Bayelsa, 3 in Rivers, 2 each in Oyo and Edo, and 1 each in Kaduna, Nasarawa, Sokoto, Gombe, Bauchi, and Ogun. Out of 36 states and the Federal Capital Territory, the participants come from 12 locations (Fig. 1.3). The southern region represents more of the participants, indicating that there are more Renewable Energy companies in these locations and women participation in the locations.

Seven percent of the total participants (3 participants) participated on behalf of their organisations, and the rest as individuals (see Fig. 1.4). The role of the participants in HR includes the MD/CEO of their companies and senior or senior HR managers (Fig. 1.4).

The gender diversity of the participants shows that 88% are female, 10% are male, and 2% (1 respondent) preferred not to say. Sixty-five percent are married, 33% are single, and 3% (1 respondent) preferred not to say (Fig. 1.5).

Forty-eight of the participants are between 30 and 40 years, 35% between 40 and 50 years, 7.5 between 50 and 60, 5% between 20 and 30 years, and 2% above 60 years (see Fig. 1.6).

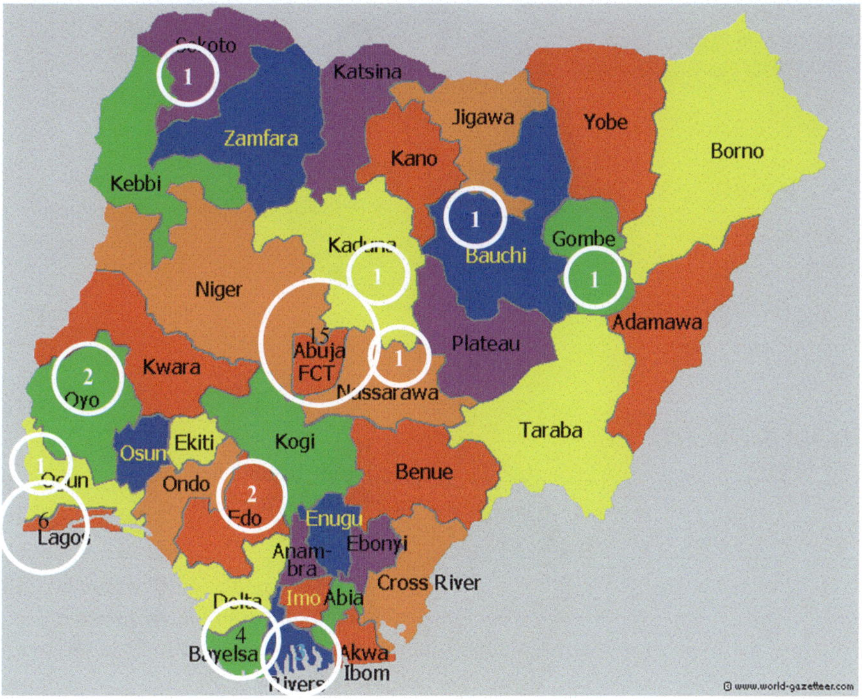

Fig. 1.3 Locations of participants in Nigeria

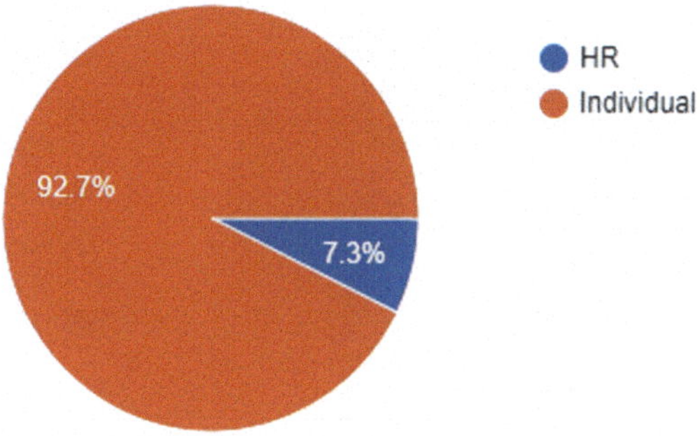

Fig. 1.4 Participants as individuals or as representatives of their companies

Educational qualification ranges from 50% of the participants having a master's degree to 18% with a diploma; and in between are 20% with a bachelor's degree and 13% with a doctorate (see Fig. 1.7).

Participants predominantly have stayed for more than 10 years in their current organisations (45%), followed by 3–5 years (25%), then 5–10 years (17%), and finally less than 3 years (15%) (Fig. 1.8). This result indicates that the participants have stayed long enough at their workplaces to have an insight into the operations of their companies.

Fig. 1.5 Gender Representation in the respondents

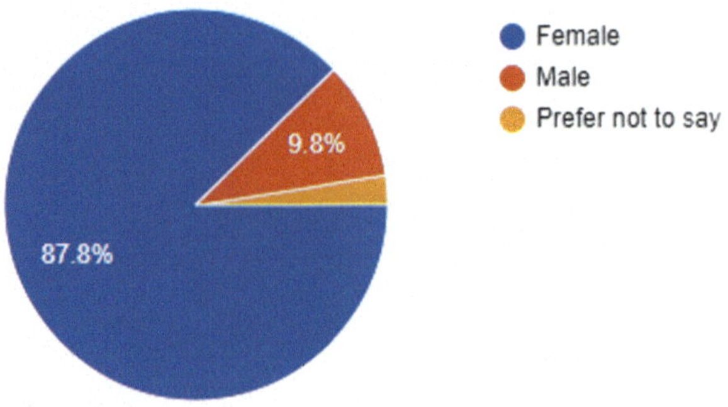

Fig. 1.6 Age range of participants

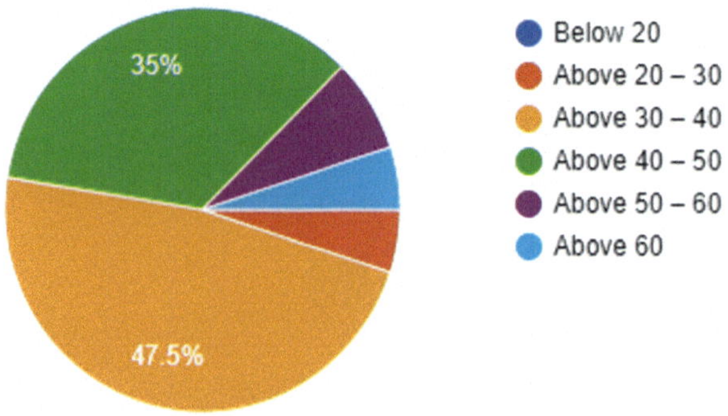

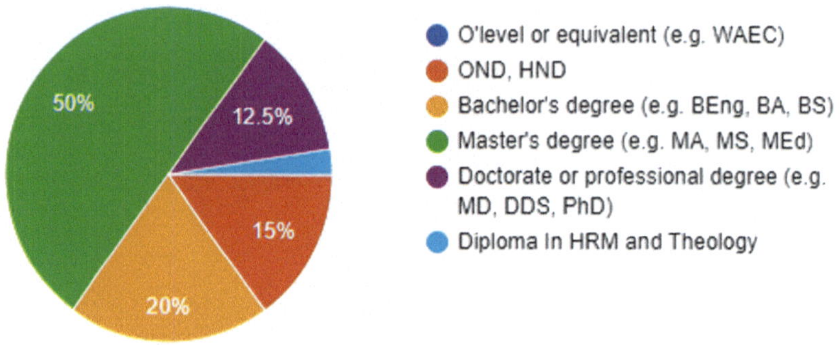

Fig. 1.7 Education level of participants

The companies dealing in Renewable Energy appear to be large companies with 59% having more than 200 employees, followed by 21% who conversely are small companies with less than 10 employees (see Fig. 1.9). Another 10% are relatively small companies still with employee numbers between 10 and 30. Cumulatively, the employee numbers indicate that Renewable Energy companies tend to be larger companies.

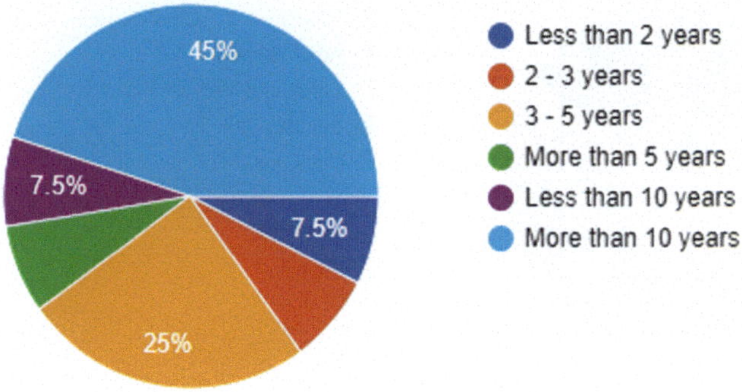

Fig. 1.8 Employment duration

Fig. 1.9 Renewable Energy Company sizes

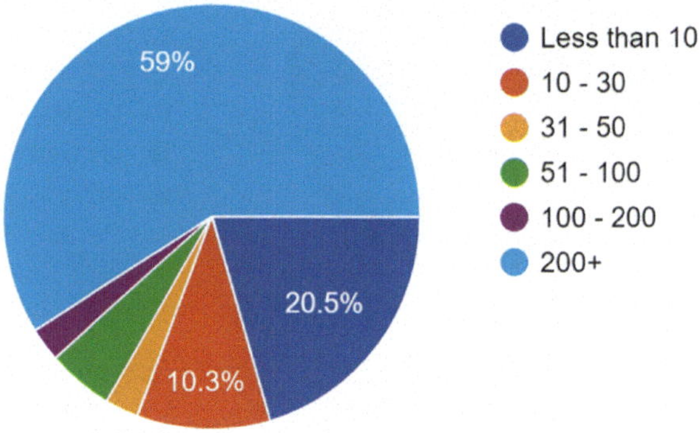

When asked what operational unit they belong to, within the Renewable Energy Company, the participants indicated a range of units (see Fig. 1.10). The most common at 25% is Research and Development, followed in a tie of 12.5% each for Sales and Marketing and Engineering. Business Development and Administration follow in a tie also, at 10%. Other units are shown in Fig. 1.10. A large amount of the employees (85%) work full-time, 10% work part-time, and the rest are contract staff (Fig. 1.9).

Similar to the operational units, the respondents indicate that they belong to various cadres as shown in Fig. 1.11. Thirty five percent work at management level, followed by junior level at 28%, senior level at 25%, and the rest of the 12% represents other cadres as shown in Fig. 1.11. The results show that there is little evidence of gender disparity in the leadership of Renewable Energy companies where the participants work (Fig. 1.11).

The participants were asked to rate the gender diversity in their organisations compared to other organisations in the industry. The results showed that 52% feel they are as diverse as others, 27% feel more diverse, and 21% feel less. Cumulatively, the participants feel the diversity is normal (Fig. 1.12).

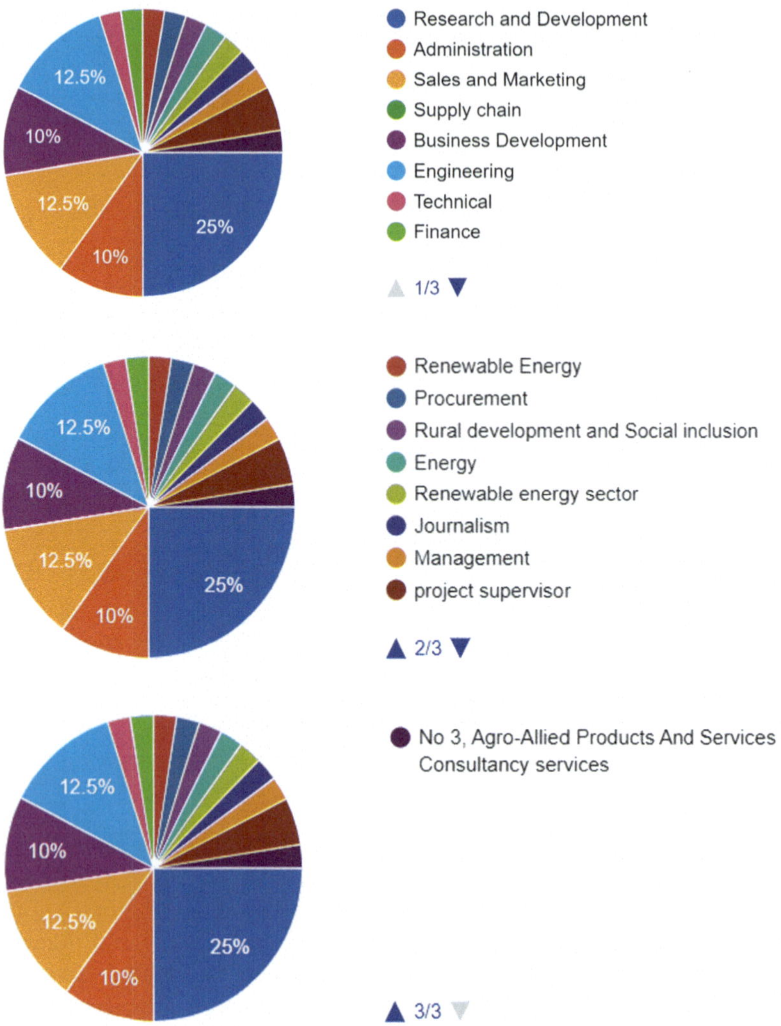

Fig. 1.10 Operational units within the Renewable Energy companies

To further explore the diversity, the participants were asked about females working at different levels in their companies (Fig 1.13). The results showed a majority of the respondents indicated that between two to five females are the most frequent employees in the junior level. However, the next frequent number of female employees is over sixty. This shows a disparity in company size. Within all the levels, the junior level has the most female employees at over sixty. In the mid-level cadre the most number is also between two and five, but having female employees of varying numbers across the level. In the senior cadre, most of the female employees fall between the and 20 female employees. This trend is not unusual in junior to senior levels of employment, i.e. more in junior level, then

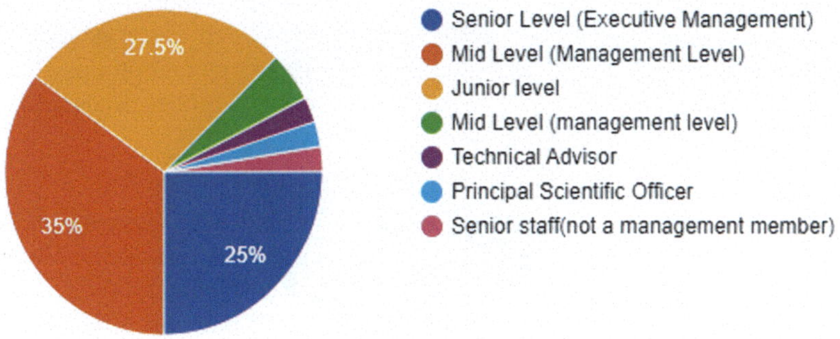

Fig. 1.11 Employment cadre

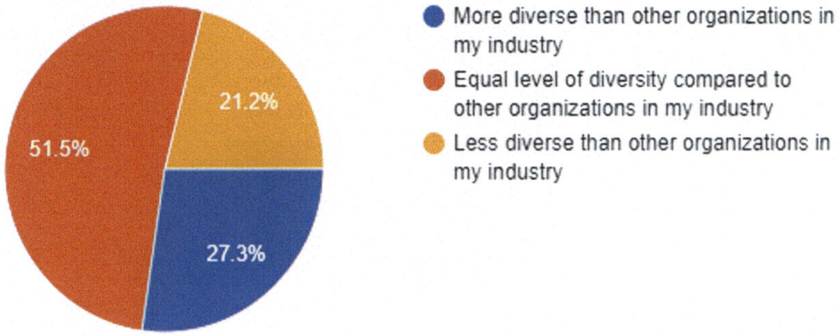

Fig. 1.12 Gender diversity in employment

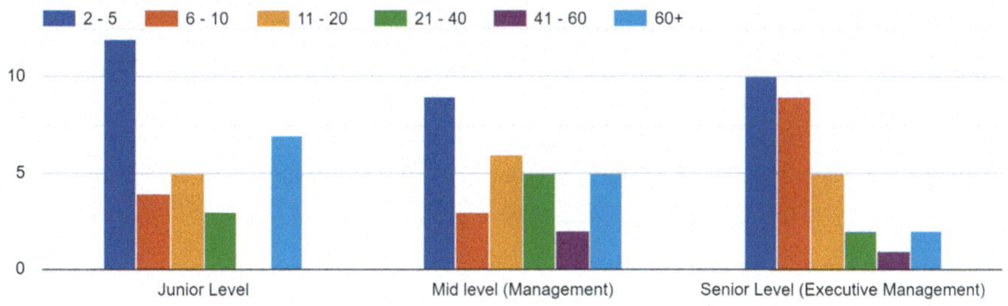

Fig. 1.13 Female employees at different levels

mid level and least in the senior level. What remains significant is the indication from the respondents that there are significantly more males than females employed in the different levels.

The participants were asked if their organisations have policies and regulations against gender pay bias, harassment, infrastructure usage, maternity leave, and other work-life balance. The results (see Fig. 1.14) show 45% agreed to having such policies and guidelines, 43% disagreed, and 13% didn't know (Fig. 1.14).

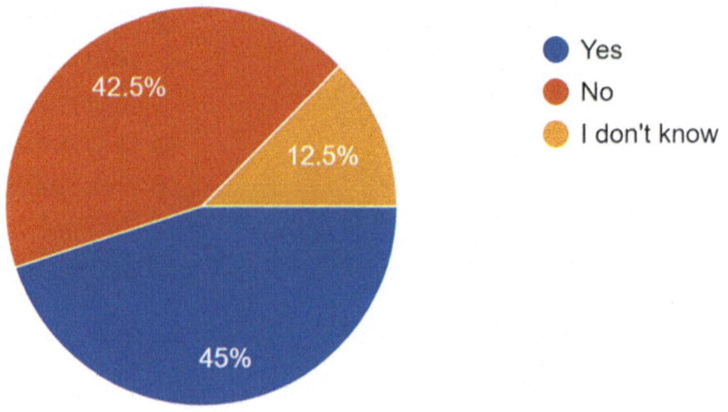

Fig. 1.14 Policies and regulations concerning gender mainstreaming

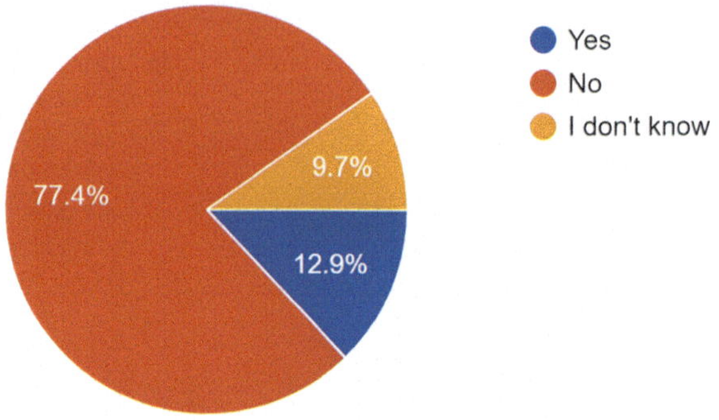

Fig. 1.15 Gender pay gap

The average annual salary is 4,165,481 (5373 USD) with a range between ₦50,000 (64.50USD) and ₦35, 000,000 (45,149 USD). When asked if they feel that there is a pay gap in their workplace, the respondents mostly (77%) indicated that there is no gap, while 13% indicated that there is a gap (Fig. 1.15). Ten percent indicated that they don't know. Further inquiries into pay gaps across the cadres also confirm that the companies with no gap in the different cadres are significantly more than those where there is a gap (see Figs. 1.14 and 1.16).

The study progressed to gender mainstreaming strategies in place to improve diversity in the companies. Some of the identified strategies include the following:

1. Early closure for nursing mothers
2. Availability of a crèche
3. Coming to work with kids
4. Harassment counselling
5. Maternity paid leave
6. No night shifts according to Factory Acts
7. Paternity leave

On the contrary, some indicated the following:

1. Very short maternity leave
2. No crèche

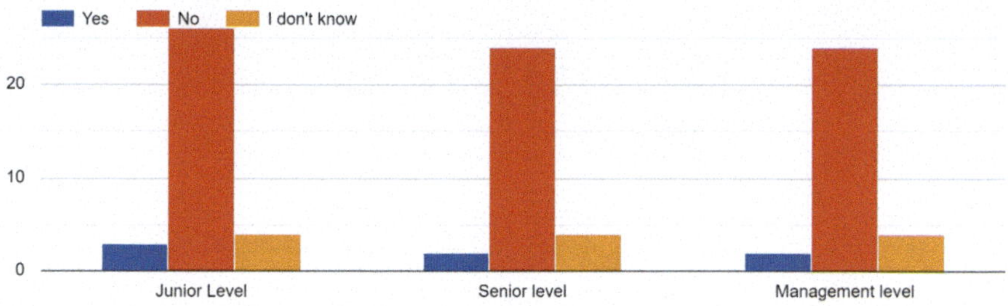

Fig. 1.16 Pay gap in different cadres

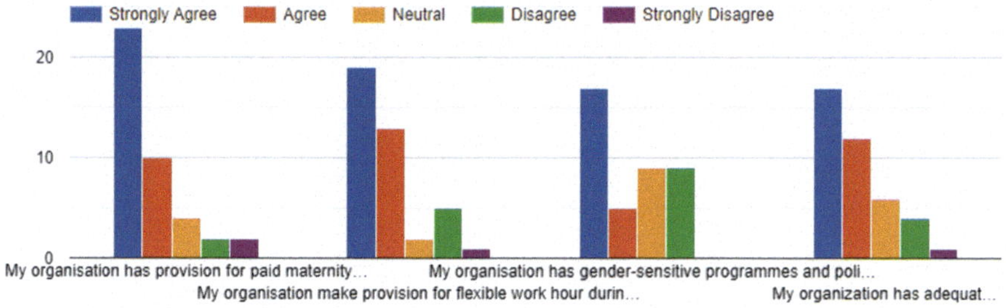

Fig. 1.17 Gender mainstreaming strategies

After further inquiry into the scale of strategies, the results (see Fig. 1.17) show that significantly more participants agree that there are strategies in place than those who feel there are not (Fig. 1.17).

My organisation has provision for paid maternity leave?	My organisation make provision for flexible work hour during post-natal	My organisation has gender-sensitive programmes and policies	My organisation has adequate physical facilities for women (such as Convenience)

The participants were asked if their organisation has identified a gender bias and is it committed to implementing a plan to reduce that gap and 51% agreed while 13% disagreed. However, a significant number of participants did not know, meaning there is an opportunity to improve communication and deliberate action in gender mainstreaming in the companies (Fig. 1.18).

When asked to rate the indicators of gender bias in Fig. 1.19, the results show that more participants agree that there are more men than women in their places of work. More of the participants strongly agree that "women in executive positions in their respective organisations are quite few compared to men" (Kemausuor et al., 2020). However, slightly more of the participants believe that the high number of men in relation to women is not as a result of preference (Fig. 1.19).

The ratio of men compared to women in my organisation/place of work is high	Women in executive position in my organisation are quite few compared to men	Men are always given preference over women in my organisation]	Men are generally preferred to women counterpart for some position in my organisation	Do you agree that women's perspectives, expertise, and leadership are critical to ensuring informed and inclusive decision-making and a larger pool of talent in the renewable energy sector?	Do you agree that the share of women in the renewable energy sector is higher than in the conventional energy sector?	Do you agree that centrality of gender equality and the increased participation of women to achieve an inclusive and fair energy transformation at the local, national, regional, and international levels should be raised?	Do you agree that motivation and preferential treatment given to men affect women's participation, leadership, and decision-making in the renewable energy sector?	Foster women's economic empowerment by increasing access to investments, aligning salary package, and capacity building in entrepreneurship and employment in the renewable energy sector	Do you agree that strengthened collaboration at all levels is a key requirement to overcoming barriers facing women's participation in the renewable energy sector?

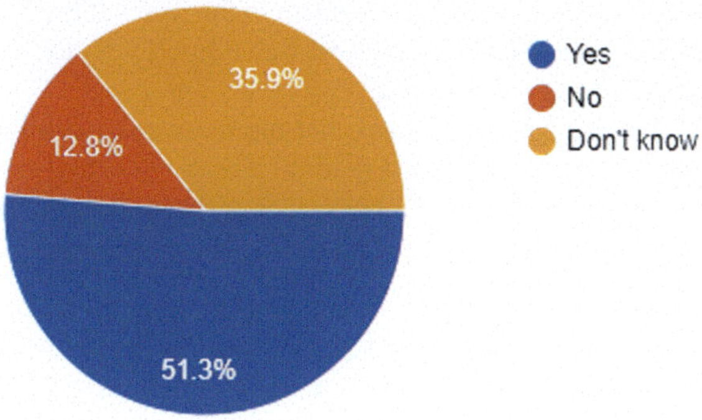

Fig. 1.18 Gender mainstreaming plan implementation. (IEA, Special Focus on Gender, 2021a, b)

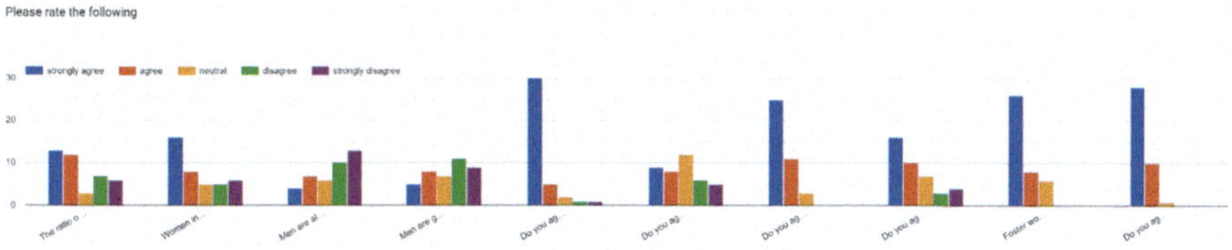

Fig. 1.19 Indicators of gender bias

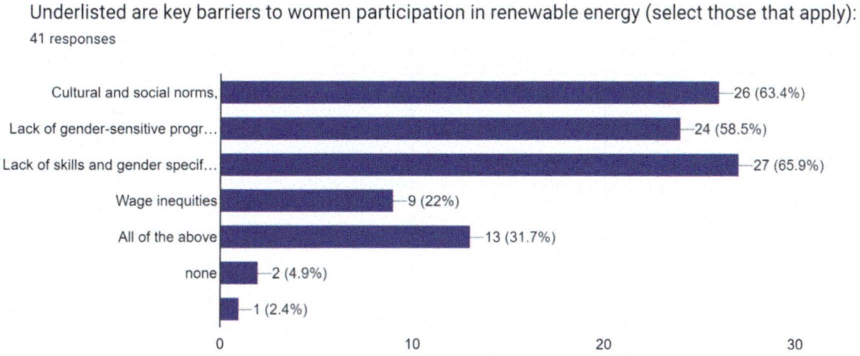

Fig. 1.20 Barriers in gender mainstreaming

Listed in Fig. 1.20 are key barriers to women participation in renewable energy. Lack of skills and gender-specific training opportunities are the biggest barrier at 66%, followed by cultural and social norms at 63% and then a lack of gender-sensitive programmes and policies at 58%. It is paramount to note that while another 32% indicated that all the barriers contributed to the gender bias, only 22% is attributed to a pay gap issue.

A striking number of the participants do believe that women's perspectives, expertise, and leadership are critical to ensuring informed and inclusive decision-making, and a larger pool of talent in the renewable energy sector and a slight margin believe that women in the Renewable Energy sector are higher than in other energy sectors.

A similar margin agrees that the motivation and preferential treatment given to men affect women's participation, leadership, and decision-making in the renewable energy sector and a striking number of the participants agree that women's economic empowerment should be fostered by the following:

1. Increasing access to investments, aligning salary package, and capacity building in entrepreneurship and employment in the renewable energy sector.
2. Strengthening collaboration at all levels is a key requirement to overcoming barriers hindering women's participation in the renewable energy sector.

Finally, the participants were asked to rank what impact the policies in Fig. 1.21 may have on promoting women participation in the renewable energy sector of Nigeria, with 1 being the first and having the most positive impact. The results show that the participants believe that gender targets and quotas with an average score of 5 have the least impact, followed with an average score of 4 for mainstreaming gender perspectives; facilitating access to finance; designing and implementing workplace policies and regulations; and providing access to education, training, and skills development programmes. The last tier of indicators, scoring an average score of 3, is creating networks and supporting mentorship; and work-life balance is identified as the indicator with the most positive impact (Fig. 1.21).

Mainstreaming gender perspectives	Creating networks and supporting mentorship	Facilitate access to finance	Gender targets and quotas	Workplace policies and regulations	Work-life balance	Access to education, training, and skills development programmes

The participants were asked to discuss gender mainstreaming in general, and the following are some of their comments.

> In our Company, women are given more opportunities except for positions like installation and maintenance that actually require more men. But we are working on getting more young ladies trained on installation and maintenance

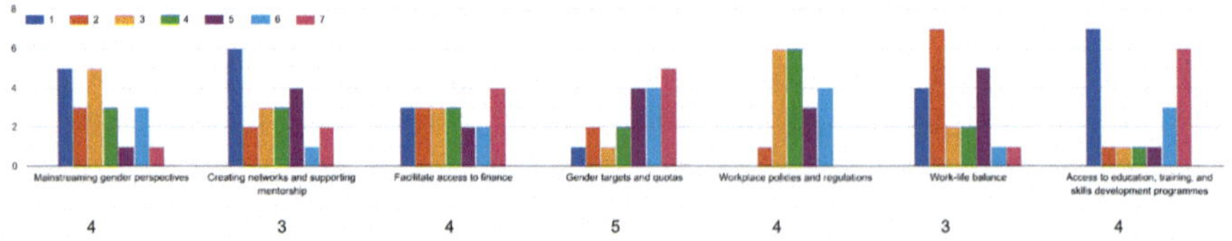

Fig. 1.21 Mainstreaming policies and their perceived impact and average impact scores, 1 being the best

If women are given the necessary platform, they will strive like they have done in other sectors

Equal access to funding opportunities will fast-track Gender participation in The Renewable Energy Sector

I implore that this survey be put to use for proper gender equality at the workplace. Thank you.

1.4 Conclusion

The objective of the study is to gather quantitative and qualitative insights on the current salary status and working conditions of women in the renewable energy sector of Nigeria. The survey received participation from mostly individual females, rather than as representatives of the companies. They are predominantly married and above the age of 20. This result, when analysed alongside their qualifications, may be an indication that people regard the sector as one requiring a higher education qualification, rather than a skilled vocation. With the growing youth population, there is an opportunity to engage more youths with skill potential in the sector.

The companies dealing in Renewable Energy appear to be large companies with 59% having more than 200 employees. This is another opportunity to drive policy in encouraging smaller-scale companies in value addition to the sector, such as maintenance and Fintech, as well as others already identified in Fig. 1.10—operational units within the Renewable Energy companies.

The study explored existing challenges in the sector including identifying a disparity in inclusion between the southern region and the north region of Nigeria, which has been historically identified as having lower rates of literacy, school enrolment, school retention, and STEM-related indicators.

The results of the inquiry on the cadres in which the participants belong show that there is little evidence of gender disparity in pay and the leadership of Renewable Energy companies where the participants work. The average annual salary is ₦4,165,481 (5373 USD) with a range between ₦50,000 (64.50 USD) and ₦35,000,000 (45,149 USD).

However, the rankings and recommendations show that while the disparities identified are expected and even normal, the participants want better inclusion, diversity, and to bridge the pay gap.

The potential solutions to improve gender diversity include creating networks and supporting mentorship; work-life balance, mainstreaming gender perspectives; facilitating access to finance; designing and implementing workplace policies and regulations; and providing access to education, training and skill development programmes and introducing gender targets and quotas. These results and analyses contribute to the ongoing dialogue on sustainable development, gender equality, and clean energy transitions in Nigeria and Africa.

References

IEA. (2021a). International Energy Agency.

IEA. (2021b). *Special focus on gender.* Africa Energy Outlook 2021.

IEA. (2022a). *Africa energy outlook 2022.* IEA. Retrieved from https://www.iea.org/reports/africa-energy-outlook-2022

IEA. (2022b). *Data and projections.* IEA. Retrieved from https://www.iea.org/reports/sdg7-data-and-projections

IRENA. (2019). *Renewable energy: A gender perspective.* International Renewable Energy Agency. Retrieved from https://www.irena.org/publications/2019/Jan/Renewable-Energy-AGender-Perspective.

IRENA. (2022). *Renewable energy targets in 2022: A guide to design.* International Renewable Energy Agency. Retrieved from https://www.irena.org/Publications/2022/Nov/Renewable-energy-targets-in-2022

Kemausuor, et al. (2020). *Gender, energy, and sustainable development: Case studies from Nigeria and Uganda.* Routledge.

News24. (2022, September 16). *US climate envoy Kerry cautions against long-term gas projects in Africa.* Retrieved from News24: https://www.news24.com/news24/africa/news/us-climate-envoy-kerry-cautions-against-long-term-gas-projects-in-africa-20220916

Paola, C., Magda, M.-G., Shehu, I. K., Sandor, S., & Godwell, N. (2023). Clean energy access as an enabler for social development: A multidimensional analysis for Sub-Saharan Africa. *Energy for Sustainable Development, 72*, 114–126. https://doi.org/10.1016/j.esd.2022.12.003

UN Women, & UNIDO. (2018). *Gender and energy in Africa.* Advancing the Sustainable Development Goals.

UN Women and UNIDO. (2023). *Gender Equality and the Sustainable Energy Transition.* United Nations Industrial Development Organization. Retrieved from https://www.uncclearn.org/wp-content/uploads/library/Gender-equality-in-the-sustainable-energy-transition-en.pdf

Chapter 2
The Viability of ECOPAVERS from Plastic Waste in Abuja, Nigeria

Amina Batagarawa and Imtissar Bashir Kurfi

2.1 Introduction

The construction industry plays a significant role in shaping the urban landscape, but it also contributes to environmental degradation and resource depletion. In Nigeria, particularly in Abuja, the capital city, there is a growing demand for building materials due to rapid urbanization which has led to a surge in construction projects. This surge further leads to an increased demand for construction elements, such as interlocking pavers, and has raised concerns about the resultant strain on the environment, sustainability and waste management (Aniekwu, 1995; Toriola-Coker et al., 2020; Business Wire, 2020).

Traditionally, cement has been the primary binding material for interlocking pavers. As the detrimental effects of cement production such as high cost, air pollution, energy consumption, and land degradation become more apparent (Etim et al., 2021; Watts, 2019), there is a pressing need to explore alternative materials that can reduce the ecological footprint of the construction sector.

One of the touted benefits of cement used in the construction industry is that its production provides employment opportunities and economic benefits (Mojekwu et al., 2013). However, alternative building materials also provide similar opportunities (Babaremu et al., 2022; Ministry of Foreign Affairs, 2020).

In a similar vein, waste plastic poses a severe environmental threat as shown in Fig. 2.1.

Plastic waste accounts for 12% of the total amount of municipal solid wastes generated globally. Globally, only 9% of plastic waste is recycled, while 22% is mismanaged as shown in Fig. 2.2 (Organisation for Economic Co-operation and

A. Batagarawa (✉)
Centre for Clean Energy and Climate Change (CLEAN), Baze University, Abuja, Nigeria

I. B. Kurfi
Ifrique Eco Solutions, Abuja, Nigeria

A. Sayigh (ed.), *Transition Towards a Carbon Free Future*, Innovative Renewable Energy, https://doi.org/10.1007/978-3-031-61660-0_2

Fig. 2.1 Global plastic waste outlook. (United Nations Environmental Programme, 2023)

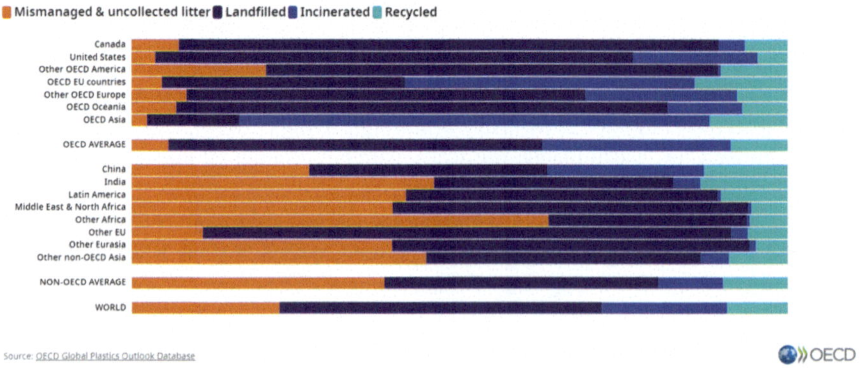

Fig. 2.2 Global plastic waste management. (Organisation for Economic Co-operation and Development, 2022)

Development, 2022). Approximately seven billion of the 9.2 billion tonnes of plastic produced from 1950 to 2017 became plastic waste, ending up in landfills or dumped. UNIDO Country Representative and Regional Director for West Africa, Jean Bakole, at the inauguration of the steering committee for the project tagged 'Promoting Sustainable Plastics Value Chains Through Circular Economy Practices', disclosed that in Nigeria, the consumption of plastics jumped by 116% within a period of 15 years to 1.25 million tonnes in 2022. Therefore, the per capita plastic consumption has grown by 5% annually from 4 kg to 6.5 kg. According to the World Bank, 14% of plastic waste is collected for recycling, but only 9% is actually recycled globally. With even more limited recycling facilities in Nigeria, plastic waste is contaminating more land ecosystems, being released into the marine environment, thereby polluting it and threatening biodiversity which alters habitats and natural processes; reducing ecosystems' ability to adapt to climate change; directly affecting millions of people's livelihoods, food production capabilities, and social well-being; and negatively impacting the blue economy.

Types of Plastic Waste Generated< Baban Bola Sorting

A significant amount of people in developing cities rely on recycled waste for their livelihoods. According to Ogwueleka and Naveen (2021), there is no formal recycling programme in Abuja, as such scavengers carry out waste segregation and identification for recyclable materials. These scavengers are predominantly male between the ages of 21 and 40 years, and they work between 10 and 12 h. They sort out 19.76 kg/c/d recyclables. The estimated quantity of recyclables generated in Abuja is 133,688 kg per day (133.688 tons per day). A waste picker earns between N1000 ($2.8) and N1500 ($4.2) per day; itinerant dealers earn between N1500 ($2.8) and N2000 ($5.6) per day, and scrap dealers earn between N10, 000 ($27.77) and N15, 000 ($41.67) per day (Figs. 2.3 and 2.4).

This research paper aims to investigate the viability of utilizing waste plastic as a binder in the production of interlocking pavers. By exploring waste plastic application in the construction industry, we seek to address two pressing issues simultaneously: the need for sustainable building materials and the challenge of plastic waste management.

Plastic waste is used in road construction, building construction, and other structural applications (Agarwal & Singhal, 2018; Bedia, 2019; Ekanayake & Ing, 2017; Siddique, 2017; Folorunsho & Amadi, 2020).

WasteAid UK and Chartered Institution of Wastes Management (Lenkiewicz & Webster, 2017; Kumi-Larbi Jnr et al., 2018) published a methodology on the production of basic concrete made with waste plastic as a binder, in place of cement. The plastic waste can be sourced from waste water sachets made of low-density polyethylene (LDPE), polyethylene terephthalate (PET), polypropylene (PP), high-density polyethylene (HDPE), and polyethylene (PE).

Siddique examines the use of waste plastic as a stabilizing additive in stone matrix asphalt. The research demonstrates the improved stability and performance of asphalt mixtures with waste plastic. The study provides insights into the

Fig. 2.3 Waste segregation in Abuja shows that plastic accounts for 15% of the total waste dumped at the landfills as shown in Fig. 2.5 (Olaide & Dias, 2020). This presents a potential for circular economy in the use of waste plastic for the construction industry

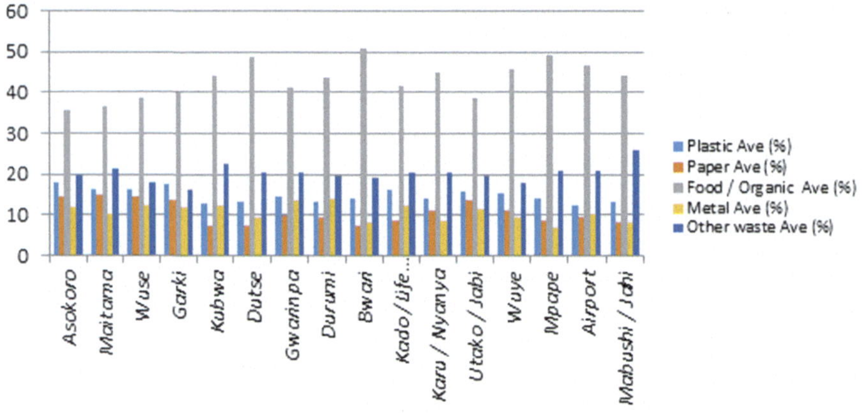

Fig. 2.4 Waste sorting in Abuja. (Olaide & Dias, 2020)

application of waste plastic in pavement materials, which can be relevant for interlocking pavers as well.

Bricks and paver blocks made from plastic waste (HDPE and PE) and sand taken in ratio 1:2, 1:3, 1:4, 1:5, and 1:6 were reported and evaluated for compressive strength, water absorption, efflorescence, hardness, and fireplace resistance (Sellakutty, 2016). A comparative study of ash bricks, plastic sand bricks, and traditional bricks showed that plastic sand bricks have the highest compressive strength of 5.12 N/mm^2 out of these bricks and paver blocks have compressive strength of 9.19 N/mm^2. The water absorption capability of plastic sand bricks was 1.10%, and that of plastic paver blocks was 1.082%. Structural property of bricks and blocks didn't show significant changes up to 180 C. All these results showed that the plastic sand bricks and paver blocks were of better quality than ash bricks and traditional clay bricks.

Despite all the benefits of using plastic waste, research on the use of plastic in concrete found that the compressive, flexural, and tensile strengths of concrete decrease as the proportion of plastic waste increases due to inadequate adhesive strength between the surface of the plastic aggregates (PA) and cement.

The above literature review provides an overview of the existing research and studies related to the utilization of waste plastic as a binder in the production of interlocking pavers. The studies provide insights into the potential use of waste plastic as a binder material. The following factors have been tested in the development of pavers using plastic waste: weight in air (kg), weight of saturated surface dry (kg), density (kg/m^3), water absorption (%), compressive strength (n/mm^2), split tensile strength, workability, flexural strength, elastic modulus, first crack strength, ductility characteristics, thermal conductivity, bulk density, abrasion resistance, fire resistance, and resistance to impact.

2.2 Methodology

In this study, firstly, an alternative approach in manufacturing pavers using waste plastic as a substitute for cement is proposed. By doing so, the aim is to reduce the dependency on cement, minimize plastic waste accumulation, and promote sustainable practices in the construction industry. The potential advantages of this alternative approach include reduced carbon emissions, conservation of natural resources, and enhanced waste management.

Secondly, this research involves an in-depth analysis of the physical, mechanical, and environmental properties of the interlocking pavers manufactured with waste plastic as a binder. However, this paper only covers the physical properties and requires further studies on the other properties.

The interlocking pavers are made out of low-density polyethylene also known as LDPE from sources such as waste pure water sachets, bin bags, plastic wrap, and other types of polythene bags. They can also be made from PET or plastic bottles like those of Coke, Pepsi, Nestle, etc. However, polythene bags are preferable as plastic bottles are reused for other purposes, whereas polythene bags are primarily single use.

The production of the pavers is based on methodology established by WasteAid UK and Chartered Institution of Wastes Management (Lenkiewicz & Webster, 2017). It is a five-stage process as shown in Fig. 2.5.

Fig. 2.5 Five-stage waste plastic paver production process

The Nigerian Building and Road Research Institute (NBRRI) conducted the following tests on the pavers:

1. Weight in air (kg)
2. Weight of saturated surface dry (kg)
3. Density (kg/m^3)
4. Water absorption (%)
5. Compressive strength (N/mm^2)

The concrete age is up to 28 days, and the sample size is:

- Length: 195 mm
- Width: 95 mm
- Thickness: 55 mm

The ratio of sand to LDPE is 60:40 or 1:0.7.

2.3 Results

The results section comprises two parts:

1. The production process of the paver
2. Properties testing

Production Process

The production of the pavers is based on methodology established by WasteAid UK and Chartered Institution of Wastes Management (2017). It is a five-stage process as shown in Fig. 2.2.

Preparation Stage

This is the very first stage, and it involves the collection or sourcing of plastic waste from sources like households, markets, supermarkets, schools, and other sources.

This process involves the separation and segregation of the plastic waste into different categories like PET, LDPE, and others for ease of production. After the sorting, the interlocking paver production begins. Only clean plastic waste is sourced commercially; therefore no pre-cleaning is done.

Plastic Melting Stage

This process requires the melting of the plastic using a heat source in a cylinder into a paste.

Plastic Mixing Stage

This process involves the mixing of the melted plastic with sand. Different ratios can be used depending on the desired outcome. Some ratios include 50:50 (50% plastic to 50% sand), 60:40 (60% plastic to 40% sand), etc.

Moulding Stage
This process involves the pouring of the mixed plastic and sand into moulds.

Setting Stage
This process involves allowing the mass poured into a mould to set. It usually ranges between 1 and 5 h depending on parameters like the weather and the type of plastic used.

Demoulding Stage
This process involves the removing of the set plastic from the mould. At this stage, the plastic would take the shape and form of the mould. When removed from the mould, it becomes an interlocking tile and it can be used immediately.

Test Results
The Nigerian Building and Road Research Institute (NBRRI) conducted the following tests on the pavers:

1. Weight in air (kg)
2. Weight of saturated surface dry (kg)
3. Density (kg/m^3)
4. Water absorption (%)
5. Compressive strength (N/mm^2)

The concrete age is up to 28 days, and the sample size is:

- Length: 195 mm
- Width: 95 mm
- Thickness: 55 mm

The results show that the average weight in air is 2.1 kg, and the range is between 2.04 and 2.18 kg, a variation of 0.14 and error margin of 6%. The average weight of saturated surface dry is 2.106 kg, and the range is between 2.047 and 2.185 kg, a variation of 0.14 and error margin of 6%. The average density is 2061 kg/m^3, and the range is between 2002 and 2139 kg/m^3, a variation of 137 and error margin of 6%. The average water absorption is 0.267%, and the range is between 0.229 and 0.34%, a variation of 0.114 and error margin of 33%. The average compressive strength is 6.607 N/mm^2, and the range is between 5.938 and 7.665 N/mm^2, a variation of 1.727 and error margin of 23%. All error margins above 10% (water absorption (%) and compressive strength (N/mm2)) indicate the need for further standardisation (Table 2.1).

Comparative Analysis with HDPE, PET, and Cement
The results obtained from the tests above are compared with those obtained by Awodiji et al. (2021). According to the American Society for Testing and Materials (1988), the dimensions of a paver shall not be greater than 160 mm in width, 240 mm in length, or 140 mm in thickness. The ECOPAVER is 95 mm in width, 195 mm in length, and 55 mm in thickness. The dimension is within acceptable standards.

Table 2.1 Physical properties of ECOPAVERS

	Weight in air (kg)	Weight of saturated surface dry (kg)	Density (kg/m^3)	Water absorption (%)	Compressive strength (N/mm^2)
1	2.080	2.086	2041	0.288	6.694
2	2.040	2.047	2002	0.343	6.262
3	2.060	2.065	2022	0.243	5.938
4	2.140	2.145	2100	0.234	6.478
5	2.180	2.185	2139	0.229	7.665
Average	**2.100**	**2.106**	**2061**	**0.267**	**6.607**
Highest	2.18	2.185	2139	0.343	7.665
Lowest	2.04	2.047	2002	0.229	5.938
Variation	0.14	0.138	137	0.114	1.727
Error margin	6%	6%	6%	33%	23%

Table 2.2 Comparative analysis on thermal properties of waste plastic PET, HDPE, LDPE, and cement pavers

	Density (kg/m^3)	Water absorption (%)	Compressive strength (N/mm^2)
Sand-PET paver 1:1	1590	7.35	1.4
Sand-PET paver 1:1.5	1545	3.06	1.0
Sand-PET paver 1:2	1478	0	0.6
Sand-HDPE paver 1:1	983	4.87	10.4
Sand-HDPE paver 1: 1.5	995	0	12.0
Sand-HDPE paver 1:2	997	0	17.0
Sand-cement paver 1:1	2282	4.04	5.7
Sand-cement paver 1: 1.5	2372	2.9	9.0
Sand-cement paver 1:2	2352	2.9	11.5
ECOPAVER 1:0.7	**2061**	**0.27**	**6.6**

Water absorption shall not be greater than 5%. The ECOPAVER has an average water absorption of 0.27% which is also within acceptable standards.

The average compressive strength shall not be less than 55 N/mm^2. The average compressive strength of the ECOPAVER is 6.6 N/mm^2, well below acceptable standard. Further additives are required to improve the compressive strength.

The results in Table 2.2 show that the highest density is sand-cement paver 1: 1.5 at 2372 kg/m^3. The decrease in density between the ECOPAVER 1:0.7 at 2061 kg/m^3 and the conventional sand-cement paver 1:1.5 at 2372 kg/m^3 is 15%.

Three of the samples absorb minimal water; sand-HDPE paver 1:2; sand-PET paver 1:1.5; and sand-PET paver 1:1. The ECOPAVER with 0.27% water absorption performs better than the conventional sand-cement paver 1:1.5 at 4.04%, an increase of 1396% as shown in Fig. 2.3 Comparing physical properties of PET, HDPE, LDPE, and cement pavers.

None of the samples perform up to standard with regards the compressive strength. However, the best results are performed by the PET samples. Further work is required to improve the compressive strength of the ECOPAVER (Fig. 2.6).

2.4 Conclusion

The findings of this study have the potential to revolutionize the building material industry in Abuja, Nigeria, and beyond. By harnessing waste plastic as a binder for interlocking pavers, we cannot only mitigate environmental degradation but also contribute to the circular economy by converting waste into a valuable resource. This research aims to provide valuable insights and recommendations for policymakers,

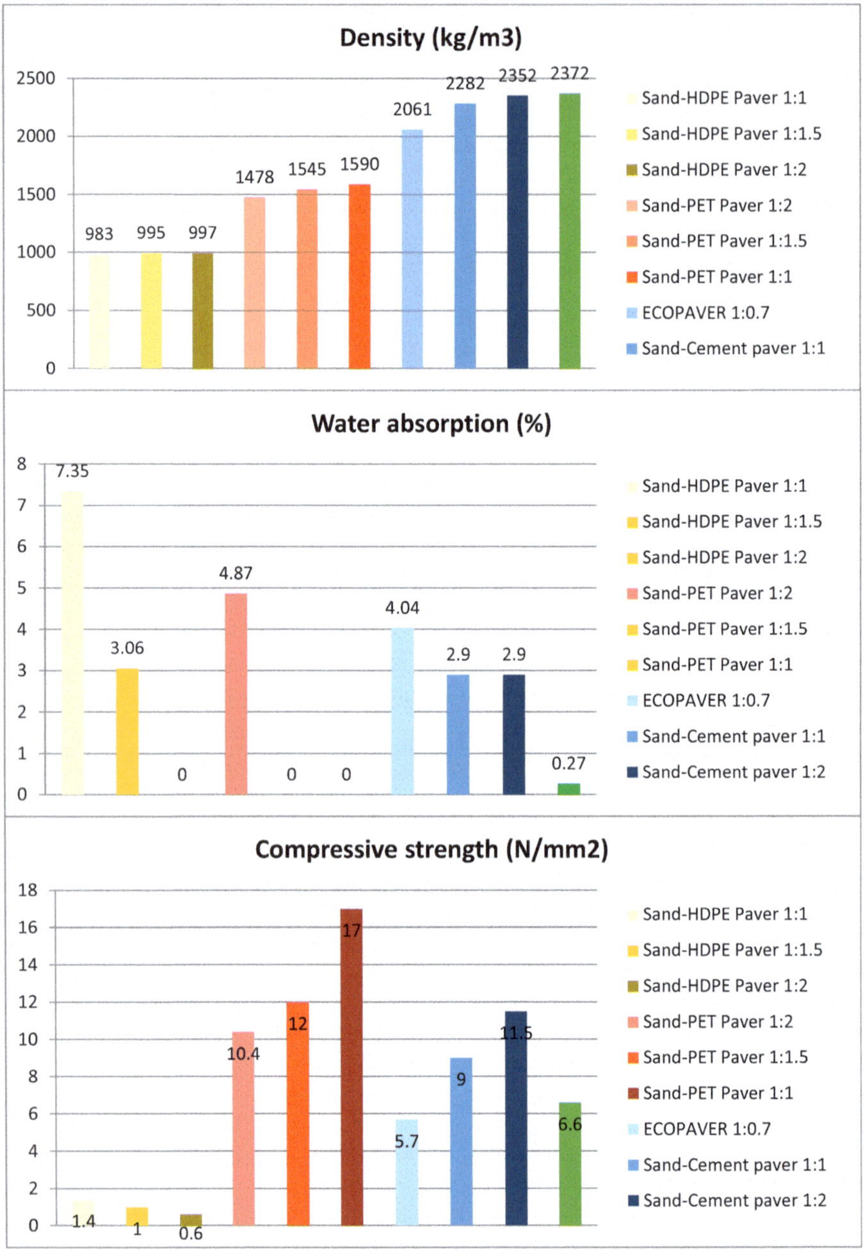

Fig. 2.6 Comparing physical properties of PET, HDPE, LDPE, and cement pavers

construction professionals, and environmentalists to facilitate the adoption of sustainable practices in the construction industry.

In conclusion, this paper sheds more light on the feasibility and benefits of utilizing waste plastic as a binder for interlocking pavers in Abuja, Nigeria. By proposing an innovative and sustainable approach to building material production, we strive to address the environmental challenges associated with the construction sector while promoting a circular economy model.

The results show that the average weight in air is 2.1 kg and the average weight of saturated surface dry is 2.106 kg. The average density is 2061 kg/m^3, while the average water absorption is 0.267%. The average compressive strength is 6.607 N/mm^2. These results demonstrate the viability of using plastic waste as a binder in pavers, especially in a developing country with plastic pollution challenges.

Future work is required in studying more of the thermo-physical properties of the paver with the aim of improving its performance. Additionally, the economic feasibility of this alternative approach, taking into account the cost of production, availability of waste plastic, and potential market acceptance, will be assessed.

References

Agarwal, S., & Singhal, D. (2018). Use of plastic waste in bituminous road construction. *International Journal of Scientific and Research Publications, 8*(3), 126–130.

Aniekwu, A. (1995). The business environment of the construction industry in Nigeria. *Construction Management and Economics, 13*, 445–455.

Awodiji, C. T., Sule, S., & Oguguo, C. V. (2021). Comparative study on the strength properties of paving blocks produced. *Nigerian Journal of Technology (NIJOTECH), 40*, 762–770.

Babaremu, K. O., Okoya, S. A., Hughes, E., Tijani, B., Teidi, D., Akpan, A., . . . Akinlabi, E. T. (2022). Sustainable plastic waste management in a circular economy. *Heliyon, 8*, e09984.

Bedia, J. P. (2019). Performance of concrete with mixed waste plastics as binder in structural applications. *Journal of Cleaner Production, 222*, 686–694.

Business Wire. (2020). *Nigeria construction market outlook 2020–2024: Opportunities & trends with city-level insights*. Business Wire. Retrieved from https://www.researchandmarkets.com/reports/5093826/nigeria-construction-industry-databook-series?utm_source=BW&utm_medium=PressRelease&utm_code=93qvwc&utm_campaign=1435544+-+Nigeria+Construction+Market+Outlook+2020-2024%3a+Opportunities+%26+Trends+with+Ci

Ekanayake, L. L., & Ing, R. K. (2017). Utilization of plastic waste as binder in the production of construction blocks. *Case Studies in Construction Materials, 7*, 280–288.

Etim, M., Babaremu, K., Lazarus, J., & Omole, D. (2021). Health risk and environmental assessment of cement production in Nigeria. *Atmosphere, 12*(9), 1111.

Folorunsho, O. W., & Amadi, A. N. (2020). Production of pavement blocks from plastic waste. *The Pacific Journal of Science and Technology, 2*(2), 33–46.

Kumi-Larbi Jnr, A., Yunana, D., Kamsouloum, P., Webster, M., Wilson, D. C., & Cheeseman, C. (2018). Recycling waste plastics in developing countries: Use of low-density polyethylene water sachets to form plastic bonded sand blocks. *Waste Management, 80*, 112–118.

Lenkiewicz, Z., & Webster, M. (2017). *Making waste work: A toolkit for community waste management in low and middle income countries*. Wasteaid UK and Chartered Institute of Wastes Management.

Ministry of Foreign Affairs. (2020). *Business opportunities in Nigeria's waste and and circular economy sector*. The Netherlands Enterprise Agency.
Mojekwu, J. N., Idowu, A., & Sode, O. (2013). Analysis of the contribution of imported and locally manufactured cement to the growth of gross domestic product (GDP) of Nigeria (1986–2011). *African Journal of Business Management, 7*(5), 360–371.
Ogwueleka, T. C., & Naveen, B. P. (2021). Activities of informal recycling sector in North-Central, Nigeria. *Energy Nexus, 1*, 100003.
Olaide, A. M., & Dias, G. A. (2020). Municipal Solid Waste Characterization as a Measure towards Sustainable Waste Management in Abuja, Nigeria. *Journal of Environmental Science and Public Health, 04*, 43–60.
Organisation for Economic Co-operation and Development. (2022, February 22). *Plastic pollution is growing relentlessly as waste management and recycling fall short*. Retrieved from Organisation for Economic Co-operation and Development: https://www.oecd.org/environment/plastic-pollution-is-growing-relentlessly-as-waste-management-and-recycling-fall-short.htm
Sellakutty, D. (2016). Utilisation of waste plastic in manufacturing of bricks and paver blocks. *International Journal of Applied Engineering Research, 11*(3), 364–366.
Siddique, R. (2017). Waste plastic as a stabilizing additive in stone matrix asphalt. *Construction and Building Materials, 156*, 69–79.
The American Society for Testing and Materials. (1988). *Standard specification for solid concrete interlocking paving units*. The American Society for Testing and Materials.
Toriola-Coker, L. O., Alaka, H., Bello, A., & Ajayi, S. (2020). Sustainability barriers in Nigeria construction practice. In *2nd international conference on sustainable infrastructural development, 3. Ota, Nigeria* (IOP Conference Series: Materials Science and Engineering).
United Nations Environmental Programme. (2023, July 14). *Plastic pollution*. Retrieved from United Nations Environmental Programme: https://www.unep.org/plastic-pollution
Watts, J. (2019, February 25). *Concrete: The most destructive material on Earth*. Retrieved from The Guardian: https://www.theguardian.com/cities/2019/feb/25/concrete-the-most-destructive-material-on-earth

Chapter 3
Photoconductive Cells Based on Type-II Conical Quantum Dots for Thermophotovoltaic and Other Mid-infrared Applications

Karen M. Gambaryan

3.1 Introduction

Besides traditional solar cells, photovoltaic (PV) elements and photoconductive cells (PCC) have found wide use in thermophotovoltaic (TPV) power converters recently. TPV power generation involves the conversion of solar or other thermal energy to electrical energy through the use of PV converters that respond to middle-wavelength infrared radiation. Traditional TPV energy conversion systems (Coutts & Ward, 1999; Wanlass et al., 1996) consist of a radiant heat source, a selective optical filter, and TPV or PCC cells, which are either semiconductor p-n junctions, or device structures based on type-II heterostructures, where the voltage generation is occurred due to spatial separation of charge carriers. TPVs and mid-infrared PCCs are made of low direct bandgap semiconductors according to the temperature of heat source (Datas & Martí, 2017; Gevorkyan et al., 2004). Theoretical studies indicate that for 1000–2000 °C black body sources, optimum TPV cell bandgaps are in the range of 0.2–0.5 eV. This range is considerably lower than almost all conventional TPV cells, therefore there is a need for significant development in both new materials used for TPVs and in processing, to produce high-performance TPV converters with low bandgaps.

Recently, many efforts have been invested in the growth of conical quantum dots (CQDs), motivated by their unique physical properties (Lin et al., 2014; Kim et al., 2012). Cone-shaped nanostructures can significantly improve the absorption capacity of photons which is widely used in PV cells. The highest conversion efficiencies for axial and radial GaAs nanocone solar cells are 20.1% and 17.4%, obtained at a

K. M. Gambaryan (✉)
Department of Physics of Semiconductors and Microelectronics, Institute of Physics, Yerevan State University, Yerevan, Armenia
e-mail: kgambaryan@ysu.am

A. Sayigh (ed.), *Transition Towards a Carbon Free Future*, Innovative Renewable Energy, https://doi.org/10.1007/978-3-031-61660-0_3

slope angle of 5° and 6°, respectively, both of which are much higher than their cylinder nanowire counterparts. In addition, CQDs can significantly improve the light transmittance and reduce the light reflection, showing superior antireflection property, as well as they have excellent superhydrophobic property and self-cleaning ability with the contact angle of around 155° (Fulmes et al., 2015). The QDs shapes, strain distributions, elastic deformation, and most importantly composition are strongly impacted on electronic structures of the QDs and therefore influenced on the output characteristics of semiconductor devices based on them. Usually CQDs are considered as QDs when the cross section can be considered as an equilateral triangle. When the angle at the top is obtuse, the nanocone looks like quantum well, due to the large diameter of nanocones in comparison with the height and quantization of energy of particles takes place only in vertical direction. At the very small sharp solid angle, CQD transforms into a nanowire with gradually decreasing diameter from the base till the tip of the cone. The last one is a unique system which has wide technical applications as a one-dimensional graded bandgap semiconductor.

Among III–V compound semiconductors, solutions in InAs-InSb-InP material system are very attractive due to their ability to cover very important mid-infrared (3–5 μm) region. In the Stranski–Krastanow (S–K) growth mode (Stranski & Krastanow, 1937), as the most useful method at the nucleation of QDs, depending on the growth conditions, the elastic strain can be relaxed by the formation of QDs with lens-shaped, conical, pyramidal, or truncated pyramidal geometries. Previously, several types of nanostructures in that material system have been successfully grown, in particularly lens-shaped and ellipsoidal QDs, quantum rings, and others (Gambaryan et al., 2012a, b, 2015). In this paper will be presented the results of fabrication and investigation of the low-bias and room temperature PCCs made of n-InAs(100) crystals with InAsSbP composition type-II CQDs on the PCC surface for TPV and other special mid-infrared applications.

3.2 Experimental Details

For III–V compound semiconductors such as InAs, InP, or InSb, one has to remark the very large difference in the respective diameters of the V element. One can therefore monitor the strain by adequate substitution between smaller/larger V element atoms. This could be used as mixed compounds such as InAsSbP adjusting its alloy composition with three V elements only. The main driving force at the nucleation of InAsSbP CQDs is the surface diffusion of antimony atoms, like in classic InAs/InSb system. But the presence of the phosphorous atoms and their surface diffusion to one or opposite directions in our quaternary material system opens up a new possibility and can be an additional tool to control the nucleation process and the type of nanostructure (Gambaryan, 2021).

A modified liquid phase epitaxy (MLPE) technique with the slide-boat crucible (Gambaryan et al., 2012a, b, 2015) was used to grow CQDs. In contrast to other growth techniques like molecular beam epitaxy (MBE) or metal organic chemical vapor deposition (MOCVD), MLPE operates comparatively closer to thermodynamically equilibrium. In our MLPE process, melted In (7N) was used as a solvent, and undoped InAs, InP, and Sb (6N) were used as solutes. Lattice mismatch of 2.5% between the n-InAs(100) substrate with thickness of 400 μm and wetting layer was chosen to provide the growth of InAsSbP composition CQDs in S–K mode. The high-resolution SEM (SEMEDXA–FEI Nova 600–Dual Beam) and AFM (TM Microscopes–Autoprobe CP) techniques are used for investigation and characterization of CQDs morphology, dimensions (size and shape), and distribution density. The IRS-21 mid-infrared spectrometer was used to investigate the photoresponse spectra of PCC structures. Au/Cr evaporation was performed to obtain ohmic contacts on the surface of the unencapsulated nanostructures. He–Ne cw laser was used as a source of mid-infrared monochromatic and coherent radiation.

3.3 Results and Discussions

We explore the resulting morphology looking at the QD size, shape, and density using high-resolution SEM and AFM microscopes. At AFM characterization, images were acquired at a constant scanning force using a silicon nitride soft cantilever probe with a pyramidal tip. In order to avoid instrumental artifacts, each investigated area is scanned several times along different directions. Figure 3.1 shows the HR-SEM and AFM (plane view) images of InAsSbP unencapsulated CQDs grown by modified LPE on an InAs(100) substrate.

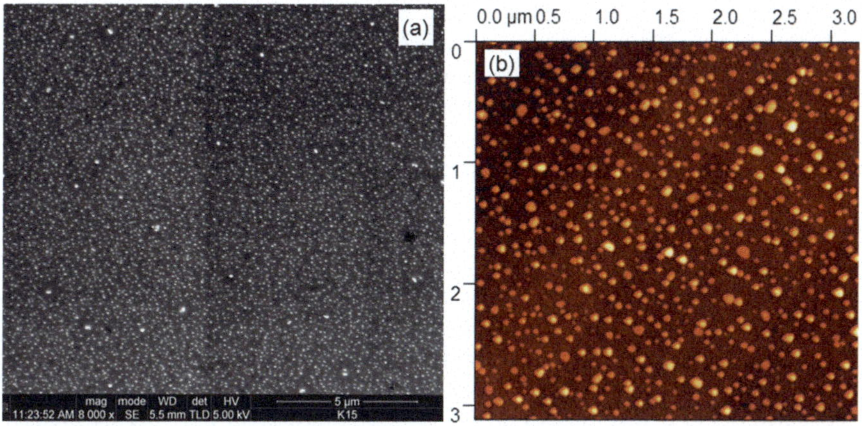

Fig. 3.1 HR-SEM (**a**) and AFM (**b**) (plane view) images of InAsSbP CQDs grown by MLPE on InAs(100) surface

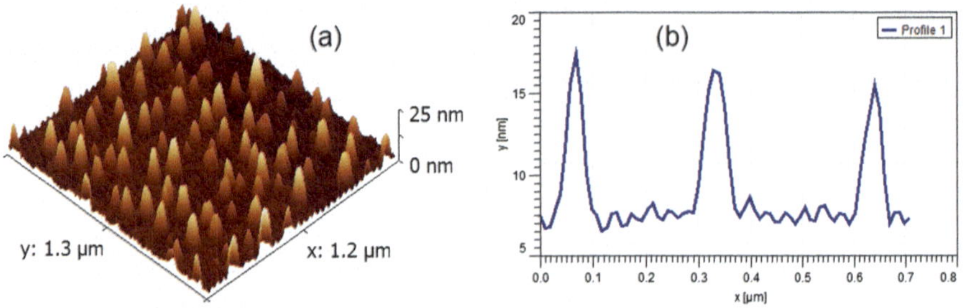

Fig. 3.2 Oblique view AFM image of CQDs (**a**) and the line-scan (**b**)

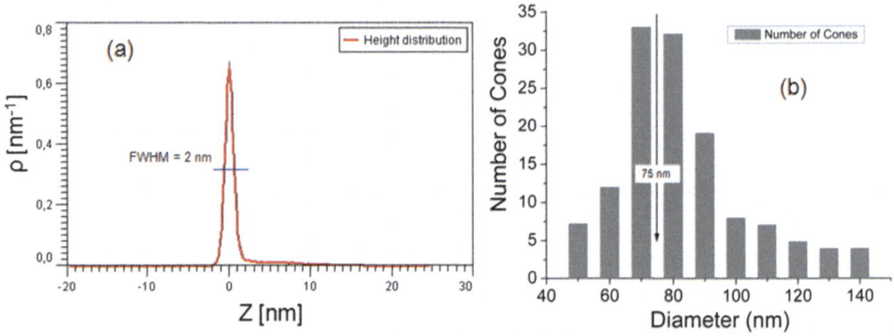

Fig. 3.3 Number of CQDs versus height (**a**) and diameter (**b**)

The InAsSbP composition CQDs are clearly visible and quite uniformly distributed over the substrate surface. The CQDs average surface density is about (2–5) × 10^9 cm^{-2} with heights from 0.5 to 20 nm and diameters ranging from 20 to 120 nm. Because the difference in lattice parameter is large enough, the growth process is consistent with the Stranski–Krastanow growth mode. The oblique view AFM image of CQDs with the line-scan is presented in Fig. 3.2. Statistical evaluation shows the Gaussian distribution of the CQDs number versus their average diameter calculated from the substrate surface $S = 10\ \mu m^2$ with the optimum size of CQDs to be at ~75 nm (Fig. 3.3b). However, from Fig. 3.3a it is quite visible that the height distribution of CQDs is much more uniform with the full-width at half maximum ~2 nm.

The room temperature photoresponse spectra (PR) of fabricated PCC structures without (Fig. 3.4a—1) and with CQDs (Fig. 3.4a—2) were measured and presented in Fig. 3.4. Figure 3.4b shows the PR spectra of CQD-based PCC structures at different low (up to 6 mV) biases. From Fig. 3.4a it is quite visible that the spectrum of CQDs-based PCC is extended to both long (up to $\lambda = 4$ µm) and short wavelength regions with the main peak at $\lambda = 3.48$ µm. This peak coincides with the energy bandgap of InAs ($E_g = 0.355$ eV). Additional peaks (especially at $U = 6$ mV applied bias) at $\lambda = 3.4$ µm, $\lambda = 3.68$ µm, and $\lambda = 3.89$ µm were also revealed on CQD-based PCC PR spectrum as a result of charge carrier's transitions through energy levels

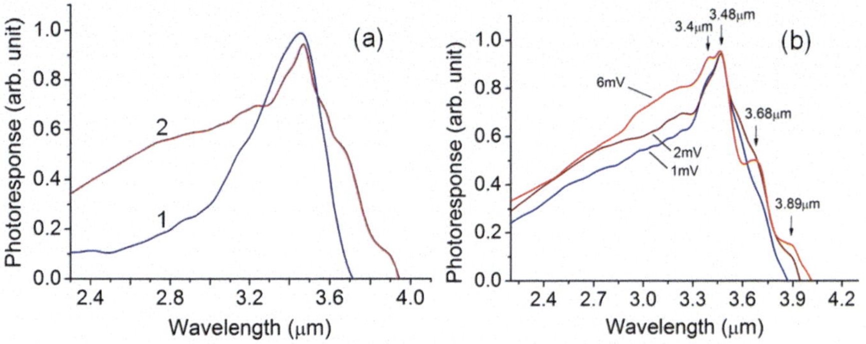

Fig. 3.4 Room temperature photoresponse spectra of fabricated PCC structures without (1) and with CQDs (2) (**a**), PR spectra of CQD-based PCC structures at different low biases (**b**)

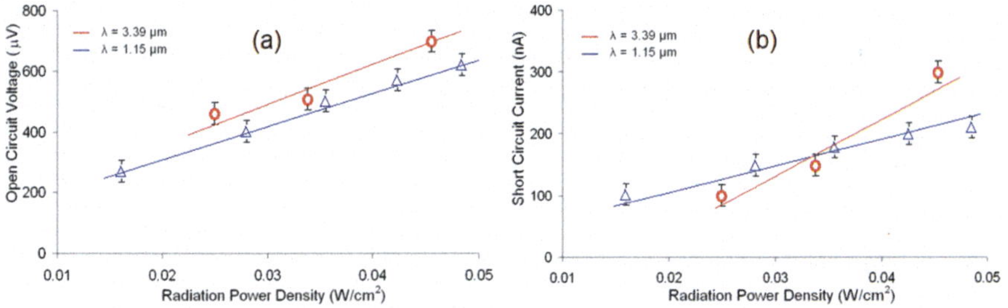

Fig. 3.5 Open-circuit voltage (**a**) and short-circuit current (**b**) of CQD-based PCC structure versus radiation power density of He–Ne laser at $\lambda = 3.39$ μm (red) and $\lambda = 1.15$ μm (blue) wavelengths

created by type-II InAsSbP CQDs grown on InAs. Actually, Fig. 3.4a shows the displacement of the absorption edge toward the long wavelength region, as well as enlargement toward the short wavelength region. Our quantitative calculation shows increasing of CQD-based PCC structure's efficiency up to 10% compared with the same structures without QDs.

The open-circuit voltage and short-circuit current of CQD-based PCC structure versus radiation power density of He–Ne laser at $\lambda = 3.39$ μm (red) and $\lambda = 1.15$ μm (blue) wavelengths are presented in Fig. 3.5a and b, respectively. The voltage and current responsivities at room temperature were equal to 1.2 V/W and 0.8 mA/W, respectively, at zero bias and $\lambda = 3.39$ μm monochromatic radiation of He–Ne laser. Those values were measured at the maximum of the spectrum and correspond to the internal quantum efficiency of ~40% (with allowance for ~30% losses for the reflection).

Additionally, investigations of electrophysical properties of the developed nano-structured PCCs showed that the formation of the CQDs on the n-InAs(100) surface leads to the increasing of the PCC's electrical sheet resistance up to one order. The results of measurements of the relative change in the surface resistance of a structure with CQDs in the presence and absence of laser radiation with $\lambda = 1.15$ and 3.39 μm

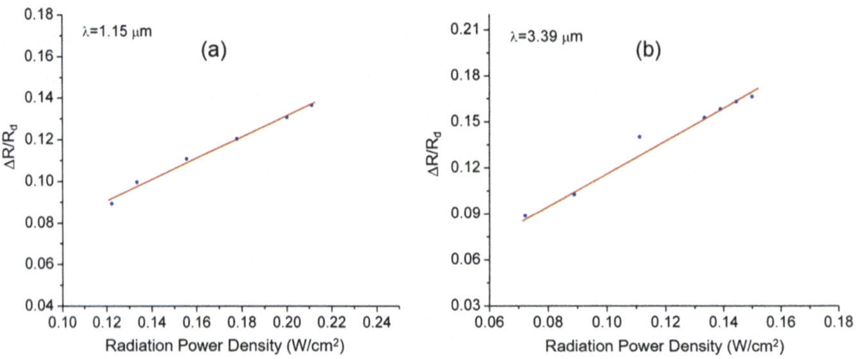

Fig. 3.6 The relative change in the surface resistance of a structure with CQDs in the presence and absence of continuously operated laser radiation with (**a**) $\lambda = 1.15$ μm and (**b**) $\lambda = 3.39$ μm

are presented in Fig. 3.6a and b, respectively. Measurements and calculations show decreasing of the surface resistance up to 12% at $\lambda = 1.15$ μm and 16% at $\lambda = 3.39$ μm of the continuously operated laser radiation.

3.4 Conclusions

Thus, we have presented the results of fabrication and investigation operated at low bias and room temperature PCCs made of n-InAs(100) crystals with InAsSbP composition type-II CQDs on the PCC surface for TPV and other special mid-infrared applications. Conical QDs were nucleated at constant temperature from quaternary In-As-Sb-P liquid phase using Stranski–Krastanow growth mode. HR-SEM and AFM techniques were used to investigate the size, shape, and distribution density of the CQDs. CQD-based PCC's open-circuit voltage and short-circuit current were measured at room temperature versus output radiation power density of the He–Ne continuously operated laser at $\lambda = 3.39$ and 1.15 μm wavelengths. It was shown that formation of the CQDs on the n-InAs(100) surface leads to the increasing of the PCC's electrical sheet resistance up to one order, as well as results in enlargement of the photoresponse spectrum toward the long wavelength region. The CQD-based PCC's voltage and current responsivity at room temperature were equal to 1.2 V/W and 0.8 mA/W, respectively, at zero bias and $\lambda = 3.39$ μm monochromatic radiation. Theoretical calculations showed increasing of CQD-based PCCs efficiency up to 10% compared with the same structures without CQDs.

The author wishes to thank to Dr. T. Boeck and his colleagues from the Leibniz Institute for Crystal Growth (IKZ, Berlin, Germany) for HR-SEM and AFM characterization, as well as to Dr. V.G. Harutyunyan from the Institute of Applied Problems of Physics (National Academy of Sciences of Armenia) for measurements of the optoelectronic properties of grown structures.

References

Coutts, T. J., & Ward, J. S. (1999). Thermophotovoltaic and photovoltaic conversion at high-flux densities. *IEEE Transactions on Electron Devices, 46*, 2145–2153.

Datas, A., & Martí, A. (2017). Thermophotovoltaic energy in space applications: Review and future potential. *Solar Energy Materials & Solar Cells, 161*, 285–296.

Fulmes, J., Jäger, R., Bräuer, A., et al. (2015). Self-aligned placement and detection of quantum dots on the tips of individual conical plasmonic nanostructures. *Nanoscale, 7*, 14691–14696.

Gambaryan, K. M. (2021). *Method for the growth of nanostructures from liquid phase* (Patent No: AM681Y). Awarded and registered: November 16, 2021; Priority: July 15, 2021; Appeal No: AM20210054Y; Entry No: 909.

Gambaryan, K. M., Aroutiounian, V. M., & Harutyunyan, V. G. (2012a). Nucleation features and energy levels of type-II InAsSbP quantum dots grown on InAs(100) substrate. *Applied Physics Letters, 101*, 093103. (5 pp).

Gambaryan, K. M., Aroutiounian, V. M., Harutyunyan, V. G., Marquardt, O., & Soukiassian, P. G. (2012b). Room temperature magnetoelectric properties of type-II InAsSbP quantum dots and nanorings. *Applied Physics Letters, 100*, 033104. (4 pp).

Gambaryan, K. M., Harutyunyan, V. G., Aroutiounian, V. M., Ai, Y., Ashalley, E., & Wang, Z. M. (2015). Magnetoresistance and capacitance oscillations and hysteresis in type-II InAsSbP ellipsoidal quantum dots. *Journal of Physics D: Applied Physics, 48*, 275302. (7 pp).

Gevorkyan, V. A., Aroutiounian, V. M., Gambaryan, K. M., Kazaryan, M. S., Touryan, K. J., & Wanlass, M. W. (2004). Liquid-phase electroepitaxial growth of low band-gap p-InAsPSbyn-InAs and p-InAsPyn-InAs diode heterostructures for thermo-photovoltaic application. *Thin Solid Films, 451–452*, 124–128.

Kim, J. S., Jeong, H. W., Lee, W., Park, B. G., Kim, B. M., & Lee, K. B. (2012). A simple and fast fabrication of a both self-cleanable and deep-UV antireflective quartz nanostructured surface. *Nanoscale Research Letters, 7*, 430–435.

Lin, Q., Leung, S.-F., Lu, L., Chen, X., Chen, Z., Tang, H., et al. (2014). Inverted nanocone-based thin film photovoltaics with omnidirectionally enhanced performance. *ACS Nano, 8*, 6484–6490.

Stranski, I., & Krastanow, L. (1937). Zur Theorie der orientierten ausscheidung von ionenkristallen aufeinander. *Monatshefte fuer Chemie, 71*, 351–364.

Wanlass, M. W., Ward, J. S., Emary, K. A., Al-Jassim, M. M., Jones, K. M., & Couts, T. J. (1996). $Ga_xIn_{1-x}As$ thermophotovoltaic converters. *Solar Energy Materials and Solar Cells, 41–42*, 405–417.

Chapter 4
100% Renewable Energy: A Stand-alone Hybrid Solar PV-Hydrogen-Battery Power Systems for Homeland Communities in Regional Western Australia

Furat Dawood, GM Shafiullah, and Martin Anda

4.1 Introduction

An emerging green renewable hydrogen industry is gaining momentum in Australia and globally, offering a promising solution for low-carbon fuel alternatives in various sectors (IEA, 2021, 2023). Hydrogen is an energy carrier, not an energy source, which means energy must be used to produce it (Yap & McLellan, 2023). The hydrogen produced from RE via water electrolysis is called renewable (green) hydrogen. Therefore, utilising hydrogen as RESS is a promising solution to decarbonise the electricity sector in the race to global zero emissions target by 2050 (IEA, 2023; Sayigh, 2022; Vivek Srinivasan et al., 2023). One of the remarkable advantages of green hydrogen is that it only generates energy and water as by-products at the used endpoint, making it a genuinely zero-carbon emissions fuel (Dawood et al., 2020a).

Australia has the potential to become a world leader in the emerging green hydrogen market, which is currently being encouraged by the Western Australian (WA) and Australian Governments (Vivek Srinivasan et al., 2023; Australian Government, 2019; DPIRD, Work Group, 2019; Dawood et al., 2020b). The WA Renewable Hydrogen Strategy and Roadmap, launched in 2019 and 2020, respectively, aim to drive WA's position as a major producer and exporter of renewable hydrogen and detail the WA Government's commitment to support and facilitate industry efforts to develop a renewable hydrogen industry in WA (WA Government, 2020; WA Government, 2021). Decentralised, distributed power generation systems have been identified as a possible solution for the significant increase in intermittent

F. Dawood (✉) · GM. Shafiullah · M. Anda
School of Engineering and Energy, College of Science, Technology, Engineering and Mathematics, Murdoch University, Murdoch, WA, Australia
e-mail: Furat.Dawood@murdoch.edu.au

A. Sayigh (ed.), *Transition Towards a Carbon Free Future*, Innovative Renewable Energy, https://doi.org/10.1007/978-3-031-61660-0_4

Renewable Energy Sources (RES)-based power production required to meet the ambitious targets at the national and state levels to achieve Net Zero by 2050 (WA Government, 2022). Renewable energy microgrids are a solution that can develop smaller and more flexible power grids that can operate in complete autonomy or even remain connected to larger central power grids and exchange power as needed (Furat et al., 2020; Dawood et al., 2019). The literature reveals significant research and demonstration projects work to use short-duration RESS to reduce the existing levels of fossil fuel usage to reduce emissions (Vivek Srinivasan et al., 2023; Horizon Power, 2022; CSIRO, 2021). Solar-PV-battery hybrid systems are typically combined with diesel generators as a backup due to a lack of long-duration energy storage using batteries only as a RESS, called BESS (Sayigh, 2022; Li et al., 2023; Karunanithi et al., 2023). Hence, hydrogen-based RESS was researched and implemented as demonstration projects worldwide to bridge the gap of long-duration RESS to improve power system reliability, quality, and autonomy (Mohammad-Alikhani et al., 2022; IRENA, 2022; Al-Buraiki & Al-Sharafi, 2022). The WA Government funded many feasibility studies and pilot projects to techno-economically assess the viability of hydrogen-based power systems (H2PS) for remote and grid-connected homeland communities (WA Government, 2020, 2021; Sandri et al., 2021). In WA, the current Remote Essential and Municipal Services (REMS) state-wide program is servicing 120 communities that are currently dependent on similar unreliable, costly, polluting diesel power stations (WA Government, 2023).

This study conducted a feasibility study to assess the practicality of establishing a self-contained microgrid in the Pilbara region in WA. This microgrid would rely entirely on renewable energy sources and incorporate a hybrid battery-hydrogen storage system. The study considers the potential social, economic, and environmental advantages. Its purpose is to provide valuable information for making an investment decision regarding implementing a pilot project in the Pilbara region. To carry out the study, Murdoch University has partnered with an indigenous mining logistics company owned by the traditional custodians of the land of two selected sites, as well as two prominent electrical contracting companies.

4.2 Methodology

Homer Pro simulation software was used to integrate the system components, and this system was simulated to optimise the sufficient sizes and capacities to meet the load (power and hydrogen) requirements. All system components were optimised for the minimum COE, resulting in a higher return on investment and minimum payback period. The simulator calculated the energy balance within the model to cover the load at all times of the year according to the site-specific RE resource availability and intensity. The RE resources and temperature profiles for site 1 and 2 locations were extracted from the NASA prediction energy resources database.

The simulator input parameters were collected from the publicly available literature reviews, technical datasheets, and the simulation software (Homer Pro) library and incorporated any necessary modifications (Energy, 2023). The CAPEX and OPEX (all in AUD) were adjusted according to the researched information and communications with the relevant existing plants in Australia, such as Horizon Power, ATCO Gas Australia, LAVO, Tempo, and Element 47 (Horizon Power, 2022; CSIRO, 2021; GKN, 2022; LAVO, 2023). Additionally, some assumptions and estimations were necessary to make.

100% RE solar-PV, hybrid battery-hydrogen energy storage systems were modelled to decarbonise the stand-alone power systems of both homeland communities. The proposed systems were designed according to the site-specific RE resources, climate, average and peak loads, as well as a sufficient period of autonomy, i.e. at least 48 hours. The system model was simulated and analysed to evaluate its feasibility techno-economically.

4.3 Site Selection and Data Collection

Murdoch University team travelled to the Pilbara region, North of WA, for site selection and data collection.

4.3.1 Site Selection Criteria

The project sites' selection criteria are to satisfy the following:

- Two sites represent small and medium sizes.
- One site is grid-connected, and the second is a stand-alone microgrid.
- Land/area availability for RE generation, in this case, a solar PV farm and/or rooftop.
- Water availability (required 10 l/kg H2, round trip useable energy about 20 kWh).
- Community licence.

The team travelled to Pilbara, WA, to select two sites fulfilling the selection criteria focusing on the power system size and orientation, i.e. grid-connected and stand-alone.

Two sites were selected as follows:

Site 1: A small homeland community of seven houses with a grid-connected power system and an average daily load profile of 157.85 kWh/d.

Site 2: A medium-sized homeland community of 35 houses, a small medical centre, and a community building. This site is a stand-alone microgrid powered by three diesel generators with an average daily load profile of 1384.5 kWh/d.

4.3.2 Load Profile

Site-specific data for site 1 revealed an average daily load profile of 157.85 kWh/d, which rounded to 158 kWh/d with an average of 6.58 kW and a peak load of 11.06 kW. The existing power system architecture and load profile are illustrated in Fig. 4.1.

Site 2 specific data revealed an average daily load profile of 1385 kWh/d, which rounded to 1.5 MWh/d supplied by three of 110 kW diesel genset power station. The average hourly load was calculated as an average between the middle of winter and summer with a 10% daily variation. The existing power system architecture and load profile are illustrated in Fig. 4.2.

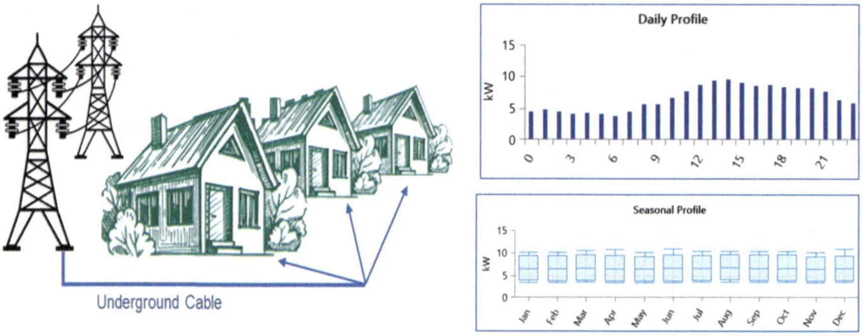

Fig. 4.1 Site 1 power system schematic diagram and extracted load profile

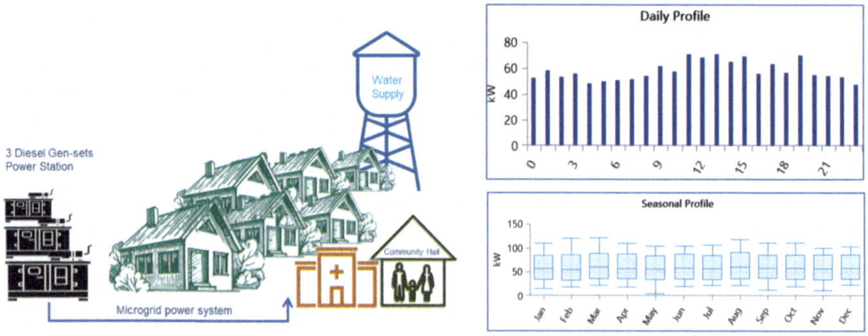

Fig. 4.2 Site 2 microgrid schematic diagram and extracted load profile

4.3.3 Modelling and Simulation

The selected sites were modelled, and the hydrogen-based power system (H2PS) was designed with a focus on the following:

- Both sites will become stand-alone microgrids with a 100% RE hybrid hydrogen-battery power system, i.e. H2PS.
- System reliability and sufficiency with at least 48 hours of autonomy.
- H2PS components have optimal sizes and costs and the least Cost of Energy (COE), i.e. $ AU/kWh.

Homer Pro software (Energy, 2023) was used for simulation with the necessary collected data. The site-specific RE resources, in this case, the peak sunshine hours and the pattern, were imported from the NASA solar irradiance database (Data, 2019).

4.3.4 H2PS Models

The engineering design of the proposed H2PSs for both selected sites comprises:

- Sufficient RE generation capacity
- Electrolyser to produce renewable (green) hydrogen
- Hydrogen storage vessel
- Fuel cell for base load power generation
- Battery bank for load fluctuation and peak following

Figure 4.3 shows a schematic of the proposed system model for both sites to convert the existing power systems to a 100% RE-H2PS.

4.3.5 Techno-economic Parameters

The techno-economic input parameters for the proposed system, i.e. including all the associated peripherals, control, and auxiliary devices, are listed in Table 4.1.

The values in Table 4.1 are collected from reliable resources for crucial input parameters (such as the price of diesel, PV panel, fuel cell, hydrogen tank, and the cost of generating electricity in the Pilbara regional). In light of this, to enhance the confidence level in the business case and business plan, a sensitivity analysis was undertaken for several key cost input values in order to determine the effect on the optimised system, as well as provide guidance around the optimal time (e.g. when hydrogen equipment prices hit trigger point $ AU per kW) or staging of a broader transition to this kind of system. The results from this sensitivity analysis were then used to inform and develop the business case.

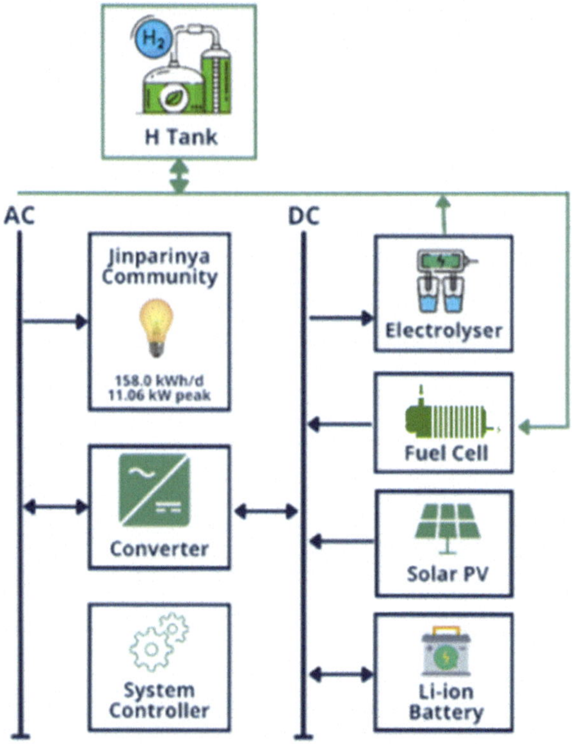

Fig. 4.3 Proposed H2PS model

4.3.6 Safety—Hazard and Operability Analysis (HAZOP)

Due to the remote locations, harsh operating environments, and social settings where the systems will be implemented, there is an enhanced safety risk, especially during the operational phase. Hydrogen-related policies and standards are developing quickly in Australia to overcome the barriers to adopting and implementing H2PS. There are existing international standards from peak organisations, such as ISO for hydrogen and hydrogen-related technologies, safety, and performance, such as ISO 14687-X, ISO 16110-X, ISO/TS 19883, and ISO 22734-X (Standardisation, 2023). Additionally, Australia, like many other countries, is adopting relevant existing national standards with revisions to determine the extent to which they are appropriate for use with hydrogen gas technologies, such as AS 2885, AS 4645, AS 4568, AS 4564, AS/NZS 5263.0, AS/NZS 5601.1, and AS 3814. However, the project must comply with WA state-specific related regulations, as well as the impact on the surroundings and the environment must be assessed during the front-end engineering design (Australian Government, 2023; HSE, 2023; Safe Work Australia, 2023). Hence, the design of all proposed systems must be assessed for hazard, safety, and handling procedures, primarily when hydrogen is in use. Standards Australia (The ME-93 Hydrogen Technologies Committee) has adopted and developed a set of hydrogen standards that must be met in the systems' design, construction, and operation (Standardisation, 2023; AS, 2023; Standards Australia, 2023).

4.3.7 Water Requirements

Typically, a quantity of 10–15 l of raw (potable) water as an electrolyser feedstock and for a cooling system is required for each kilogram of the produced hydrogen. The electrolyser's input must be purified and de-ionised water at 9.9 l for each kg H_2. The energy requirements for this process can be conducted during the excess RE availability (peak sunshine hours). Additionally, a water treatment unit is required, and its cost was included in the electrolyser package, estimated at $ AU 2–3 k for 10–20 l per hour of capacity.

4.4 Results and Discussion

The model was fed into the Homer Pro software with the case-specific data for each site separately, as shown in Table 4.1, to simulate the system functionality and optimise the component sizes for lease COE over the project lifetime of 25 years. The simulation results for the three most techno-economically feasible options are shown in Table 4.2.

4.4.1 Site 1

Table 4.2 shows the simulation results of the proposed H2PS model for site 1. The three most feasible options (scenarios) were tabulated for comparison.

The least COE scenario is option 1, which uses a large battery capacity of 128 kWh and less hydrogen system output capacity of 6 kW. However, this scenario shows that FC capacity is insufficient to cover the load in the case of a RE intermittency, i.e. no sufficient autonomy. Therefore, we recommended option 2 as an optimal configuration for site 1, which has an initial total plant capital cost of $ AU 437,105 and an annual operating cost of $ AU 41,094.82. The net present cost for the system is $ AU $ AU968,359 with a COE of $ AU1.299/kWh and a simple payback of 6.75 years (discounted payback period of 8.1 years), as plotted in Fig. 4.4. The $ AU1.299 per kWh COE compared to a conventional diesel-powered mini-grid or grid COE at $ AU1.65 per kWh before subsidy (Parkinson, 2017) presents a reduction of 21.3% in the COE and eliminates 33,448.6 kg CO2-e per year (Scope 2 emissions factor of 0.58 kg CO2-e/kWh for the NWIS, WA) (DISER, 2021). Also, option 2 offers 408 kWh of stored energy in the form of hydrogen,

Table 4.1 The RE model techno-economic input parameters

System component	Type	CAPEX/OPEX (AUD)
Solar PV system	Generic, no tracking (installed)	$ AU 1000 kW / $ AU 10 kW/year
Li-Ion battery bank	Generic Li-Ion package (installed) @ eff.	$ AU 550 kWh / $ AU 5 kWh
H2 electrolyser	Generic PEM package (installed) @ eff. 55%	$ AU 2500 kW / 10 kW/year
H2 FC.	Generic PEMFC package (installed) @ eff. 50%	$ AU 3350 kW /$ AU 1 Op hr
H2 tank low-pressure	Generic package (<30 bars) tank @ eff. 95%	$ AU 1000 kg H2 / $ AU 5 kg H2
Converter	Generic (DC/AC inverter) (installed) @ eff. 95%	$ AU 300 kW / 0
System controller	Generic load following (Smart PLC & SCADA)	$ AU 30,000 Complete System/0
Fixed capital	Engineering design, tendering, approvals, and civil work	$ AU 220,000[a] $ AU 781,000[b]
Yearly operational	Supervising and remote monitoring	$ AU 24,000[a] $ AU 104,000[b]
H2 fuelling station	Package[c]	$ AU 2.8 million[c]
Discount rate (%)	Assumed average over the project's lifetime	8
Inflation rate (%)	Assumed average over the project's lifetime	2
Project lifetime (year)	All projects' models	25

[a]Site 2 H2PS
[b]Site 1 H2PS
[c]Hydrogen fuelling station, ACT, Australia (Riotact, 2023)

which has a longer discharge duration than batteries and is sufficient for 48 hours of autonomy with 9 hours of battery autonomy at 65% FC efficiency. Hence, the optimal H2PS for site 1 consists of the following:

- 12 kW electrolyser
- 25 kg H_2 storage system
- 7 kW fuel cell system
- 80 kWh battery usable capacity

It was found that the potable water requirement for the H2PS in site 1 is 7147.5 l per year at an average of 15 l/kg H_2 (9 l of pure de-ionised water per kg H_2), which is a relatively small percentage of the average water consumption in the Pilbara region.

Site 1 H2PS model functionality was simulated, and the results revealed that the hydrogen tank of 20 kg H_2 capacity is full most of the time during the good-resources session (summer); thus, the controller stops the electrolyser from overfilling the storage tank. Hence, the hydrogen storage capacity can be increased at any time to store more hydrogen, adding more autonomy hours or use as a direct fuel or for heat energy requirements. The battery and the 20% oversized converter

Table 4.2 Techno-economic results for the three selected options

System component	Option (1)	Option (2)	Option (3)
Architecture/solar PV (kW)	58	84	104
Architecture/Li-Ion battery (kWh) nominal	128	80	64
Architecture/FC capacity (kW)	6	7	8
Architecture/electrolyser capacity (kW)	4	12	15
Architecture/H2 tank capacity (kg)	20	25	30
Architecture/converter (kW)	14	14	14
Architecture/dispatch controller type	LF.	LF.	LF.
Cost/net present cost (NPC) ($ AU)	$ AU 777,917	$ AU 968,359	$ AU 1,094,572
Cost/cost of energy (COE) ($ AU)	$ AU 1.044	$ AU 1.299	$ AU 1.469
Cost/operating cost ($ AU/year)	$ AU 28,547.50	$ AU 41,094.82	$ AU 48,727.70
Cost/initial capital ($ AU)	$ AU 408,869	$ AU 437,105	$ AU 464,644
Cost/fuel cost ($ AU/year) (hydrogen)	$ AU 0	$ AU 0	$ AU 0
Cost/O&M ($ AU/year)	$ AU26,613.69	$ AU38,750.55	$ AU46,303.44
System/Ren Frac (%)	100	100	100
System/excess elec (%)	53.39	46.03	50.83
System/total H2 fuel consumption (kg/year)	53.4	452	683
System/cap short (%)	0.09	0.06	0.05
System/elec prod (kWh/year)	109,592	165,150	207,040
System/elec cons (kWh/year)	61,209	83,789	96,924
Electrolyser/H2 production (kg/year)	73.4	476.5	712.7
FC/operating hours (hr)	209	1895	2580
FC/production (kWh/year)	890	7532	11,376
FC/hydrogen fuel consumption (kg)	53.39	451.93	682.55
FC/O&M cost ($ AU/year)	$ AU1,254	$ AU13,265	$ AU20,640
Hydrogen tank autonomy (kg)	20	24.5	29.7
Autonomy in the form of stored hydrogen (kWh)	333	408	495
Li-Ion batt/autonomy (hr)	15.55	9.72	7.78
Li-Ion batt/annual throughput (kWh/year)	30,212	22,404	17,929
Li-Ion batt/nominal capacity (kWh)	128	80	64
Li-Ion batt/usable nominal capacity (kWh)	102	64	51
System/CO2-e (kg/year)	0	0	0
Water requirement potable (l/year) @ 15 l/kg H2	1101	7147.5	10,690.5

(14 kW) will cover the start-up currents, such as electric machinery or pumps. The particular type of stand-alone inverter/converter stabilises the microgrid voltage and frequency while continuously powered by the battery. Figure 4.5a shows site 1 option 2 H2PS component functionality plot over a year. The hydrogen storage tank is almost full in good RE resources seasons in the Pilbara, from mid-February to

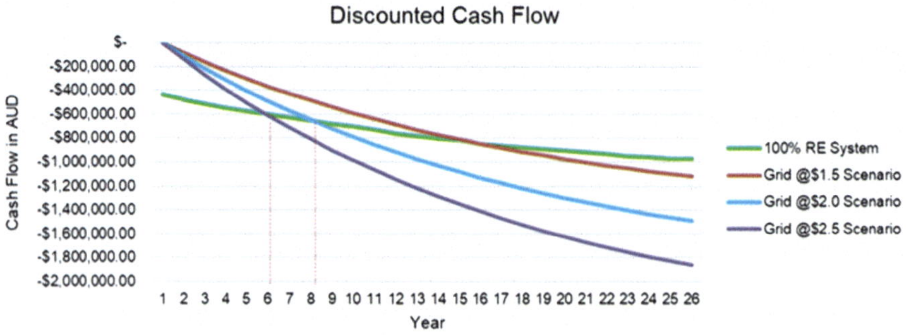

Fig. 4.4 Site 1 nominated scenario (option 2) discounted payback period

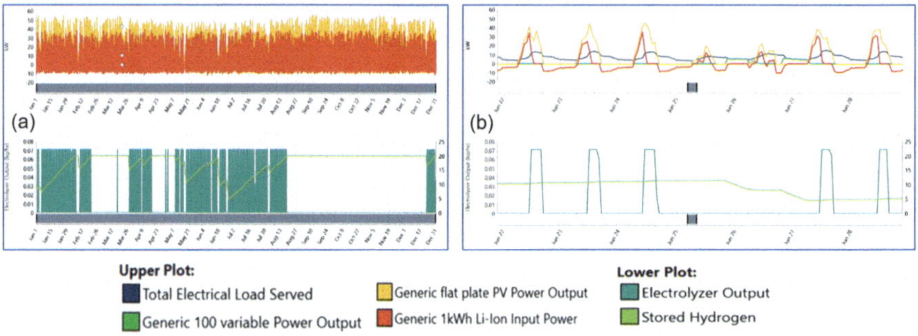

Fig. 4.5 Site 1 H2PS components' functioning option 2. (**a**) yearly profile, (**b**) daily profile (example of the worst period of the year)

the end of March and from mid-August till mid of December. This demonstrates the hydrogen merit of long-duration seasonal RE storage. Figure 4.5b focuses on the worst-case scenario when RE is intermittent or unavailable for a few days, i.e. demonstrates the autonomy viability of the recommended scenario.

4.4.2 Site 2

Table 4.3 shows the simulation results of the proposed H2PS model for site 2. The 100% RE H2PS most feasible is option 1 simulation results which were tabulated for comparison with a proposed 98.5% RE penetration in option 2. Both Options were compared with the base case scenario, i.e. the existing three 110 kW diesel gensets. Option 2 was recommended for two reasons as follows:

1. Using one of the diesel genset to contribute 1.5% of the energy balance reduces the COE by about 28.4% compared to the base case COE. However, option 2 will incur less than 2% of emissions.

Table 4.3 Techno-economic results for the three selected options

System component	Base case	Option 1	Option 2
Architecture/genset 1 (kW)	110	0	0
Architecture/genset 2 (kW)	110	0	0
Architecture/genset 3 (kW)	110	0	110
Architecture/genset working hours (hr)	8802	0	220
Architecture/solar PV (kW)	0	966	966
Architecture/Li-Ion battery (kWh) nominal	0	600	480
Architecture/FC capacity (kW)	0	70	60
Architecture/electrolyser capacity (kW)	0	300	200
Architecture/H2 tank capacity (kg)	0	300	200
Architecture/converter (kW)	0	150	150
Architecture/dispatch controller type	LF.	LF.	LF.
Cost/net present cost (NPC) ($ AU)	$ AU7,702,048	$ AU5,869,335	$ AU5,402,124
Cost/cost of energy (COE) ($ AU)	$ AU1.18	$ AU0.902	$ AU0.827
Cost/operating cost ($ AU/year)	$ AU593,467	$ AU168,968	$ AU169,106
Cost/initial capital ($ AU.)	$ AU30,000	$ AU3,712,000	$ AU3,216,000
Cost/fuel cost ($ AU/year) (hydrogen)	0	0	0
Cost/fuel cost ($ AU/year) (diesel)	$ AU434,094.2	0	$ AU7,254.52
Compare economics/IRR (%)	Base case	14,5	17.7
Compare return on investment (%)	Base case	10.5	13.5
Compare economics/simple payback (year)	Base case	6.3	5.4
Compare economics/discounted payback (year)	Base case	8.2	6.6
System/Ren Frac (%)	0	100	98.50
System/excess elec (%)	0	42.4	38.9
System/cap short (%)	0	0	0
System/elec prod (kWh/year)	505,626	1,797,167	1,968,298
System/elec cons (kWh/year)	505,525	1,281,839	1,165,995
Electrolyser/H2 production (kg/year)	0	8260	9259.2
FC/operating hours (hr)	0	2106	2718
FC/production (kWh/year)	0	132,874	153,454
FC/hydrogen fuel consumption (kg)	0	7972	9207.21
FC/O&M cost ($ AU/year)	0	$ AU14,742	$ AU16,308
Hydrogen tank autonomy (kg)	0	288	52.2
Autonomy in the form of stored hydrogen (kWh)	0	4800	870
Li-Ion batt/autonomy (hr)	0	8.32	6.65
Li-Ion batt/annual throughput (kWh/year)	0	131,338	102,795
Li-Ion batt/nominal capacity (kWh)	0	600	480
Li-Ion batt/usable nominal capacity (kWh)	0	480	384
System/CO2-e (kg/year)	405,503.1	0	6776.7
Water requirement potable (Kl/year) @ 15 l/kg H2	Cooling top-up	123.90	138.89

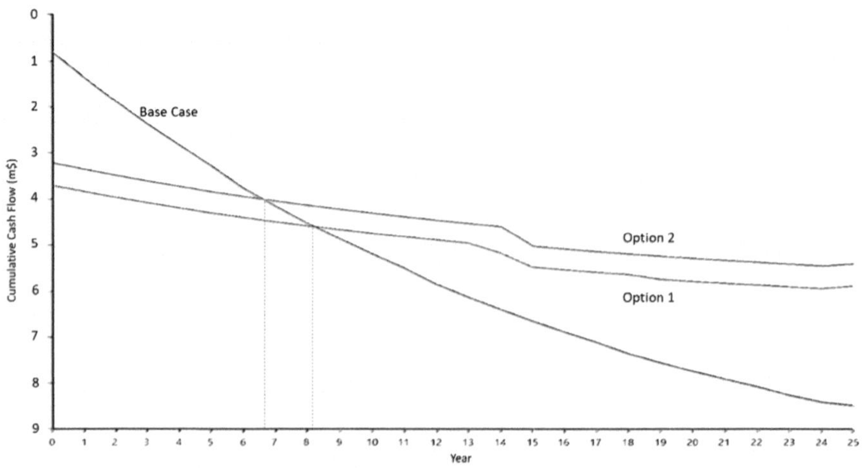

Fig. 4.6 Site 2 proposed H2PS (options 1 and 2) cash flow compared to the base case scenario

2. Utilising one of the gensets as a backup for the proposed H2PS can ensure the system's reliability and increase autonomy without increasing the hydrogen system capacity.

The proposed model's simulation and calculation for site 2 revealed that both proposed scenarios (options 1 and 2) are techno-economically feasible and will reduce the COE (over 25 years project lifetime) compared with the base case scenario by 23.6% and 28.4%, respectively. The optimal configuration for site 2 H2PS, option 2, has an initial total plant capital cost of $ AU 3,216,000 and an annual operating cost of $ AU169,106 (retaining one of the existing diesel gensets of 110 kW). The net present cost for the system is $ AU5,402,124 with a COE of $ AU0.827/kWh and a simple payback of 5.4 years (discounted payback period of 6.6 years) compared with the existing (diesel genset) base case, as plotted in Fig. 4.6. The optimal H2PS configuration for site 2 H2PS consists of the following:

- 200 kW electrolyser
- 200 kg H_2 storage system
- 60 kW fuel cell system
- 480 kWh battery usable capacity.

The potable water requirement for the H2PS option 2 in site 2 is 138.89 Kl/year at an average of 15 l per kg H_2 (9 l of pure de-ionised water per kg H_2), which is comparably small percentage of the average water consumption in the Pilbara region, which is relatively a small percentage of the community water consumption in the Pilbara region.

Site 2 and options 1 and 2 models' functionality was simulated, and the yearly profile results plotted in Fig. 4.7a revealed in option 1 that the hydrogen tank is mostly full during the good session (summer). However, the 100% RE system (option 1) stored hydrogen level goes down to a minimum of 60% of its capacity

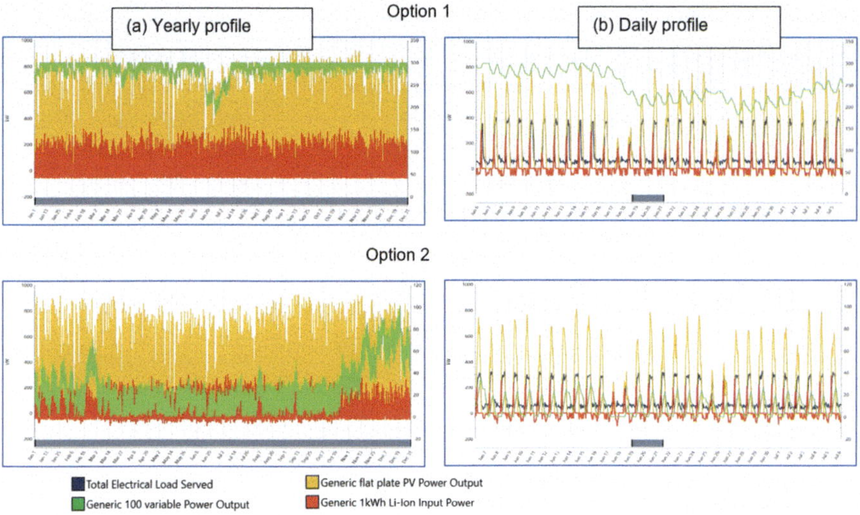

Fig. 4.7 Site 2 H2PS models for options 1 and 2 components' functioning: daily and yearly profiles

during the low solar irradiance season (winter) while falling to near zero in the option 2 scenario. In comparison, the option 2 functioning plot shows stable hydrogen autonomy as the diesel genset covers the load when hydrogen levels go too low. Figure 4.7b zooms on the predicted worst solar irradiance period in the year for both options 1 and 2 to demonstrate accurately how each system component is predicted to function. The hydrogen production and storage capacities can be increased to add more system autonomy hours, but that comes at a cost, i.e. CAPEX and OPEX of bigger solar PV, electrolyser, and the hydrogen tank.

4.4.3 Financial and Project Cost Risk

Some economic/cost values in the modelling listed in Table 4.1 were default values for key input parameters (such as the price of diesel, PV panel, fuel cell, hydrogen tank, etc.) and the cost of generating electricity at regional Pilbara. These may not reflect the actual values, especially for implementation in remote areas. It is also well known that implementing projects in remote regions and sites is inevitably more expensive than initially thought. There is a risk of financial and project cost (e.g. CAPEX for the system and site works) increases/changes when undertaking the project, which could affect the viability and financial benefits of the project. To quantify this risk's size, the project undertook a sensitivity analysis around key financial values and assumptions used in the modelling. The variations in key economic factors for the two projects are shown in Table 4.4, and COE sensitivity results are plotted in Fig. 4.8.

Table 4.4 Variation in the hybrid hydrogen system model techno-economic input parameters

System component	CAPEX/OPEX (AUD)	+/− % on sensitivity
Solar PV system	$ A U1,000 kW / $10 kW/year	−10%, −20%, +10%, +20%
Li-Ion battery bank	$ A U550 kWh / $5 kWh	−10%, −20% and − 30%
H2 electrolyser	$ AU 2500 kW / 10 kW/year	−50%, −20%, +20%
H2 FC	$ AU 3350 kW /$1 Op hr	−50%, −20%, +20%
H2 tank low-pressure	$ AU 1000 kg H2 / $ AU 5 kg H2	−20%, +20%
Fixed capital	$ AU 220,000[a] $ AU 781,000[b]	−20%, +20%, +50%, +100%
Yearly operational	$ AU 24,000[a] $ AU104,000[b]	−20%, +20%, +50%
Diesel price	$ AU 2.80 per litre	−20%, +20%, +50%

[a]Site 1 H2PS
[b]Site 2 H2PS

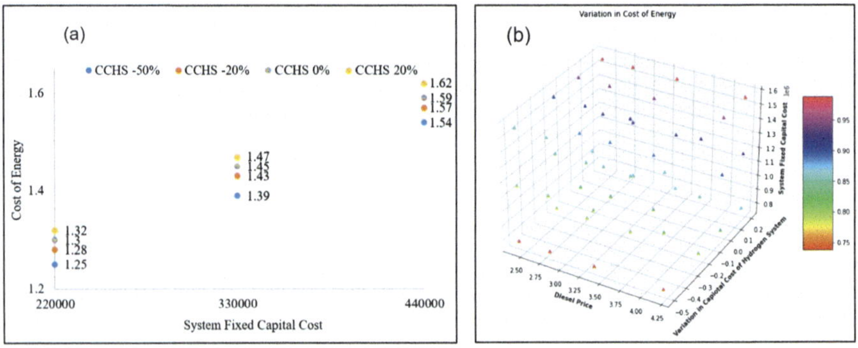

Fig. 4.8 Sensitivity analysis for (**a**) site 1 project, (**b**) site 2 Project

From the modelling, the costs that are most likely to lead to financial risk for the projects are increases in:

- Fixed capital costs: engineering design, tendering, approvals, and civil work.
- Capital costs of the system components (e.g. Solar PV system, H2 fuel cell, etc.)

For the optimal site 1 system design, even with a doubling (100% increase) of the system fixed capital cost and a 20% increase in capital cost of the hydrogen system, the COE for the optimal system design only increases by 24% as shown in Fig. 4.8a, which is still below the lower range of the diesel-only mini-grid price of $1.65/kWh.

Site 2 optimal system design, assuming the base case diesel price ($2.80/l) even with a doubling (100% increase) of the system fixed capital cost and 20% increase in capital cost of the hydrogen system, the COE for the optimal system design only (base case) increases by 19% as shown in Fig. 4.8b. This means an increase in COE from $0.827/kWh to 0.981/kWh, which is still well below the lower range of the diesel-only system price of $1.18/kWh. Even with a diesel price increase of 20%, i.e. $3.36/l, the COE only increases to $0.924/kWh, which is still below the diesel-only system, whose COE will also increase due to the diesel price increase.

4.5 Conclusions

This feasibility study successfully developed the optimised technical designs and financial analyses from modelling the two stand-alone hybrid hydrogen-based power systems (H2PS) for a small-scale homeland community and a medium-scale community in the Pilbara regional WA. Both sites were modelled as a 100% RE stand-alone hybrid H2PS for their actual load profile to become pilot net zero-emission projects and provide the basis for a larger capacity system. The proposed model comprises solar-PV, electrolyser, hybrid battery-hydrogen storage system, and a fuel cell. Several scenarios were considered in the modelling at each site and simulated to techno-economically evaluate the energy balance and cost of energy (COE) reduction using optimal component sizes. Simulation analyses revealed that a larger hydrogen system capacity could add more robustness and energy autonomy. Therefore, the recommended option 2 for site 1 is to use an optimal balance between the battery and hydrogen system capacities for reliability and sufficient autonomy. The options for site 2 were simulated for the optimal energy balance, autonomy, and COE. This option was compared with a new proposed scenario (option 2) of including one of the three existing diesel generators in the model as a backup power generator and autonomy assurance. Option 2 can reduce the COE further by 8.5% due to reducing the H2PS capacity at a penalty of 1.5% fossil fuel penetration; therefore, it is 1.5% emissions compared to the base case scenario.

About 120 remote homeland communities in WA are served by the current Remote Essential and Municipal Services (REMS) state-wide program. These communities are dependent on unreliable, costly, polluting diesel power stations. The present study has demonstrated a potential transition model to sustainable energy and improved outcomes for these communities.

Acknowledgement The feasibility study (Study) received grant funding from the Western Australian Government's Renewable Hydrogen Fund, administered by the Department of Jobs, Tourism, Science, and Innovation.

References

Al-Buraiki, A. S., & Al-Sharafi, A. (2022). Hydrogen production via using excess electric energy of an off-grid hybrid solar/wind system based on a novel performance indicator. *Energy Conversion and Management, 254*, 115270.

AS. (2023). *Standards Australia portal.* Available from: https://www.standards.org.au/

Australian Government. (2019). CEC, National Hydrogen Strategy, Clean Energy Council, Editor. Australian Government.

Australian Government. (2023). *Regulations and standards*. Available from: https://www.industry.gov.au/regulations-and-standards

CSIRO. (2021). *HyResource* [cited 2023; Projects Database]. Available from: https://research.csiro.au/hyresource/

Data, N. G.. (2019). *NASA surface meteorology and solar energy*. Available from: https://data.nasa.gov/Earth-Science/Prediction-Of-Worldwide-Energy-Resources-POWER-/wn3p-qsan/data

Dawood, F., Shafiullah, G. M., & Anda, M. (2019). *Power to gas energy storage system for energy self-sufficient smart cities development*. Springer International Publishing.

Dawood, F., Anda, M., & Shafiullah, G. M. (2020a). Hydrogen production for energy: An overview. *International Journal of Hydrogen Energy, 45*(7), 3847.

Dawood, F., Shafiullah, G. M., & Anda, M. (2020b). A hover view over Australia's Hydrogen Industry in recent history: The necessity for a Hydrogen Industry Knowledge-Sharing Platform. *International Journal of Hydrogen Energy, 45*(58), 32916–32939.

DISER. (2021). *National greenhouse accounts factors*. Australian Government, Department of Industry, Science, Energy and Resources.

DPIRD, Work Group. (2019). *Western Australian renewable hydrogen strategy*. Department of Primary Industries and Regional Development.

Energy. (2023). *H. HOMER Pro Version 3.14.5* [cited 2023 1 June]. Available from: https://www.homerenergy.com/products/pro/version-history.html.

Furat, D., Shafiullah, G. M., & Martin, A. (2020). Stand-alone microgrid with 100% renewable energy: A case study with hybrid solar PV-battery-hydrogen. *Sustainability, 12*(5), 2047.

GKN. (2022). *HY2 energy systems*. Available from: https://www.gknhydrogen.com/

Horizon Power. (2022). *Denham hydrogen demonstration plant*. Available from: https://www.horizonpower.com.au/your-community/getting-future-ready/denham-hydrogen-demonstration-plant/

HSE. (2023). *Health safety environment Australia*. Available from: https://hseaustralia.com.au/

IEA. (2021). *Global hydrogen review 2021*. IEA Publications.

IEA. (2023). *Credible pathways to 1.5 °C*. International Energy Agency.

IRENA. (2022). *Global hydrogen trade to meet the 1.5°C climate goal: Part I—Trade outlook for 2050 and way forward*. International Renewable Energy Agency.

Karunanithi, K., et al. (2023). Investigations of stand-alone PV system with battery-supercapacitor hybrid energy storage system for household applications. *International Journal of Information Technology, 15*(1), 279–287.

LAVO. (2023). LAVO website 2023. Available from: https://www.lavo.com.au/

Li, R., et al. (2023). Techno-economic analysis of a wind-photovoltaic-electrolysis-battery hybrid energy system for power and hydrogen generation. *Energy Conversion and Management, 281*, 116854.

Mohammad-Alikhani, A., et al. (2022). Multiobjective optimization of system configuration and component capacity in an AC minigrid hybrid power system. *IEEE Transactions on Industry Applications, 58*(3), 4158–4170.

Parkinson, G. (2017). *How Horizon Power plans to remove world's biggest fossil fuel subsidy*. Renew Economy. Available from: https://reneweconomy.com.au/how-horizon-power-plans-to-remove-worlds-biggest-fossil-fuel-subsidy-58724/

Riotact. (2023). *Australia's first public hydrogen refuelling station opens in Canberra*. Available from: https://the-riotact.com/australias-first-public-hydrogen-refuelling-station-opens-in-canberra/449845

Safe Work Australia. (2023). *Codes of practice*. Available from: https://www.safeworkaustralia.gov.au/law-and-regulation/codes-practice.

Sandri, O., et al. (2021). Hydrogen for all? Household energy vulnerability and the transition to hydrogen in Australia. *Energy Research & Social Science, 79*, 102179.

Sayigh, A. (2022). *Sustainable energy development and innovation: Selected papers from the World Renewable Energy Congress (WREC) 2020. Innovative renewable energy*. Springer International Publishing AG.

Standardisation. (2023). I.-I.O.f. ISO/TC 197 hydrogen technologies. Available from: https://www.iso.org/committee/54560.html

Standards Australia. (2023). *Committee ME-093, Hydrogen Technologies*. Available from: https://www.standards.org.au/engagement-events/strategic-initiatives/hydrogen

Vivek Srinivasan, B. D., Dollman, R., Towns, A., Charnock, S., Palfreyman, D., Hayward, J., Graham, P., Foster, J., Reedman, L., & Tourbier, D. (2023). *Renewable energy storage roadmap*. CSIRO.

WA Government. (2020). *Western Australian Renewable Hydrogen Roadmap*. WA Government, Department of Jobs, Tourism, Science and Innovation (JTSI).

WA Government. (2021). *Western Australian Renewable Hydrogen Strategy*. WA Government, Department of Jobs, Tourism, Science and Innovation (JTSI).

WA Government. (2022). *Ambitious interim target set for State Government emissions*. Western Australia Government WA, Australia.

WA Government. (2023). *Remote essential and municipal services*. Available from: https://www.wa.gov.au/organisation/department-of-communities/remote-essential-and-municipal-services

Yap, J., & McLellan, B. (2023). A historical analysis of hydrogen economy research, development, and expectations, 1972 to 2020. *Environments, 10*. https://doi.org/10.3390/environments10010011

Chapter 5
Amazing Photovoltaics: From Research Curiosity to Technology Reality

Lawrence L. Kazmerski

5.1 Introduction and Background

Photovoltaic technology has emerged to be a major part of electrical power markets worldwide (REN21 2023a, b; IEA, 2023; Mints, 2023). Its recent growth has been amazing—with both technological improvements and dramatic price declines. PV is meeting its long-anticipated status for economic competitiveness, reliability, environmental compatibility, and application versatility. In early 2022, the world cumulative installations surpassed 1-TWp. In 2023 (PV Magazine, 2022), the PV industry is expected to ship one-third terawatt or more product into the world markets. Prices for PV modules are hovering near \$0.40 USD currently in most markets (Mints, 2023). Worldwide, PV accounted for more than 70% of the electric power added to our grids the past 2 years. Factories have production capacities exceeding 10 MW/year. New country markets continue to emerge, including Brazil ranking *fourth* and Poland *eighth* in installed PV in 2022—the first time for both in the "top ten" listings (REN21, 2023a, b). PV is a real part of emerging and developed countries' energy roadmaps—and is being counted on as a major player toward climate change mitigation. And PV is not stagnant, knowing there are still limitations and barriers to overcome and opportunities to address. Research investments on materials and devices are at an all-time high—with time gaps from laboratory discovery to commercialization shrinking from several decades to less than a few years. PV module reliabilities are entering the markets backed by 30-year warranties—and those modules have efficiencies converting 20% or more of the sun's power into useful electricity (NREL, 2023a).

Amazing—but much remains to be accomplished. This paper examines where we are with photovoltaics—technology, applications, performance, cost, and

L. L. Kazmerski (✉)
Renewable and Sustainable Energy Institute (RASEI), University of Colorado Boulder, Boulder, CO, USA

A. Sayigh (ed.), *Transition Towards a Carbon Free Future*, Innovative Renewable Energy, https://doi.org/10.1007/978-3-031-61660-0_5

production. The present and future build on the past and the origins of modern photovoltaics are highlighted, with the evolution, then revolution, over the past seven decades. Finally, some expectations on where PV is going and what the future holds for the technologies. And what can the coming generations of the PV technology community and consumers anticipate. Much of this is a brief reflection of the author's own experiences and observations living and being part of these years of amazing technology realization.

5.2 Modern PV Beginnings

Certainly, the history of the photovoltaic effect and suitable materials goes back centuries (PVResources, 2023). With the report by Charles Fritts of the first solid-state device in 1883, PV lingered as a research curiosity with scientists and research engineers for the next half century. The modern tipping point for this technology was somewhat modest, starting with the announcement of the first viable research solar cell by Bell Telephone Laboratories in 1954 having 6% solar conversion efficiency, ~2 cm^2 in area, producing about 10 mW of power (Kazmerski, 2017). The team leaders were Gerald Pearson (the physicist), Daryl Chapin (the engineer), and Calvin Fuller (the chemist). Figure 5.1 shows the Bell Labs notebook cross-sectional graphic of the cell and the photos of the solar cell itself. However, the New York Times published the vision of what this small device it would eventually become for the humanity:

> New York Times, April 16, 1954: MURRAY HILL, N.J.—A solar battery, the first of its kind, which converts useful amounts of the sun's radiation directly and efficiently into electricity, has been constructed here by Bell Telephone Laboratories ... It may mark the beginning of a new era, leading eventually to the realization of one of mankind's most cherished dreams—the harnessing of the almost limitless energy of the sun for the uses of civilization.

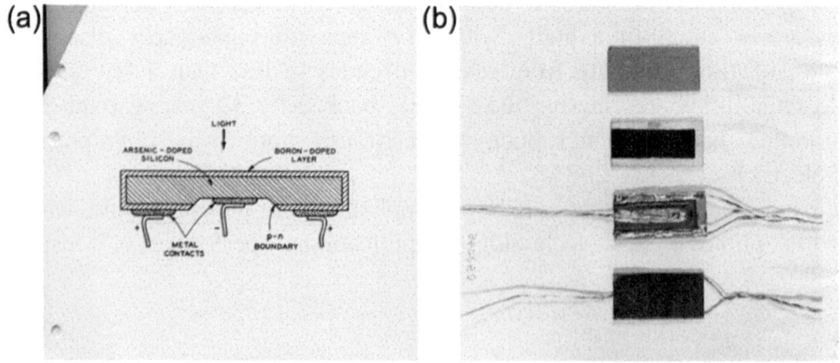

Fig. 5.1 Bell Telephone Laboratories Si solar cell: (**a**) cross section of device from Daryl Chapin's laboratory notebook (1953); (**b**) photo of actual device

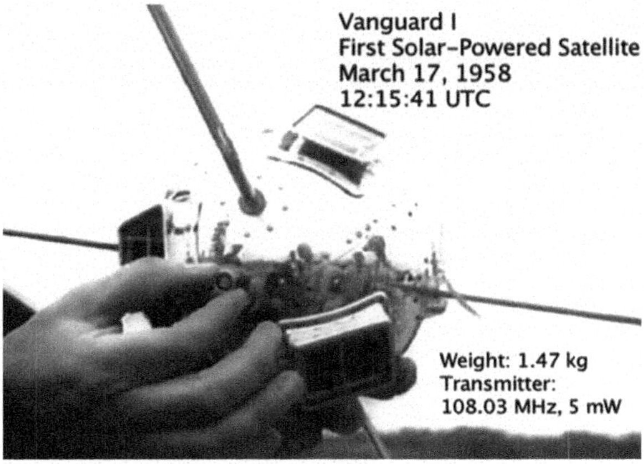

Fig. 5.2 Vanguard I, the first PV-powered satellite launched on March 17, 1958

The application that Bell Labs had for this new development was to power telephone repeaters in the southeastern USA. The batteries used at the time were costly and short-lived, and the PV alternative proved to be a technologically good alternative, but not one that was yet cost competitive. PV was not yet ready for earth-bound markets. But an opportunity presented itself with the interest in space with the plans for the launching of an artificial earth satellite as part of an international effort led by the USA in the International Geophysical Year (Nicolet, 1983). The satellite was Vanguard—but the Soviet Union shocked the world with the *Sputnik*—the first to successfully orbit the earth. Vanguard finally made it to space—after two failures—but had an important distinction. Vanguard I became the first solar-powered satellite (Fig. 5.2). The solar cells for this satellite were developed and manufactured by Hoffman Electronics, which had fortunately decided to invest in the Bell development and coaxed away a key, young scientist, Dr. Morton Prince (ISES, 2020) from the Bell Lab solar team, to lead the effort. The Hoffman Electronics team was a "who's who" of PV pioneers—Prince, Martin Wolf, Eugene Ralph, Peter Iles, Bernd Ross, Maurice Paradise, and later Don Bickler and Hans Rauschenbach. Their six small "modules" on Vanguard each provided enough power for the 50 mW transmitter that emitted it "beep-beep-beep," initially by the "trustworthy" batteries that the project management team mandated as the prime source, then by the solar cells which took over after about 2 weeks when the batteries failed. That management team did not have confidence in this new technology, and the solar cells continued to power that signal, tying up the needed frequency for the next ~7 years because they provided no mechanism to switch off the power. Finally Vanguard intercepted the Van Allen radiation belt (which was not discovered yet at the time of launch) and caused the electronics to fail. This was the first demonstration of the *reliability* of PV and heralded the opportunity for the first real markets for PV—*space power.*

The down-to-earth break for PV came in the early 1970s, when the fossil fuel-hungry world was hit by the Arab oil embargo (US Department of State, 2023). In the resulting energy shortage storm, "alternative energy" sources were sought to provide some relief from the overwhelming fossil dependence. Many countries

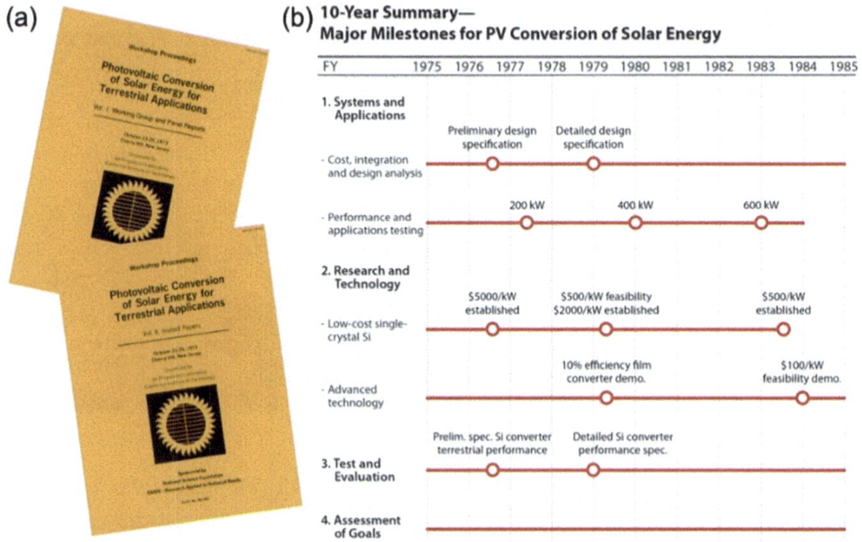

Fig. 5.3 Cherry Hill Workshop that established the US terrestrial PV program in 1973. (**a**) Workshop proceedings; (**b**) implementation plans and expectations

started to consider wind, hydro, biofuels, thermal, wave, tidal, and solar energy resources. The USA, for example, brought pioneered an invited gathering of 134 scientists, engineers, and decision-makers in Cherry Hill, New Jersey, in October 1973 to formulate a US solar PV plan and program (Cherry Hill Workshop, 1972). The *Photovoltaic Conversion of Solar Energy for Terrestrial Applications* workshop (Fig. 5.3) produced a plan for a US terrestrial PV program, highlighted by objectives for research and systems development through the year 2000—as well as some specific design plans for predicted early applications. The goal (though not defined as cost or price) was 0.50 USD/watt (though not defined for a solar cell or a module). A table showing the objectives is shown in Fig. 5.8, with depictions of some of the applications. This was the financial level proposed for PV to be competitive with existing utility-scale power generation sources. As this paper will cover later, these expectations have now been realized and exceeded—though some taking a longer time span than hoped for.

The 1980s through early 2000s marked the time for research, development, and moderate proof-of-concept deployments. PV was marginally in a commercial phase—but still too pricey except for some specialized markets. But the materials and device landscape, scientific understanding, and manufacturing understanding were in a stage of expansion. The USA led with early research and development contributions and technology improvements; Japan showed how to manufacture and established the basis for profitability; Germany understood the value and implemented policy and consumer incentives that attracted and educated the public. But it was finally China that showed how to scale up production and rapidly drive down prices in the period 2011 to present. This was the real "vanguard" for PV expansion into the world energy framework.

5.3 PV Technology Progress and Status

The realization of PV as a reliable electricity source progresses through evolutionary, disruptive, and revolutionary phases from the 1980s through the present time. The development had barriers that ranged from technical through political and competing well-funded non-renewable energy businesses. But eventually even these could not overcome the true viability and need for this clean power source.

5.3.1 Worldwide Products, Markets, Sources, and Users

Perhaps the most compelling illustration of how far PV has advanced is to first look at the prices for the PV module—from near the start of the research on terrestrial PV to now—shown in Fig. 5.4. In 1977, a modest power module (30–75 W) costs about 77 USD/W (in 1977 dollars or ~350 USD/W in 2022 USD). As shown, in 2022 a 700 W module can be purchased for around 0.40 USD/W! Amazing! Currently, module prices range from about 0.36 to 0.44 USD/W worldwide (Mints, 2023). The year 2011 marked another tipping point for PV—that is when China took over and prices were from about 4USD/W to current levels.

One benchmark to evaluate growth is through what the manufacturers have *shipped* each year into the world markets. Figure 5.5 shows recent data and

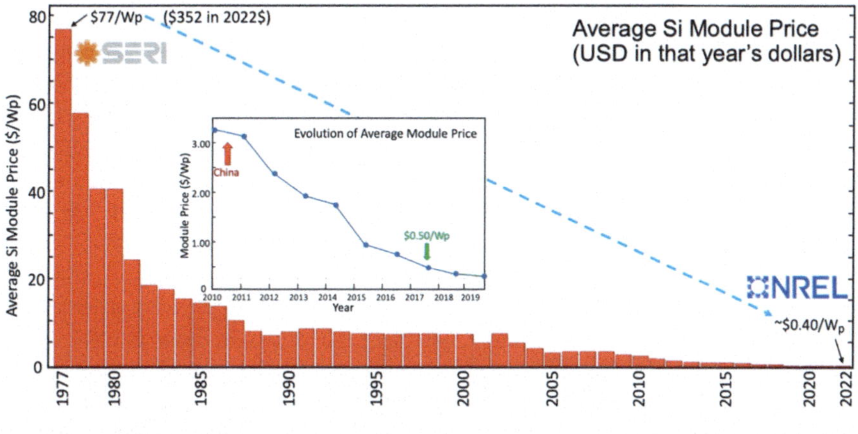

Fig. 5.4 PV module price evolution from 1977 to 2022. The tipping point that brought PV prices to competitiveness for consumers is in the timeframe 2010–2011 when China entered commercialization and expansion of the PV technology

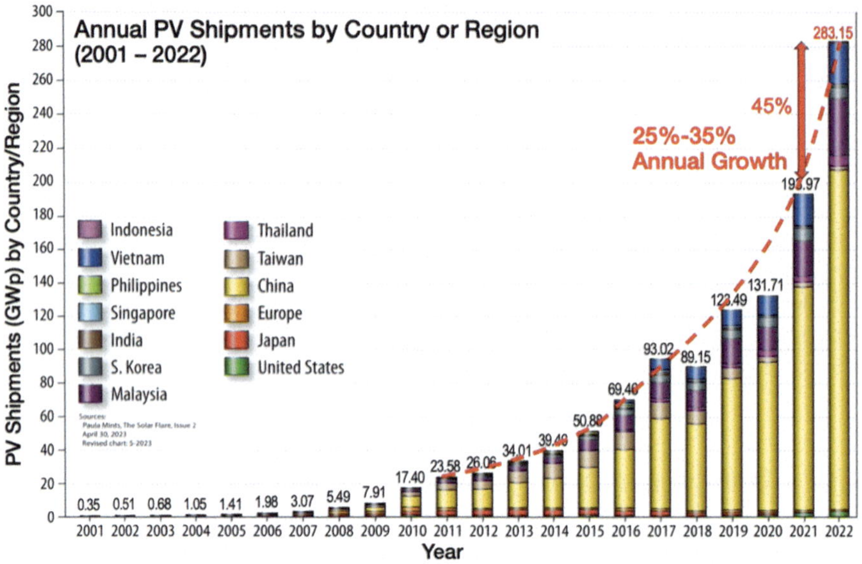

Fig. 5.5 Annual PV module shipments into world markets by country or region for the period 2001 to 2022 (after Mints, 2023)

information provided by Paula Mints (Mints, 2023), showing the shipments over the period 2001 through 2022 originating from manufacturing facilities in various world geographical regions. Actually, since the 1990s, these shipments have experienced fairly consistent and positive growths in the 20–35% range (though 1 year was hit by some interruptions in policy incentive in some countries and use of existing inventories). This past year showed an incredible 45% growth over 2021. Amazing.

But who is supplying this PV? The answer is shown in Fig. 5.6—China, accounting for 71% of the shipments, and Asia for >97% (Mints, 2023). This is impressive but may herald a sign of caution. There are analyses that warn of significant increases in prices if the world continues to depend on mainly a *single source* for its product. This single sourcing can be hit by geopolitical and resource issues very rapidly in our tumultuous world situations. There is now a major effort to diversify the manufacturing—especially noted by programs initiated in the USA, India, Europe, and Brazil (Mints, 2023; Shiradkar et al., 2022). (Note: 2023 shipments exceeded 500 GW, surpassing expectations).

And who is using all these photovoltaics? Again, as shown in Fig. 5.7, China is #1. Although the USA and Japan rank 2 and 3, India, with its aggressive programs, is closing in. It is interesting that in 2022, emerging markets have come into focus. As already noted, Brazil ranks *fourth* and Poland *eighth* in that year. The total cumulative installations for the top 10 countries are highlighted in Fig. 5.3. It is the performance, but REN21 has put the number at 1185 GW installed (currently estimates are 1100 and 1250 GWp, depending on the particular report) (REN21, 2023a, b). PV topped the 1TWp installed in 2022—and is likely to reach *1 TWp/year* shipments into the world markets by 2030.

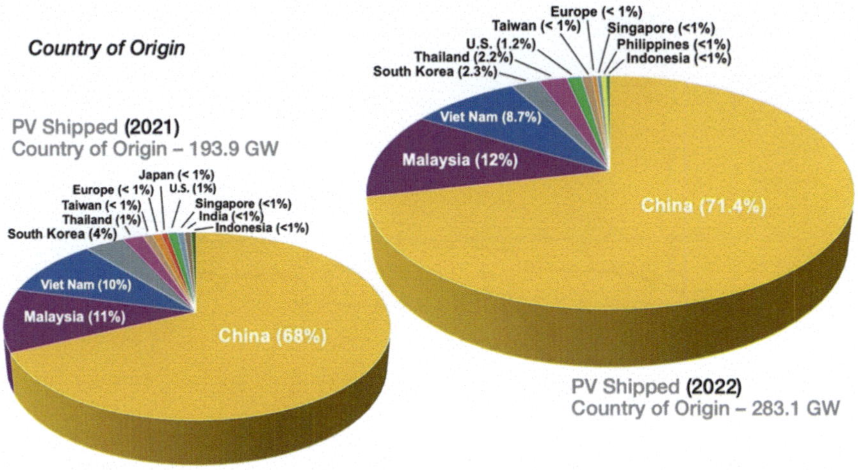

Fig. 5.6 Sources of PV shipments in 2022 by country

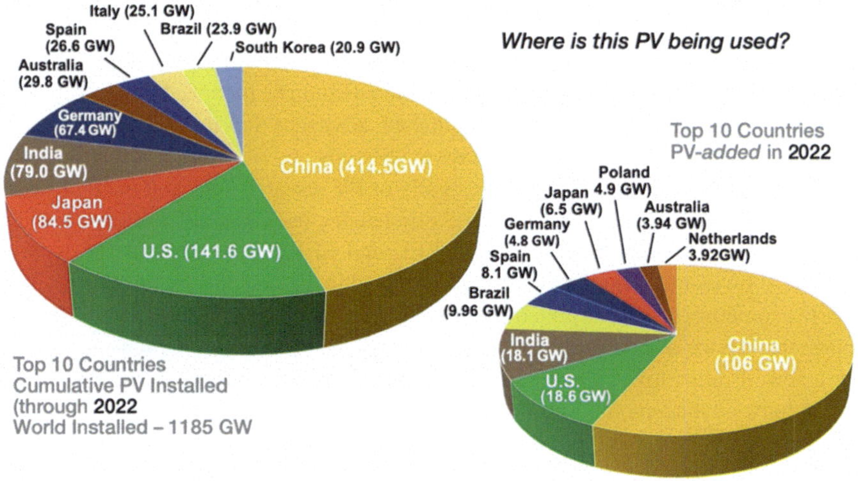

Fig. 5.7 Top 10 world users for PV: (**a**) cumulative installations through 2022; (**b**) 2022 installations (after REN21, 2023a, b). Cumulative installations topped 1TW in February 2022 and 1.5TW in 2023. China accounted for more than 600GW in 2023, adding 235 GW in that single year

5.3.2 *Devices, Material, and Technologies*

Commercial PV is a silicon stronghold, accounting for >94% of shipments in 2022. Thin-film CdTe represented about 5%, thin-film CIGS about 1%, and a Si:H-based <1% of the total (Mints, 2023). These are still respective numbers in "GW" shipped but represent a small market share. The new, higher-efficiency (<19%) and higher power (>400 W) thin-film CdTe modules certainly have gained some respect—

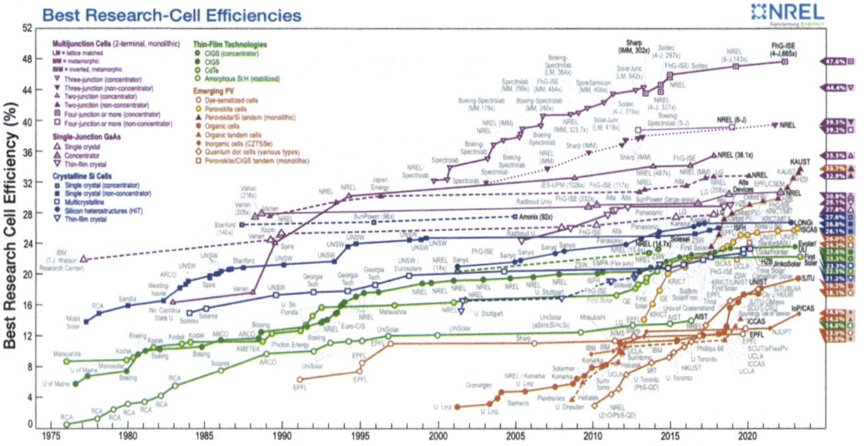

Fig. 5.8 Best in class and confirmed PV research cell efficiencies by year (NREL, 2023b)

especially for better performances (kWh/kW) over the course of a day and in higher-temperature climate zones. The CdTe technology also claims better environmental factors—as well as effective existing program for recycling. The thin-film technologies also hold potential manufacturing advantages for speed, flexibility, and material utilization. Despite these significant advances for the thin film sector, Si-based PV continues to make impressive advances.

Figure 5.8 represents the incredible family of PV research cells and materials that have been recorded with "best-in-class" efficiencies over the half century (NREL, 2023b). This chart has been kept by NREL and SERI since the mid-1980s, and Fig. 5.9 shows one of the first charts documented in 1985–1986 by the author. The chart is sometimes misinterpreted as only "evolutionary" in performance. In fact, it represents device technology—and in many cases the progress from point-to-point is for a new or innovative cell structure. Disruptive, but not evolutionary or Edisonian technology advancement. Similar progress is seen for the module technologies, represented by the "champion device" chart in Fig. 5.10.

5.3.2.1 The World of Silicon PV

For Si device configurations, some 59% of the devices produced in 2022 were higher-efficiency *passivated-emitter and rear-contact* (PERC) monocrystalline solar cells (Mints, 2023; Blakers, 2019; Min et al., 2019; Tous et al., 2018). The choice of materials and the PERC device (Fig. 5.11) reflect the importance of performance (efficiency) in continuing to lower the cost of the module technologies. The PERC device structure is shown in Fig. 5.12a. Though still the greatest portion of the market, there is a trend toward the tunnel-oxide passivated contact or TOPCon PV cell (Fig. 5.12c) (Gifford, 2020; Feldmann et al., 2014; Messmer et al., 2019).

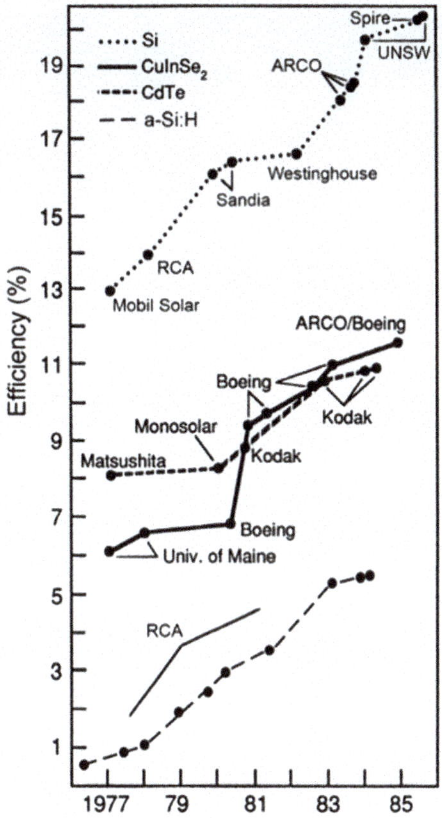

Fig. 5.9 One of first versions of SERI chart (1985–86) by the author

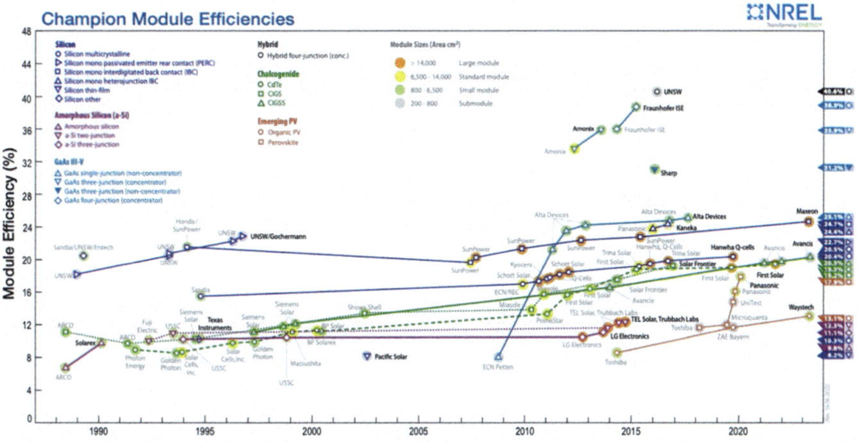

Fig. 5.10 NREL champion module chart (NREL, 2023a)

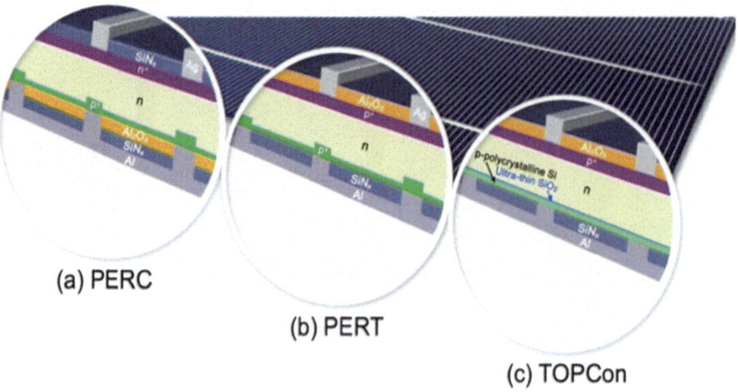

Fig. 5.11 Si device cross sections: (**a**) passivated emitter rear contact (PERC), (**b**) passivated emitter rear totally diffused (PERT), and (**c**) tunnel oxide passivated contact (TOPCon solar cells)

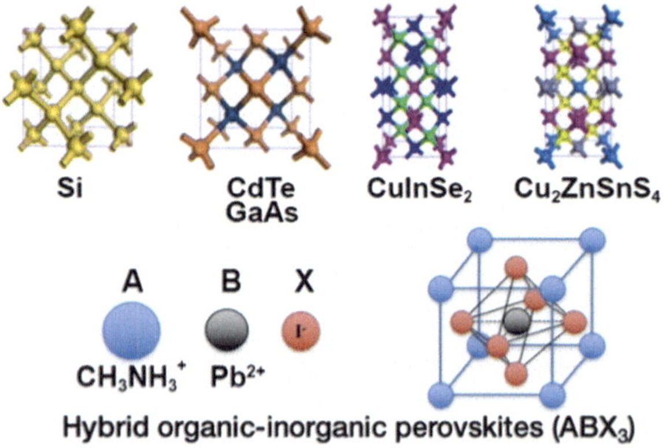

Fig. 5.12 Materials "playing field" from elemental and binary through ternary, multinary, and complex hybrid binary inorganic-organic perovskites

The cell structure shown in Fig. 5.10b is similar to the PERC, but with the tunnel oxide providing better performance levels. This cell type captured about 12% of the markets in 2022—and continues to expand in production. This cell has attracted the attention and adoption by the industry because (1) it can be manufactured in current PERC lines and (2) it provides a gain of about 1% absolute in efficiency. The remainder of the Si market is filled by monocrystalline interdigitated back contact cells (Swanson, 1990) (1% share) and Si heterojunction (6% share) or "HIT" cells (Taguchi et al., 2013). These are both high efficiency options (Fig. 5.8) that first penetrated the 20% and 25% barriers for Si commercial cells early on and provided the pathway toward the world of super performance. Progress for modules has complemented the cell advancements, with bifacial modules (collecting energy from both front and back sides of the module) becoming the product of choice for many manufacturers and power plants (especially in single-axis tracking installations).

5.3.2.2 Thin-Film PV

Thin films have always been considered a serious contender through the early 2000s. The commercial thin-film landscape was dominated in the 1980s and into the 1990s by amorphous Si:hydrogen (a-Si:H) technologies (Fig. 5.8) (Dalal, 1983; Kabir et al., 2011). The prominent technology was the triple-junction structure, with the bottom (low-bandgap) cell a-SiGe:H or microcrystalline-Si (μc-Si). Though these cells reached stabilized 13% efficiencies, they dropped out of major commercial markets due to the rise of crystalline Si and lingering uncertainties about their stability under sunlight. Two other contenders entered the commercial scene in the 1990s and early 2000s: $Cu(In,Ga)(S,Se)_2$ or CIGSSe (Ramanujam & Singh, 2017; Wang et al., 2019; Mohan & Paulose, 2019) and CdTe (Lee & Ebong, 2017; Sinha et al., 2019), with the latter technology far more successful as a commercial product. Both of these heterojunction solar cells demonstrated high efficiencies, reaching 23.6% (Evolar and Uppsala University) (NREL, 2023b) and 22.3% (First Solar) (NREL, 2023b), respectively, in 2023. The most common device architecture is the formation of the heterostructure with CdS (or alloy). The perceived concern with "Cd" as a toxic material provided a path for other junction partners for CIGSSe (e.g., ZnSe, InSe, InP. . .). Modules for both technologies have exceeded 19% efficiency as shown on the NREL module chart (Fig. 5.10) (NREL, 2023a). The thin-film PV technology is still dominated by one manufacturer, First Solar, which has expanded its capacities in the USA, India, as well as Southeast Asia. They have brought innovative design to their cells and modules—enabling them to boast better performances under high temperature conditions, better kWh/kW, lowest warrantied degradation rates, and proven environmental manufacturing and a PV module with among the lowest carbon footprints and fastest energy payback times of the entire menu of large-scale PV products.

5.3.2.3 The Upstart-Searching for the Holy Grail

Hybrid-perovskites: PV history has shown that evolution, innovation, and creativity have dominated technology development toward the next generations of solar conversion for the next generations of electricity consumers. Certainly, an examination of Fig. 5.8 overviews many advances, which can be complemented by the increasing complexity of the material investigation for PV over time (Fig. 5.12). The PV research and commercialization communities recently have been consumed with the rapid introduction and development of the hybrid (organic-inorganic) perovskites (Fig. 5.12). Although there was some research on these materials in the 1990s and early 2000s, the precipitous progress from 10% to the >25% regime for laboratory devices in ~6–7 years (Figs. 5.8 and 5.13) has spiked now interest of R&D (and funding) in almost every PV laboratory in the world (KAUST, 2023; Wang, 2019; Davis & Yu, 2020; Oxford PV, 2020). Now closing in on commercialization, the evolution is taking this material into two areas: (1) single-junction

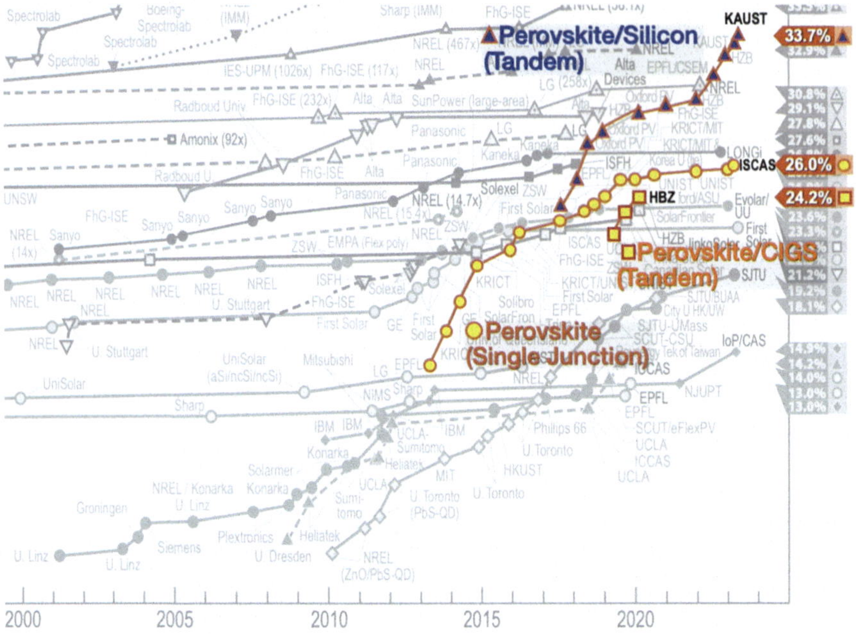

Fig. 5.13 Rapid advancement of efficiencies of research hybrid-perovskite solar cells shown in enlarged portion of Fig. 5.8 for single junction, tandems with CIGS, and tandems with Si

perovskites and (2) tandem or multijunction cells, primarily with Si but also considered recently with other lower bandgap partners (e.g., CIGS) and all-perovskite layered multijunctions. The Si tandem, in particular, is probably the closest to being a viable commercial product (Oxford PV, 2020)—and has demonstrated not only the highest efficiencies but improved stabilities.

Generic hybrid-perovskite solar cell cross sections are presented in Fig. 5.14. The hole transport layer performs two functions—to decrease charge recombination and to prevent direct contact between the perovskite and the metal contact. Though no longer a complex structure for a solar cell, the device does pose several regions in which "shades of stability" have to be considered: (1) contact materials that can be a source of doping; (2) interfaces that have defects, chemical interactions, and band adjustments; (3) perovskite material (absorber) that can phase segregate, exhibit ion migration, and decompose; and finally, (4) encapsulation (module) to protect from the environment, especially water vapor and oxygen (and sometimes, sunlight) which adversely affect the perovskite material (Mints, 2023; Davis & Yu, 2020). A major focus is on the hybrid-perovskite itself, since relatively little is known about the extent and nature of the decomposition and related effects that cause possible material instability. Each of these areas is critical for this new technology and is a massive topic of ongoing research—spanning experimental through theoretical evaluations. [Note: Longi has had perovskite/Si tandems at 34.6% in 2024].

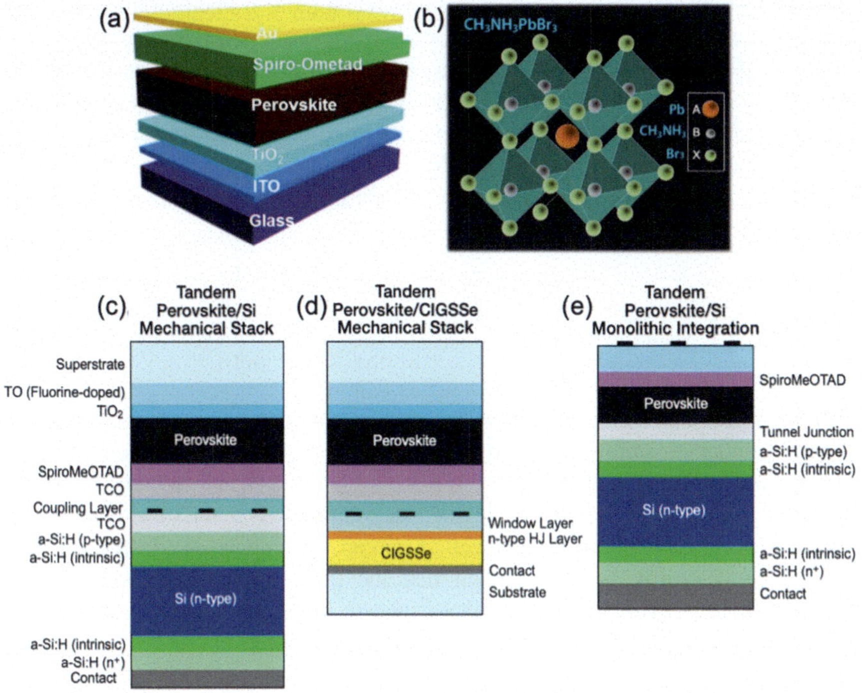

Fig. 5.14 Cross sections of various perovskite solar cell configurations: (**a**) generic cell; (**b**) ABX3 hybrid perovskite structure; (**c**) mechanical stack tandem on Si; (**d**) mechanical stack tandem on CIGSSe; (**e**) monolithic tandem

Organic photovoltaics (OPV): Though there was early research on organic approaches, this technology was really introduced at the turn of the twenty-first century in the search for more inexpensive approaches. Some successful applications for electronics unpinned the research for PV. OPV did exhibit good efficiencies through the 2010–2012 timeframe, but the stability was always the inhibitor (Rwenyagila, 2017; Cui et al., 2020). Commercial ventures were started, but most closed because of lack of funding and interest from venture capitalists. This "plastic" solar cell offers flexibility in structure and ability to manufacture continuously by roll-to-roll processing, is perhaps the least expensive materials, and provides benefits for other markets, such as architectural design (Sunew, 2023; Wijewardane & Kazmerski, 2023). However, with the rise of the Si cell, the sources for support lost interest because of issues with being stuck at lower efficiencies at the time—and those issues with lifetime. But interest has been "reborn" by the recent jump in performance from the 10% to beyond the 19% level (NREL, 2023b; Zhu et al., 2022) (Fig. 5.9). This may be the example that we do not have sufficient patience with technology development in our PV sector—and the worry is that we might prematurely throw out potential "killer technologies." The message is to keep tuned and have an open mind. What the future might hold is not what we have today.

Innovation is moving up for a future passing rectangular framed modules that are slapped on our rooftops or in our fields, toward a blend of energy-producing PV that is integrated into our living environment, architecturally beautiful and more than meeting our electricity requirements with exceptional performances. It should also be pointed out that many of the OPV experts have provided foundational understanding and experience underlying the rapid related hybrid-perovskite solar device—and the recent advances in performance may have benefited from knowledge gain on that hybrid technology.

5.4 The Future

The PV research and commercial have been consistent in their innovation, creativity, thinking—and their ability to surprise even the most conservative doubters. This spirit of advancement will continue. We are well on the path toward the 100% renewable future, but PV has to go further. The good news is that PV, whether on the system, module, cell, or material level, is driven by innovation, dedication, and competence. Recently, it was thought that the pandemic would severely slow progress. Many political sectors said these solar technologies were not needed. But the importance of this electricity source and the dedication of the PV community have kept the momentum. PV is too big, too advanced, too important to not succeed.

Can a "100% renewable electricity" future become a reality? The answer is yes, but most of the renewable energy technologies have current limiting situation—in that they are not "24-hour" producers. There is a TWhr challenge that requires more innovation, investment, and technology commitment. Already artificial intelligence (AI) (or machine learning (ML)) permeates the entire value chain from materials discovery through system operation and reliability—and will continue to stimulate solar PV progress and expansion. The key is the development of adequate storage technologies and efficient appliances that can couple with PV to supply a constant, reliable electricity system. Renewable electricity will rule the future world's clean energy sector by mid-century. PV will be the major segment of any sizeable renewable electricity future. Innovation will take "evolution" to successful "revolution"—and our assured 100% renewable energy future.

For decades now, Dr. Ali Sayigh (WREN Director General) has devoted his life and organization to this clean energy future. So many times, he has entreated us about our renewable energy future: "We have to do it ,. . and we will." Amazingly, he has been steadfastly correct in his vision.

Acknowledgements The author acknowledges his association with the US National Renewable Energy Laboratory since 1977 for the knowledge and opportunities the laboratory provided to his career development. The author also thanks the Fulbright Foundation for furthering his understanding and experience with real technology deployments over the past year. And the author would especially like to thank Dr. Ali Sayigh of WREN for being a model for unflinching support of and dedication to renewable energy for all of us.

References

Blakers, A. (2019). Development of the PERC solar cell. *IEEE Journal of Photovoltaics, 9*, 629–635. https://doi.org/10.1109/JPHOTOV.2019.2899460

Breyer, C., Bogdanov, D., Khalili, S., & Keiner, D. (2021). *Solar photovoltaics in 100% renewable energy system. Encyclopedia of sustainability science and technology*. Springer Science+Business Media, LLC. https://doi.org/10.1007/978-1-4939-2493-6_1071-1

Cherry Hill Workshop. (1972). *Workshop proceedings: Photovoltaic conversion of solar energy for terrestrial applications, Vol 1 and 2* (National Science Foundation, Research Applied to National Needs, AG-485), Cherry Hill, New Jersey, October 23–25, 1973. https://ntrl.ntis.gov/NTRL/dashboard/searchResults/titleDetail/PB242529.xhtml

Cui, Y., Yao, H., Hong, L., Zhang, T., Tag, Y., Lin, B., Xian, K., Gao, B., An, C., & Bi, P. (2020). Organic photovoltaic cell with 17% efficiency and superior processability. *National Science Review, 7*, 1239–1246. https://doi.org/10.1093/nsr/nwz200

Dalal, V. (1983). *Tandem junction amorphous semiconductor photovoltaic cell*. U.S. Patent 4,387,265 (1983); Also, Dalal, V. (1981). US patent 4,253,882. Multiple gap photovoltaic device. https://uspto.report/patent/grant/4253882

Davis, M., & Yu, A. (2020). A review of flexible halide perovskite solar cells towards scalable manufacturing and environmental sustainability. *Journal of Semiconductors, 41*, 041603. https://doi.org/10.1088/1674-4926/41/4/041603

Feldmann, F., Bivour, M., Reichel, C., Steinkemper, H., Hermle, M., & Glunz, S. W. (2014). Tunnel oxide passivated contacts as an alternative to partial rear contacts. *Solar Energy Materials & Solar Cells, 131*, 46–50. https://doi.org/10.1016/j.solmat.2014.06.015

Gifford, J. (2020, March 7). TOPCon n-type solar cell technology could be a rival to mono PERC. *PV Magazine*. https://pv-magazine-usa.com/2020/03/07/topcon-n-type-solar-cell-technology-could-be-a-rival-to-mono-perc/

IEA. (2023). *World energy outlook 2022*. International Energy Agency, Paris, France October 2022. https://iea.blob.core.windows.net/assets/830fe099-5530-48f2-a7c1-11f35d510983/WorldEnergyOutlook2022.pdf

ISES SWC50. (2020). *International solar energy society on-line archive: The century of solar, stories and visions, compiled by G. Stapleton, Morton Prince and Eugene Ralph biographies*. https://www.swc50.org/pageflip/files/basic-html/page44.html

Kabir, M. I., Ibarahim, Z., Sopian, K., & Amin, N. (2011). A review on progress of amorphous and microcrystalline silicon thin-film solar cells. *Recent Patents on Electrical Engineering, 4*, 50–62. https://doi.org/10.2174/1874476111104010050

KAUST. (2023, April). *KAUST team sets new world record for tandem solar cell efficiency*. https://www.kaust.edu.sa/en/news/kaust-team-sets-world-record-for-tandem-solar-cell-efficiency#:~:text=Researchers%20in%20the%20KAUST%20Photovoltaics%20Laboratory%20(KPV%2DLab)%20of,(HZB)%20record%20at%2032.5%25

Kazmerski, L. L. (2017). The bell telephone laboratories discovery: Ushering in our modern age of solar photovoltaics. In A. A. A. S. Sayigh (Ed.), *Photovoltaics for sustainable electricity and buildings* (pp. 15–23). Springer. https://doi.org/10.1007/978-3-319-39280-6_2

Lee, T. D., & Ebong, A. U. (2017). A review of thin film solar cell technologies and challenges. *Renewable and Sustainable Energy Reviews, 70*, 1286–1297. https://doi.org/10.1016/j.rser.2016.12.028

Messmer, C., Fell, A., Felmann, F., Wöhrle, N., Schön, J., & Hermie, M. (2019). Efficiency roadmap for evolutionary upgrades of PERC solar cells by TOPCon: Impact of parasitic absorption. *IEEE Journal of Photovoltaics, 10*, 335–342. https://doi.org/10.1109/jphotov.2019.2957642

Min, B., Muller, M., Wagner, H., Fischer, G., Brendel, R., Altermatt, P. P., & Neuhaus, H. (2019). A roadmap toward 24% efficient PERC solar cells in industrial mass production. *IEEE Journal of Photovoltaics, 7*, 1541–1550. https://doi.org/10.1109/JPHOTOV.2017.2749007

Mints, P. (2023). PV market report the solar flare, Issue 2, SF-22023, SPV Market Research, April 30, 2023; See also, SPV Market Research, Photovoltaics Manufacturer Capacity, Shipments, & Revenues 2022/2023, SPV-Suppl 11, April 2023. https://www.spvmarketresearch.com/services.html

Mohan, R., & Paulose, R. (2019). Brief review on copper indium gallium diselenide (CIGS) solar cells. In X.-Y. Wang (Ed.), *Photoenergy and thin film materials*. Wiley Online Library. https://doi.org/10.1002/9781119580546.ch4

Nicolet, M. (1983). Geophysical Year (1957–1958): Great achievements and minor obstacles. *GeoJournal, 8*, 303–320. https://www.jstor.org/stable/41143298

NREL. (2023a, July). NREL champion module efficiency chart, courtesy of the National Renewable Energy Laboratory. https://www.nrel.gov/pv/module-efficiency.html

NREL. (2023b, July). NREL research cell efficiency chart, courtesy of the National Renewable Energy Laboratory. https://www.nrel.gov/pv/cell-efficiency.html

Oxford PV. (2020). *The dance to deliver the world's most efficient solar panel*. https://commercialsolarguy.com/2020/04/06/perovskite-silicon-tandem-solar-cells/

PV Magazine. (2022, March 15). *World has installed 1TW of solar capacity*. https://www.pv-magazine.com/2022/03/15/humans-have-installed-1-terawatt-of-solar-capacity/

PVResources. (2023). *Photovoltaics: Historical development*. https://www.pvresources.com/en/introduction/history.php

Ramanujam, J., & Singh, U. P. (2017). Copper-indium-selenide based solar cells—Review. *Energy and Environmental Science, 10*, 1306. https://doi.org/10.1039/c7ee00826k

REN21. (2023a). Renewables 2023 global status report: Demand module report, REN21 Secretariat, UN Environment Programme. https://www.ren21.net/wp-content/uploads/2019/05/GSR2023_Demand_Modules.pdf

REN21. (2023b). Renewables 2023 global status report: Supply module report, REN21 Secretariat, UN Environment Programme. https://www.ren21.net/wp-content/uploads/2019/05/GSR-2023_Energy-Supply-Module.pdf

Renne, D. (2022). Progress, opportunities, and challenges for achieving net-zero emissions and 100% renewables. *Solar Compass, 1*, 100007. https://doi.org/10.1016/j.solcom.2022.100007

Rwenyagila, E. R. (2017). A review of organic photovoltaic energy source and its technological designs. *International Journal of Photoenergy, 2017*, 2656512. https://doi.org/10.1155/2017/1656512

Shiradkar, N., Arya, R., Chaubal, A., et al. (2022). Recent developments in solar manufacturing in India. *Solar Compass, 1*, 100009. https://doi.org/10.1016/j.solcom.2022.100009

Sinha, T., Lilhare, D., & Khare, A. (2019). A review on the improvement in performance of CdTe/CdS thin-film solar cells through optimization of structural parameters. *Journal of Materials Science, 54*, 12189–12205. https://doi.org/10.1007/s10853-019-03651-0

Sunew. (2023). *Greenest energy everywhere* (Accessed 2021). https://sunew.com.br

Swanson, R. M. (1990). *Method of fabricating back surface point contact solar cells*. US Patent US4927770A. https://patents.google.com/patent/US4927770A/en

Taguchi, M., Yano, A., Tohoda, S., Matsuyama, K., Nakamura, Y., Nishiwaki, T., & Fujita, K. (2013). 24.7% record efficiency HIT solar cell on thin silicon wafer. *IEEE Journal of Photovoltaics, 4*, 96–99. https://doi.org/10.1109/JPHOTOV.2013.2282737

Tous, L., Choulat, P., Singh, S., John, J., Aleman, M., Firat, M., Duerlinckx, F., & Szlufcik, J. (2018). Efficiency roadmaps for industrial bifacial pPERC and nPERT cells. *AIP Conference Proceedings, 2147*, 120001. https://doi.org/10.1063/1.5123883

U.S. Department of State. (2023). *US DOS office of the historian, Oil Embargo, Oil Embargo, 1973–1974*. https://history.state.gov/milestones/1969-1976/oil-embargo

Wang, Y.-C., Wu, T-T., & Chueh, Y.-L. (2019). A critical review on flexible Cu(In,Ga)Se2 (CIGS) solar cells. *Materials Chemistry and Physics, 234*. https://doi.org/10.1016/j.matchemphys.2019.04.066

Wijewardane, S., & Kazmerski, L. L. (2023). Inventions, innovations, and new technologies: Flexible and lightweight thin film solar PV based on CIGS, CdTe, and a-Si:H. *Solar Compass, 7*, 100053. https://doi.org/10.1016/j.solcom.2023.100053

Zhu, L., Zhang, M., Xu, J., Li, C., Yan, J., et al. (2022). Single-junction organic solar cells with over 19% efficiency enabled by refined double-fibril network morphology. *Nature Materials, 21*, 656–663. https://doi.org/10.1038/s41563-022-01244-y

Chapter 6
The Compatibility of Water/Core-Shell $Ag\text{-}SiO_2$ Nanofluid as a Spectral Splitting Optical Filtration Fluid to Six Types of Photovoltaic Solar Cells Under Concentrated Solar Conditions

A. S. Abdelrazik

6.1 Introduction

Many research investigations on various solar systems have been undertaken during the previous decade to maximize the utilization of incident sun radiation. The hybrid photovoltaic/thermal system is one of the technologies that has recently gained much attention. The scientists wanted to increase the amount of electricity generated and heat collected. One way to improve the hybrid system's overall performance is spectrally selective optical filtration (SSOF). However, the partial obstruction to incident solar light remained a barrier to the complete adoption of the new approach (SSOF).

Solving the PV cell overheating was the primary goal that drove the investigations on the SSOF. The cells use only 20% of the broad spectrum of incident solar energy to create electricity, with the majority of the remaining solar spectrum being absorbed as heat. As a result, the temperature of the cells rises, which reduces the production of electrical energy at a rate of around 0.5%/°C. The primary objective of the SSOF is to divide the incident sun spectrum into two wavelength bands (Han et al., 2019), therefore transmitting only the solar spectrum that is effective for photovoltaics while blocking the rest from reaching the PV cells by absorption. Theoretically, this should lead to generating the necessary electricity without causing the PV cell to overheat. According to (Chemisana et al., 2018), the liquid filters' spectral transmittance needs should be the following two factors: (1) transmit the largest amount of light to maximize PV electrical output, and (2) absorb the most amount of light to maximize heat production and dissipation.

A. S. Abdelrazik (✉)
Interdisciplinary Research Center for Sustainable Energy Systems, King Fahd University of Petroleum & Minerals (KFUPM), Dhahran, Saudi Arabia
e-mail: ahmeds@kfupm.edu.sa

A. Sayigh (ed.), *Transition Towards a Carbon Free Future*, Innovative Renewable Energy, https://doi.org/10.1007/978-3-031-61660-0_6

The SSOF has been the subject of dozens of studies to attain the aforementioned parameters. Hassani et al. (2016a, b) conducted numerical analyses to determine the impact of an Ag/water SSOF on the performance of CPV/T and PV/T systems. It was claimed that the NF did not prevent sun rays from reaching the PV module, which produced power. They also showed that PV/T with SSOF generated high-grade exergy. According to the authors, the best NF transmitted 82% of the sun's spectrum. Finally, they showed that nanofluid absorbed all of the required UV spectrum as well as some IR spectrum. A PV/T system's overall performance was evaluated by Crisostomo et al. (2017) employing water/core-shell Ag-SiO2 NF for SSOF. The system was tested outdoors, and the thermal receiver channel was disconnected from the PV panel. This solution added 12% more weighted energy output compared to the standalone PV system. The generated NF samples' transmission spectra nearly matched the PV panel's necessary spectra, with out-of-spectrum transmission occurring in the infrared and short wavelength ranges. While heat output and overall system efficiency recorded increases, electrical output was reported to be decreasing. In their study of the water/Ag NF, Abdelrazik et al. (2019b) found that larger nanoparticle loadings and path lengths had a detrimental impact on the NF's transmittance, which led to more obstructions for solar radiation to reach the PV module. At lower path length and load values, the hybrid PV/T with an SSOF performed better. The hybrid system's equivalent electrical efficiency was higher than the standard PV system's; however, it declined at higher loadings.

Han et al. (2019) examined how the optical fluid type affected the performance of the hybrid CPV/T system when it was paired with an SSOF. Testing has been done on several liquids. Only the Valvoline oil and copper sulfate aqueous testing liquids had limited transmittance in the Si and GaAs cells' electrically efficient spectrum region. The remaining liquids exhibit high transmittance within the desired band. Utilizing oil filters, moderate NIR absorption has been seen. In another study, Han et al. (2019) reported that both the copper sulfate aqueous salt and Valvoline oil resulted in low electrical output for the Si and GaAs cells. Apart from that, excellent transmittance has been accomplished. Most oil filters, except for Valvoline and glycols, have been used to achieve maximum cell efficiencies. Ag-TiO2 suspended in water has been used by Li et al. (2019) as an optical filter for PVT systems. High transmittance in the electrically efficient spectral range and high absorptance in the visible range were reported. Higher nanoparticle loadings (0–200 ppm) led to lower PV conversion efficiencies. The scientists came to the conclusion that higher nanoparticle loadings increased thermal energy efficiency but decreased electrical efficiency, and vice versa, suggesting that the loading can be adjusted depending on the hybrid system's intended use.

Adam et al.'s (2020) investigation looked at the optical performance of water/SiO2 NF employed as SSOF with hybrid CPV/T systems at extremely high temperatures. Only at short wavelengths has the existence of various spectral transmittances been identified. Furthermore, at exposure temperatures ranging from 25 to 90 °C, the nanofluid's absorptance has been decreased. Huang et al. (2020) conducted a numerical study to evaluate the performance of a SSOF-integrated concentrated PV/T (CPV/T) system at a solar concentration ratio of 50. According to their findings, the CdTe and Si PV cells could absorb 39% and 27% of inefficient solar energy, respectively. The cells become less heated as a result. Han et al. (2020) investigated the [$CoSO_4$+propylene glycol]/Ag hybrid NF's optical capabilities. The hybrid NF displayed substantial absorption beyond solar wavelengths (325–670 nm). When used on two different types of cell materials (Si and GaAs), the NF demonstrated good transmittance in those wavelength electrically efficient ranges (615–970 nm). The best filtering performance was attained for the Si cells at an Ag nanoparticle loading of 31.8 ppm and for the GaAs cells at a 44.6% efficiency. More than 500% more energy was generated by the hybrid system as a whole than by a single PV system, according to reports.

Our team has done multiple research on SSOF-integrated PV-based systems during the past 3 years (Abdelrazik, 2023a, b; Abdelrazik et al., 2021, 2022). When evaluated at a 5-sun solar concentration, the scientists found that the electrical and thermal efficiencies for the water/Ag NF for SSOF were 8.4% and 66.7%, respectively, compared to only 3.7% for the standalone PV system. Additionally, compared to the standard PV and PV/T systems, integrating the SSOF with the PV/T system has decreased the average PV temperature by 6.3% and 42.1%, respectively. Deionized water was found to be the best SSOF fluid, and water/core-shell silver-silica was the best nanofluid with the highest agreement with the spectrum response of the monocrystalline silicon PV cells. The SSOF-integrated PV-based systems were advised for use at room temperature and high sun concentrations. SSOF was regarded as a ground-breaking method that could control and manage the spectral sunlight distribution ranges received by the PV module in PV/T systems for effective electrical and thermal productivity.

From the studies mentioned above, the water basefluid and water/core-shell Ag-SiO_2 NF SSOF have shown high potential with the hybrid CPV/T systems. Their integration resulted in better overall performance for the studied hybrid systems; however, no study has been proposed to the moment considering the effective spectral response of the different PV cell types and how the same SSOF fluid can act in different ways depending on the cell type.

In this study, we propose a computational evaluation of the compatibility of water/Ag-SiO2 NF as an SSOF fluid when integrated with six distinct PV cell types: mSi, pSi, aSi, CdTe, CIGS, and perovskite PV cells. Investigations are being conducted into the SSOF's integration with various PV cell types in conventional PV and PV/thermal (PV/T) systems. Each cell type's four systems are evaluated in comparison. The investigation was carried out under various solar concentrations, ranging from a single sun to 50 suns, and the governing differential equations for each system were derived. In addition to comparing the electrical efficiency and average PV cell temperature, the functionality of the systems are examined in terms of their outputs of thermal and electrical energy. The evaluation has been applied for a whole day in mid of July in Dhahran, Saudi Arabia.

6.2 Methodology

6.2.1 Model Geometry

This work investigates four solar collectors: a single PV system, a hybrid SSOF/PV system, a hybrid PV/cooling fluid (PV/CF) system, and a hybrid SSOF/PV/CF system (see Fig. 6.1). The second system (SSOF/PV) has an OF channel on the front side of the PV panel, while the third system (PV/CF) has a CF channel on the backside of the PV panel. The SSOF and CF channels are present in the final system (SSOF/PV/CF) in their suggested placements on the front and back sides of the PV panel, respectively. The glass covers of the SSOF channel are made of high transmittance glass in SSOF systems, and the sidewalls are presumed to be insulated in all systems to prevent heat loss from their sides and to direct all heat transfer to the fluids in motion. None of the systems are insulated from the bottom to allow a fair comparison between their performance and that of a standalone PV system. The upper surfaces of all systems exchange heat with their environment by convection to/from the atmosphere and radiation to space. Convection and radiation exchange heat with the surrounding environment at the bottom surface.

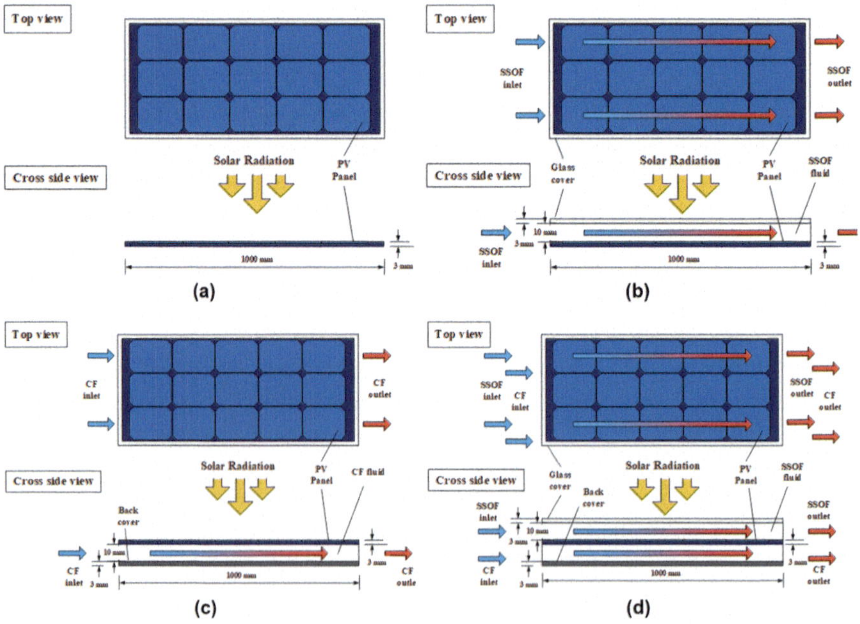

Fig. 6.1 A schematic diagram for the studied PV-based systems as (**a**) standalone PV system, (**b**) SSOF/PV, (**c**) PV/CF, and (**d**) SSOF/PV/CF

6.2.2 Operation of the SSOF-Based Systems

The incident solar energy strikes the glass cover, where the majority of it is passed to the water/Ag-SiO_2 SSOF, and the glass absorbs the remainder. A certain portion of the received light is transmitted and arrives at the PV module based on the spectral distribution of the SSOF transmittance, while the remaining component is absorbed in the SSOF. The received sun spectrum is separated into two wavelength bands based on the type of integrated PV cell. The first range of radiation received is converted to electrical energy by the cell, but the second range is absorbed in the cell, heating it and increasing its temperature.

6.2.3 Evaluation Method

The finite difference method was used to build and analyze the transient governing energy equations for the four separate systems. Matlab 2022b was used to solve the numerical problems. For each cell type, the spectral distribution of conversion efficiency was examined. The spectral distribution of the conversion efficiency was analyzed using the spectral ratios for the PV cells described in (Micheli et al., 2019) and the reported conversion efficiency for the mSi, pSi, aSi, CdTe, CIGS, and perovskite cells (Belyakov, 2019), as presented in Fig. 6.2. In mid-July, the performance of each of the four systems was examined for a full day using the six proposed PV cell types (see Fig. 6.3a). The daily electrical and thermal energy outputs, as well as the electrical efficiency and average PV temperatures, were all considered in the evaluation.

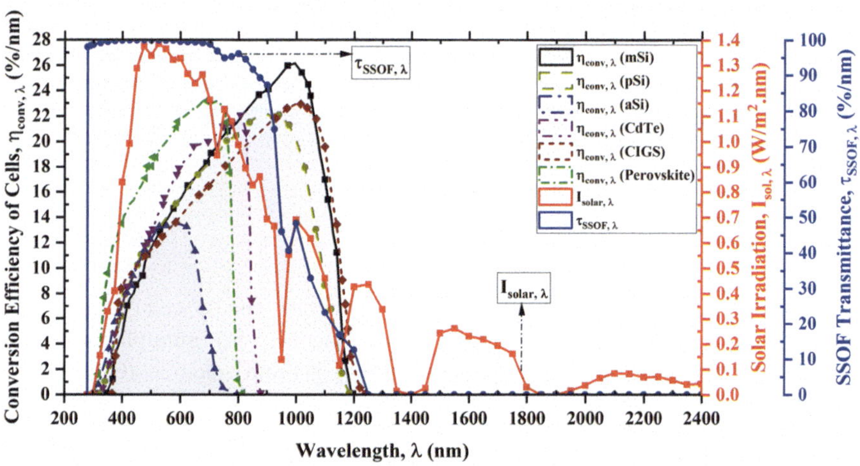

Fig. 6.2 The spectral distribution of the conversion efficiency of the six PV cell types, the proposed solar irradiation (according to ASTM G-173-03), and the transmittance of the SSOF fluid

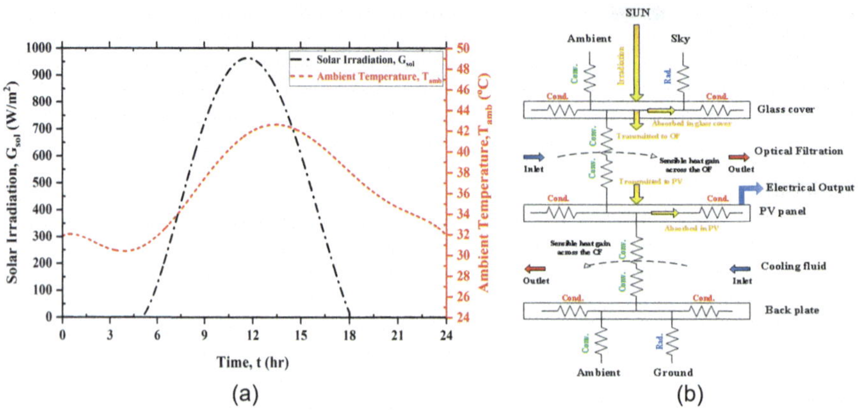

Fig. 6.3 (**a**) Thermal circuit for the energy distribution along the different layers of the hybrid SSOPV/PV/CF system and (**b**) weather data in mid-July in Dhahran, Saudi Arabia (Weather Spark, n.d.)

6.2.4 Mathematical Models

The differential equations regulating energy flow across each of the PV, SSOF/PV, PV/CF, and SSOF/PV/CF systems were obtained from the energy balance along each layer of the system. The time rate of internal energy change for each layer due to heat exchange with adjoining layers was calculated. Figure 6.3b depicts the thermal circuit for the hybrid SSOF/PV/CF system. The water/Ag-SiO_2 NF was regarded as the flowing fluid with constant characteristics in both the SSOF and the CF. PV cell thermophysical characteristics were presumed constant.

Table 6.1 summarizes the finite difference formulations of the differential equations regulating the energy transfers between the layers of the hybrid SSOF/PV/CF system, while Table 6.2 summarizes the system's performance evaluation parameters. The correlations needed to evaluate the heat transfer coefficients shown in Table 6.1 were previously discussed in (Abdelrazik et al., 2020a, b, c, d, 2021).

6.2.5 Parametric Study

The four systems use PV cells from six different varieties. The PV cell types include monocrystalline silicon (m-Si), poly-crystalline silicon (p-Si), amorphous silicon (a-Si), cadmium telluride (CdTe), copper indium gallium selenide (CIGS), and perovskite cells. Additionally, the SSOF was typically advised to be employed at greater solar concentrations to protect the PV cells from direct exposure to intense solar radiation (Abdelrazik et al., 2021; Hassani et al., 2016b). As a result, the functionality of the 24 models (4 systems × 6 PV cell types/single system) was assessed at six different solar concentrations: 1, 2, 5, 10, 20, and 50 suns.

Table 6.1 The finite difference formulations for the governing differential equations of the hybrid SSOF/PV/CF system

Layer	Finite difference formulation	
Glass cover	$t_g \rho_g c p_g \frac{T_{g_i}^{\ n} - T_{g_i}^{\ n-1}}{\Delta t} = \alpha_g C \int_{0.25\ \mu m}^{2.5\ \mu m} G_\lambda d\lambda + h_{\text{rad},g} \left(T_{\text{sky}} - T_{g_i}^{\ n}\right)$ $+ h_{\text{conv},g}\left(T_{\text{amb}} - T_{g_i}^{\ n}\right) + h_{\text{conv,of}}\left(T_{of_i}^{\ n} - T_{g_i}^{\ n}\right) + \frac{t_g k_g}{\Delta x^2}\left(T_{g_{i+1}}^{\ n} - 2T_{g_i}^{\ n} + T_{g_{i-1}}^{\ n}\right)$	(1)
SSOF	$t_{of} u_{of} \rho_{of} C_{p_{of}} \frac{T_{of_i}^{\ n} - T_{of_{i-1}}^{\ n}}{\Delta x} = \tau_g C \int_{0.25\ \mu m}^{2.5\ \mu m} \alpha_{of,\lambda} G_\lambda d\lambda + h_{\text{conv,of}}\left(T_{g_i}^{\ n} - T_{of_i}^{\ n}\right) + h_{\text{conv,of}}\left(T_{pv_i}^{\ n} - T_{of_i}^{\ n}\right)$	(2)
PV panel	$t_{pv} \rho_{pv} c p_{pv} \frac{T_{pv_i}^{\ n} - T_{pv_i}^{\ n-1}}{\Delta t} = C\tau_g \left\{ \alpha_{pv} \int_{0.25\ \mu m}^{2.5\ \mu m} \tau_{of,\lambda} G_\lambda d\lambda - \left[1 - \beta_{pv}\left(T_{pv_i}^{\ n} - T_{\text{ref}}\right)\right] \int_{0.25\ \mu m}^{2.5\ \mu m} \tau_{of,\lambda} \eta_{\text{ref},\lambda} G_\lambda d\lambda \right\}$ $+ h_{\text{conv,of}}\left(T_{of_i}^{\ n} - T_{pv_i}^{\ n}\right) + h_{\text{conv,cf}}\left(T_{cf_i}^{\ n} - T_{pv_i}^{\ n}\right) + \frac{t_{pv} k_{pv}}{\Delta x^2}\left(T_{pv_{i+1}}^{\ n} - 2T_{pv_i}^{\ n} + T_{pv_{i-1}}^{\ n}\right)$	(3)
CF	$t_{cf} u_{cf} \rho_{cf} C_{p_{cf}} \frac{T_{cf_i}^{\ n} - T_{cf_{i-1}}^{\ n}}{\Delta x} = h_{\text{conv,cf}}\left(T_{pv_i}^{\ n} - T_{cf_i}^{\ n}\right) + h_{\text{conv,cf}}\left(T_{bp_i}^{\ n} - T_{cf_i}^{\ n}\right)$	(4)
Back cover	$t_{bp} \rho_{bp} c p_{bp} \frac{T_{bp_i}^{\ n} - T_{bp_i}^{\ n-1}}{\Delta t} = h_{\text{conv,cf}}\left(T_{cf_i}^{\ n} - T_{bp_i}^{\ n}\right) + h_{rad,bp}\left(T_{\text{ground}} - T_{bp_i}^{\ n}\right)$ $+ h_{\text{conv,bp}}\left(T_{\text{amb}} - T_{bp_i}^{\ n}\right) + \frac{t_{bp} k_{bp}}{\Delta x^2}\left(T_{bp_{i+1}}^{\ n} - 2T_{bp_i}^{\ n} + T_{bp_{i-1}}^{\ n}\right)$	(5)

Table 6.2 Performance evaluation parameters of the different systems

Parameter	Symbol	Formula	Equation	Valid for
Input solar power	$\dot{P}_{\text{in}}$	$A_s \int_{0.25\ \mu\text{m}}^{2.5\ \mu\text{m}} G_\lambda d\lambda$	(6)	All systems
Electrical power output (from the PV cells)	$\dot{P}_{\text{el}}$	$A_s C \int_{0.25\ \mu\text{m}}^{2.5\ \mu\text{m}} G_\lambda \eta_{\text{ref},\lambda}\left[1-\beta\left(T_{pv}-T_{\text{ref}}\right)\right] d\lambda$	(7)	PV and PV/CF
		$A_s C \tau_g \int_{0.25\ \mu\text{m}}^{2.5\ \mu\text{m}} \tau_{of,\lambda} G_\lambda \eta_{\text{ref},\lambda}\left[1-\beta\left(T_{pv}-T_{\text{ref}}\right)\right] d\lambda$	(8)	SSOF/PV and SSOF/PV/CF
Thermal power output (collected by the OF and CF)	$\dot{Q}_{\text{th}}$	Zero	(9)	PV
		$\dot{Q}_{\text{th,of}} = \dot{m}_{of} c_{p,of}\left(T_{\text{of,out}} - T_{\text{of,in}}\right)$		SSOF/PV
		$\dot{Q}_{\text{th,cf}} = \dot{m}_{cf} c_{p,cf}\left(T_{\text{of,out}} - T_{\text{of,in}}\right)$		PV/CF
		$\dot{Q}_{\text{th,of}} + \dot{Q}_{\text{th,cf}}$		SSOF/PV/CF
Input solar energy/day	E_{in}	$\int_0^{\text{One day=24 hrs}} \dot{P}_{\text{in}}\, dt$	(10)	All systems
Electrical energy/day	E_{el}	$\int_0^{\text{One day=24 hrs}} \dot{P}_{\text{el}}\, dt$	(11)	
Thermal energy/day	E_{th}	$\int_0^{\text{One day=24 hrs}} \dot{Q}_{th}\, dt$	(12)	
Electrical efficiency	η_{el}	$E_{\text{el}}/E_{\text{in}}$	(13)	
Thermal efficiency	η_{th}	$E_{\text{th}}/E_{\text{in}}$	(14)	

Table 6.3 Thermophysical and optical properties of the different systems' layers

Layer	Density, ρ (kg/m^3)	Specific heat, c_p (J/kg.K)	Thermal conductivity, k (W/m.K)	Absorptivity, α	Transmissivity, τ	Emissivity, ε
Glass	2600	700	1.05	0.05	0.9	0.9
PV	2330	710	150	0.945	–	0.9
Back cover	2700	910	205	–	–	0.09

Table 6.4 The employed constant parameters in this study

Parameter	Value	Parameter	Value
Length of the collector, L	0.7 m	Spatial horizontal step, Δx	0.05 m
Width of the collector, W	0.3 m	Spatial vertical step, Δy	0.002 m
PV panel thickness, t_{pv}	3 mm	Time step, Δt	10 s
Back cover thickness, t_{bc}	3 mm	The reference temperature, T_{ref}	25 °C
Optical filtration channel height, t_{of}	10 mm	The inlet temperature of the OF fluid, $T_{of,\ in}$	25 °C
Cooling fluid channel height, t_{cf}	10 mm	The inlet temperature of the CF fluid, $Tc_{f,\ in}$	25 °C
Wind speed, u_a	0 m s^{-1}		

Tables 6.3 and 6.4 provide a list of the layers' thermophysical and optical characteristics as well as the constant parameters currently used according to (Abdelrazik et al., 2019a, b, 2020a, b; Abd-Elrazik et al. 2017). Besides, Fig. 6.3a shows the seasonal weather distribution of atmospheric temperature and solar radiation in the middle of July, while Fig. 6.2 shows the spectral variations of solar radiation intensity, reference conversion efficiencies of the six PV cell types, and the transmittance of the SSOF (measured experimentally and reported in (Abdelrazik et al., 2019b).

6.3 Results

6.3.1 Model Validation

The numerical performance findings were checked against experimental results before analyzing the hybrid system's performance using differential equations. Our previous study (Abdelrazik et al., 2021) validated the performance findings of a single-channel hybrid PV/SSOF system against Erdil et al. (2008). Erdil and his team used water for optical filtration and measured the output water temperature under particular conditions. The same parameters for estimating the output water temperature were used for the numerical work in this study. The analyzed examples revealed a 3.4% average disparity between numerical results and experimental data.

6.3.2 Compatibility and Performance Assessment

Figures 6.4a, b show the effect of connecting the SSOF channel to conventional PV and hybrid PV/CF systems on the generated electrical energy when different PV cells and solar concentration ratios are used. The results demonstrated a general trend for the various systems and cells studied. The generated electrical energy grew with solar concentration to a peak value before decreasing and becoming inefficient at a specific solar concentration value.

As shown in Fig. 6.4a, incorporating the water/Ag-SiO_2 SSOF into the standalone PV system enhanced the electrical energy production at similar investigated solar concentration values. The hybrid SSOF/PV system's electrical energy output peaked at an average solar ratio of 14.5 suns, compared to only 5 suns for the solo PV system, after which the energy generated began to degrade with solar concentration. Furthermore, the system remained efficient with the SSOF for a greater range of solar concentrations up to 28.8 suns, compared to only 10.4 without the SSOF. The increasing sinusoidal relationship between electrical energy and solar concentration can be attributed to the direct proportionality between increasing the number of suns and the matching electrical energy production. However, increasing the solar concentration meant increasing the PV module's stored thermal energy, which had a detrimental impact on its electrical productivity. As a result, increasing the concentration ratio for a certain system and up to a specific solar concentration ratio had a stronger beneficial impact on electrical productivity than on heat absorption. After this point, the negative influence of thermal absorption takes precedence, resulting in a drop in overall electrical energy output. This promise is confirmed and assured when examining electrical productivity following SSOF channel integration. The system was more efficient at greater solar concentration ratios due to better cooling and decreased exposure to solar radiation (see Fig. 6.4a).

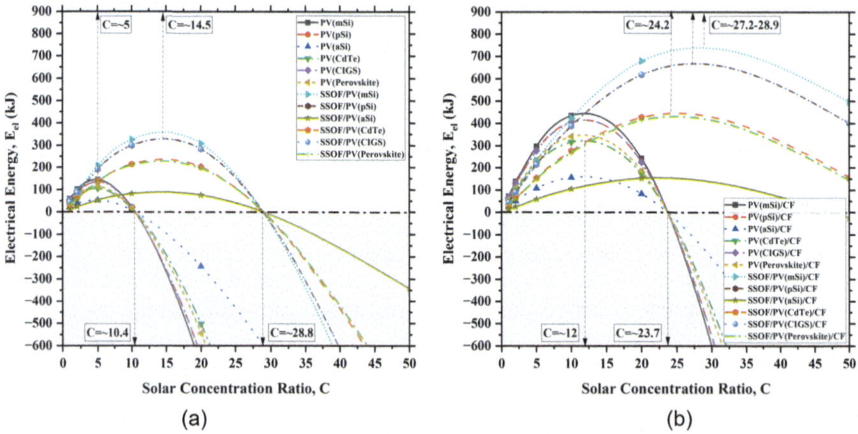

Fig. 6.4 The electrical energy of the (**a**) standalone PV and hybrid SSOF/PV and (**b**) hybrid PV/CF and SSOF/PV/CF systems for six PV cell types under different solar concentrations

Figure 6.4b shows similar results, but at a greater order of magnitude, due to bottom cooling to the PV module via the CF channel. The hybrid PV/CF and SSOF/PV/CF systems produced the most electricity at 12 suns and 24.2–28.9 suns, respectively (depending on the cell type). Again, with the sinusoidally shaped behavior, the systems remained efficient at solar concentrations of up to 23.7 and 50 suns, respectively (depending on cell type).

When the electrical energy production of the various systems was compared, the hybrid PV/CF system had the highest electrical productivity at low solar concentrations due to the twofold cooling impact from the front and back sides of the PV module. The SSOF/PV/CF combination produced the maximum electricity values at high solar concentrations (>10). At high sun concentrations, the standalone PV system was the least productive, while at unity solar concentrations, it outperformed the hybrid SSOF/PV system (see Fig. 6.4a). The SSOF/PV switched to be more efficient at high solar concentration ratios depending on the PV cell used, which is ascribed to SSOF's superior suitability to higher solar concentration situations since it prevents direct exposure to concentrated solar radiation. This makes the hybrid SSOF/PV system capable of covering a higher solar concentration range than the conventional hybrid PV/CF system.

Different electrical energy outputs were observed depending on the cell type, as shown in Fig. 6.4. In the standard PV and hybrid PV/CF systems, these variances were caused by the varied spectrum responses of the cells, which allow for different reactions in terms of their directly produced energy and the amount of heat absorbed. These distinctions are essential when integrating SSOF into traditional systems because the resultant combined effect of the SSOF liquid's spectral transmittance range and the various spectral response ranges of the cells adds further order of variety.

Using water/Ag-SiO2 NF as the SSOF liquid, m-Si PV cells produced the most electrical energy, followed by p-Si PV cells. The poorest values were obtained using a-Si PV cells, which were initially expected due to the low spectral response efficiency and restricted spectral response range relative to the SSOF fluid's spectral transmittance, as shown in Fig. 6.2. Figure 6.4 depicts the results, which reveal that the solar concentration ratio at which the systems hit that peak in energy generation was nearly the same for all types of PV cells, with only the system itself varying. On the other hand, with the hybrid SSOF/PV/CF system shown in Fig. 6.4b, the use of different cells resulted in maximum electricity output at various solar ratios. This was ascribed to the combination double-cooling impact of the CF and SSOF, as well as the variable spectrum responses of the cells and the application of SSOF, which generated apparent variation in the system's responsiveness to greater solar concentrations based on the attached PV cell type. Maximum electricity production was attained using a-Si PV cells at a minimum of 24 suns and a maximum of 28.9 suns. Because the tested PV cells have similar thermophysical properties, there is not much difference between the systems.

As a function of the used PV cell and solar concentration ratio, Fig. 6.5a shows the change in the thermal energy produced by each of the four systems under study. The results shown in Fig. 6.5a give the first impression that the hybrid SSOF/PV/CF

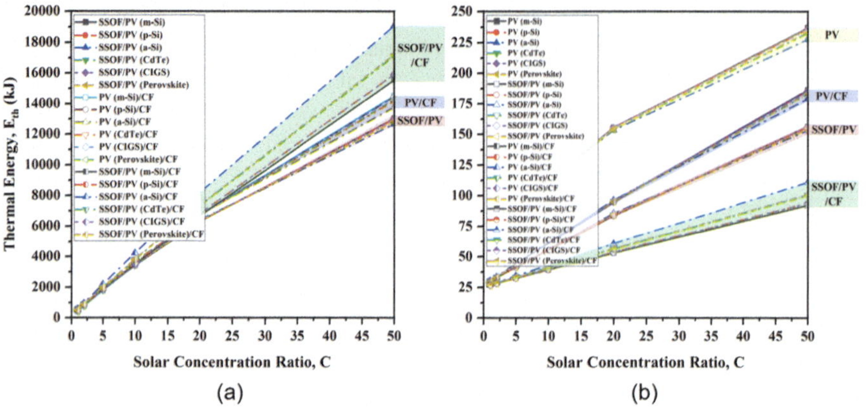

Fig. 6.5 (**a**) The thermal energy and (**b**) average PV temperature of the different systems, for six PV cell types and under different solar concentrations

system, which double-cools the PV module from both sides, can collect the most thermal energy. However, the worst decision regarding the thermal energy collected was sticking solely to front-side cooling (SSOF/PV system).

In terms of PV cell type, different thermal energy were gathered for the same system depending on the installed PV cell. When integrated into the hybrid SSOF/PV/CF system, the a-Si cells collected the most total thermal energy, but when integrated into the SSOF/PV system with no below PV cooling, they offered the worst scenario. At 50 sun solar concentration ratio, the first system collected 50.6% more heat. This was attributed to the a-Si cells' narrow spectral response range compared to the water/Ag-SiO2 SSOF's spectral transmittance, enabling more non-electrically-efficient solar energy to be transferred and absorbed as heat in the a-Si cells. This heat is subsequently collected by the SSOF and CF fluids in the first system and lost via the PV backside in the second.

Integrating the SSOF with the traditional PV/CF at greater sun intensities became more suggested. Using a-Si cells in the SSOF/PV/CF system at 20 suns increased collected thermal energy by 27.8% and 21%, compared to the hybrid SSOF/PV and PV/CF systems, respectively. At 50 suns, these percentages grew to 50.6% and 38.6%, respectively. However, obtaining such high sun concentration values was not recommended because less electrical energy could be captured (see Fig. 6.4). At five suns, the differences in collected thermal energy between all systems with different cells were within 8%, except for the hybrid SSOF/PV/CF system with the a-Si, which was 18.3% greater than the lowest collected energy. At significantly lower sun intensities, the difference in collected thermal energy between the various alternatives became almost negligible.

Figure 6.5b depicts the variance in average PV temperature for each of the four systems investigated as a function of the PV cell and solar concentration ratio used. There is no question that the assumption of comparable thermal properties between the different cells, as a result of the scarcity of information in the literature, influenced the results. However, the primary goal was to assess the impact of the

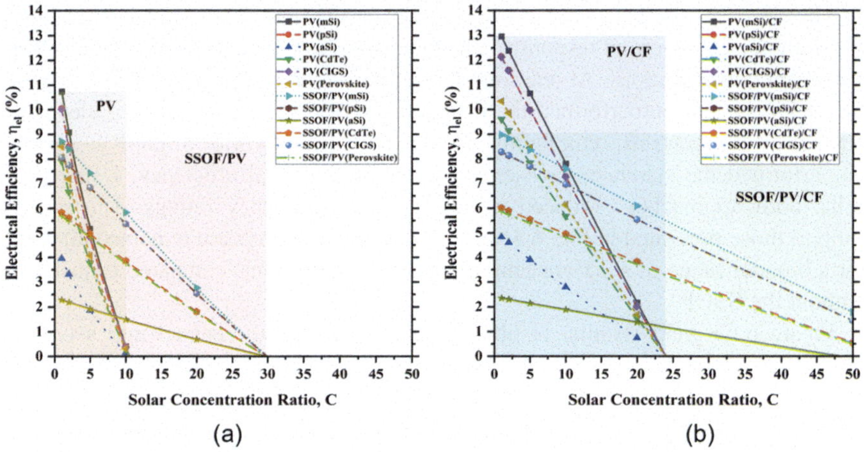

Fig. 6.6 The electrical efficiency of the (**a**) standalone PV and hybrid SSOF/PV (**b**) hybrid PV/CF and SSOF/PV/CF systems for six PV cell types and under different solar concentrations

various spectral response spectrum ranges on the optical characteristics of water, as shown in Figs. 6.4 and 6.5a. Lower PV temperatures can be attained by integrating SSOF into both conventional PV and PV/CF systems, as shown in Fig. 6.5b. The hybrid SSOF/PV/CF system had the lowest PV temperatures, followed by the SSOF/PV system, and the standalone PV system had the highest temperatures. Minor variances in average PV temperature have been documented at unity sun concentration due to the impact of the cell type in each system. At greater solar concentrations, the disparities in PV temperatures across the four systems were more noticeable. As concentrations increase, incorporating the SSOF channel becomes more recommended to protect the solar panels from direct solar radiation exposure. Going to a unity solar concentration, the cost of the system will be the deciding factor between the three hybrid systems, as they have essentially identical PV temperatures, which are lower than the single PV system.

Figure 6.6 depicts the change in electrical efficiency of the four tested systems as a function of the used PV cell and solar concentration ratio. The inverse relationship between the four systems' electrical efficiency and the rise in solar concentration ratio is clearly shown. The rise in incident radiation with concentration was more prominent in magnitude than the equivalent increase in output electrical energy. However, Fig. 6.4 showed an increase in electrical output with solar concentration up to a point before a decrease; the significant increase in input energy with concentrations led to an overall decrease in the electrical efficiency of the systems with the solar concentration ratio. According to the results shown in Fig. 6.6a, the standalone PV system with m-Si cells produced the highest electrical efficiency at unity solar concentration, while the standalone PV system with a-Si cells produced the lowest, compared to the rest of the options depicted in the exact figure. Compared to a standalone PV system, integrating the SSOF with the PV module boosted system efficiency at specific points and decreased the deterioration rate with solar concentration.

The electrical efficiency was strongly reliant on the kind of integrated PV cell due to differences in spectrum responses, and this variation was much more noticeable at lower sun concentrations. At one solar concentration and employing m-Si cells, the single PV system outperformed the hybrid SSOF/PV system in terms of electrical efficiency. In contrast, employing CdTe cells, the hybrid SSOF/PV system outperformed the standalone system in terms of electrical efficiency. Using p-Si cells, both approaches produced nearly identical efficiency ratings. The results support those presented in Fig. 6.4, as with the SSOF, the system remained efficient for a broader range of solar concentrations up to ~29.5 suns, compared to just ~10 without the SSOF.

Figure 6.6b yields similar results. Integrating an SSOF channel into a typical PV/CF system has lowered the rate of degradation in electrical efficiency, allowing it to operate over a broader range of solar concentration ratios. Again, the electrical efficiency was strongly reliant on the type of integrated PV cell, and this variation was much more noticeable at lower sun intensities. However, nearly parallel relations were found for the hybrid SSOF/PV/CF system, depending on the integrated cell, reflecting a broader concentration range of efficiency dependency on cell type. In contrast to the findings shown in Fig. 6.6a, the hybrid PV/CF system was always advised above the hybrid SSOF/PV/CF system at unity solar concentration due to its better electrical efficiency for all cell types. Once more, SSOF was effective at higher solar concentrations because it shielded panels from concentrated sun radiation. The hybrid SSOF/PV/CF system with an a-Si cell had the lowest electrical efficiency at unity solar concentration, whereas the PV/CF system with m-Si cells had the maximum electrical efficiency.

To summarize the present study's findings, the water/Ag-SiO_2 SSOF was more compatible with the m-Si, p-Si, and CIGS cells for higher electrical outputs, while it was less compatible with the rest of the cells. Table 6.5 provides a preference for the best-recommended system and integrated cell at each examined solar concentration from an electrical efficiency point of view. The hybrid PV/CF system, with m-Si cell, was the most recommended at low concentrations (up to 10). The integration of p-Si, followed by CIGS cells, was recommended as first and second alternative to the m-Si cells. The hybrid SSOF/PV/CF, with m-Si, was the best suited at higher concentrations, with the CIGS and p-Si cells as the first and second alternatives.

Table 6.5 Descending ordering of systems and cells according to the highest electrical efficiency at each examined solar concentration

Pref.	SC = 1	SC = 2	SC = 5	SC = 10	SC = 20	SC = 50
#1	PV/CF (m-Si)	PV/CF (m-Si)	PV/CF (m-Si)	PV/CF (m-Si)	SSOF/PV/CF (m-Si)	SSOF/PV/CF (m-Si)
#2	PV/CF (p-Si)	PV/CF (p-Si)	PV/CF (p-Si)	SSOF/PV/CF (m-Si)	SSOF/PV/CF (CIGS)	SSOF/PV/CF (CIGS)
#3	PV/CF (CIGS)	PV/CF (CIGS)	PV/CF (CIGS)	PV/CF (p-Si)	SSOF/PV/CF (p-Si)	SSOF/PV/CF (p-Si)

Similarly, in descending order, Table 6.6 lists, for each system, the potential cells to achieve the highest electrical and thermal energy outputs and lowest PV temperatures. The m-Si cell was the best choice for all systems to achieve the highest electrical energy. However, it was the worst option regarding low PV temperatures, except for the hybrid SSOF/PV/CF system, as m-Si cells remained the best option. On the other side, using the a-Si cell was the worst option for the highest electrical energy and the best for the lowest PV temperature, except for the SSOF/PV/CF system. It can be noted that the p-Si and CIGS are frequently switching their orders due to close spectral responses. Similarly, it occurs with the CdTe and perovskite cells, with a bit of change in the SSOF/PV/CF case.

6.4 Conclusions

The water/Ag-SiO_2 nanofluid was used as the incident radiation SSOF fluid. The SSOF's compatibility with six distinct types of PV cells with differing spectral response ranges has been numerically examined. A transient study was conducted for a full day in mid-July in Dhahran, Saudi Arabia, comparing the performance of SSOF-integrated PV and PV/T systems to their traditional counterparts. The following critical conclusions have been released:

- SSOF-integrated PV and hybrid PV/CF systems produced higher electrical energies at lower PV temperatures when operating at high solar concentrations. SSOF/PV/CF produced the most electricity, whereas standalone PV produced the least.
- When conventional systems are combined with SSOF, the spectral transmittance range of the liquid and the cells' independent spectral response ranges extend the number of possible combinations.
- The water/Ag-SiO_2 SSOF was more compatible with the m-Si, p-Si, and CIGS cells, but less compatible with the others.
- Due to the CF and SSOF's double-cooling effect and the cells' different spectrum responses, the hybrid SSOF/PV/CF system produced the maximum electricity at various higher solar ratios.
- m-Si cells were the most electrically efficient, followed by p-Si and CIGS.
- a-Si cells were the most thermally efficient when combined with SSOF/PV/CF.
- An SSOF/PV/CF system attained the lowest PV temperature by employing m-Si.
- SSOF/PV was not the best option when solar concentration ratios were low.
- The SC reduced the electrical efficiency of the four systems in a cell-type-dependent fashion.
- At low SCs, cell type had less impact on average PV temperature.
- The type of integrated PV cell utilized and the solar concentration all impacted the electrical and thermal output and the average PV temperature.

Table 6.6 Descending ordering of cells, for the different PV systems, according to the (a) highest electrical energy output, (b) thermal energy output, and (c) lowest average PV temperature

System/cell	m-Si			p-Si			a-Si			CdTe			CIGS			Perovskite		
Evaluation	(a)	(b)	(c)	(a)	(b)	(c)	(a)	(b)	(c)	(a)	(b)	(c)	(a)	(b)	(c)	(a)	(b)	(c)
PV	1		6	2		5	6		1	5		2	3		4	4		3
SSOF/PV	1	1	6	3	3	4	6	6	1	4	4	3	2	2	5	5	5	2
PV/CF	1	1	6	2	2	5	6	6	1	5	5	2	3	3	4	4	4	3
SSOF/PV/CF	1	6	1	3	4	3	6	1	6	4	3	4	2	5	2	5	2	5

Finally, utilizing different types of PV cells and creating more optimal SSOF fluids for various heat control applications should receive greater attention. Additionally, various filtrations can be used on the PV glazing to match the spectrums required by various cell types.

References

Abdelrazik, A. S. (2023a). The potential of liquid-based spectrally-selective optical filtration and its use in hybrid photovoltaic/thermal solar systems. In *Solar energy* (Vol. 249, pp. 569–605). https://doi.org/10.1016/j.solener.2022.12.010

Abdelrazik, A.S., (2023b). Water liquid compatibility as a spectral splitting optical filtration fluid to six types of photovoltaic solar cells under high solar concentrations. *Energy Conversion and Management, 294*, p.117557. https://doi.org/10.1016/j.enconman.2023.117557

Abdelrazik, A. S. S., Al-Sulaiman, F. A. A., & Saidur, R. (2019a). Optical behavior of a water/silver nanofluid and their influence on the performance of a photovoltaic-thermal collector. *Solar Energy Materials and Solar Cells, 201*(March), 110054. https://doi.org/10.1016/j.solmat.2019.110054

Abdelrazik, A. S. A. S. A. S., Al-sulaiman, F. A. A., Saidur, R., & Ben-mansour, R. (2019b). Evaluation of the effects of optical filtration and nanoPCM on the performance of a hybrid photovoltaic-thermal solar collector. *Energy Conversion and Management, 195*(November 2018), 139–156. https://doi.org/10.1016/j.enconman.2019.04.083

Abdelrazik, A. S., Al-Sulaiman, F. A., & Saidur, R. (2020a). Numerical investigation of the effects of the nano-enhanced phase change materials on the thermal and electrical performance of hybrid PV/thermal systems. *Energy Conversion and Management, 205*(December 2019), 112449. https://doi.org/10.1016/j.enconman.2019.112449

Abdelrazik, A. S., Saidur, R., & Al-Sulaiman, F. A. (2020b). Thermal regulation and performance assessment of a hybrid photovoltaic/thermal system using different combinations of nano-enhanced phase change materials. *Solar Energy Materials and Solar Cells, 215*(October 2019), 110645. https://doi.org/10.1016/j.solmat.2020.110645

Abdelrazik, A. S., Tan, K. H., Aslfattahi, N., Arifutzzaman, A., Saidur, R., & Al-sulaiman, F. A. (2020c). Optical, stability and energy performance of water-based MXene nanofluids in hybrid PV/thermal solar systems. *Solar Energy, 204*(April), 32–47. https://doi.org/10.1016/j.solener.2020.04.063

Abdelrazik, A. S. S., Tan, K. H. H., Aslfattahi, N., Saidur, R., Al-Sulaiman, F. A., & Al-Sulaiman, F. A. (2020d, February). Optical properties and stability of water-based nanofluids mixed with reduced graphene oxide decorated with silver and energy performance investigation in hybrid photovoltaic/thermal solar systems. *International Journal of Energy Research*, er.5770. https://doi.org/10.1002/er.5770

Abdelrazik, A. S., Saidur, R., & Al-Sulaiman, F. A. (2021). Investigation of the performance of a hybrid PV/thermal system using water/silver nanofluid-based optical filter. *Energy, 215*, 119172. https://doi.org/10.1016/j.energy.2020.119172. Elsevier.

Abdelrazik, A. S., Al-Sulaiman, F. A., & Saidur, R. (2022). Feasibility study for the integration of optical filtration and nano-enhanced phase change materials to the conventional PV-based solar systems. *Renewable Energy, 187*, 463–483. https://doi.org/10.1016/j.renene.2022.01.095

Abd-Elrazik, A. S., Al-Sulaiman, F. A., Saidur, R., & Ben-Mansour, R. (2017). Analytical study on the effect of optical filtration and Nano-PCM on The Performance of PV/T Solar Collectors. In *ISES Solar World Congress 2017 – IEA SHC international conference on solar heating and cooling for buildings and industry 2017, proceedings* (pp. 1038–1047). https://doi.org/10.18086/swc.2017.18.01

Adam, S. A., Ju, X., Zhang, Z., Lin, J., Abd El-Samie, M. M. M. M. M., & Xu, C. (2020). Effect of temperature on the stability and optical properties of SiO2-water nanofluids for hybrid photovoltaic/thermal applications. *Applied Thermal Engineering, 175*(December 2019), 1–11. https://doi.org/10.1016/j.applthermaleng.2020.115394

Belyakov, N. (2019). Solar energy. *Sustainable Power Generation*, 417–438. https://doi.org/10.1016/B978-0-12-817012-0.00031-1

Chemisana, D., Fernandez, E. F., Riverola, A., & Moreno, A. (2018). Fluid-based spectrally selective fi lters for direct immersed PVT solar systems in building applications. *Renewable Energy, 123*, 263–272. https://doi.org/10.1016/j.renene.2018.02.018

Crisostomo, F., Hjerrild, N., Mesgari, S., Li, Q., & Taylor, R. A. (2017). A hybrid PV/T collector using spectrally selective absorbing nanofluids. *Applied Energy, 193*, 1–14. https://doi.org/10.1016/j.apenergy.2017.02.028

Erdil, E., Ilkan, M., & Egelioglu, F. (2008). An experimental study on energy generation with a photovoltaic (PV)-solar thermal hybrid system. *Energy, 33*(8), 1241–1245. https://doi.org/10.1016/j.energy.2008.03.005

Han, X., Xue, D., Zheng, J., Alelyani, S. M., & Chen, X. (2019). Spectral characterization of spectrally selective liquid absorption filters and exploring their effects on concentrator solar cells. *Renewable Energy, 131*, 938–945. https://doi.org/10.1016/j.renene.2018.07.125

Han, X., Chen, X., Sun, Y., & Qu, J. (2020). Performance improvement of a PV/T system utilizing Ag/CoSO4-propylene glycol nanofluid optical filter. *Energy, 192*, 116611. https://doi.org/10.1016/j.energy.2019.116611

Hassani, S., Saidur, R., Mekhilef, S., & Taylor, R. A. (2016a). Environmental and exergy benefit of nanofluid-based hybrid PV/T systems. *Energy Conversion and Management, 123*, 431–444. https://doi.org/10.1016/j.enconman.2016.06.061

Hassani, S., Taylor, R. A., Mekhilef, S., & Saidur, R. (2016b). A cascade nanofluid-based PV/T system with optimized optical and thermal properties. *Energy, 112*, 963–975. https://doi.org/10.1016/j.energy.2016.06.142

Huang, G., Wang, K., & Riera, S. et al. (2020, *July*). Performance analysis of fluid-based spectral-splitting hybrid photovoltaic-thermal solar collectors. *Proceedings international conference on efficiency, cost, optimization, simulation, and environment impact of energy systems*

Li, H., He, Y., Wang, C., Wang, X., & Hu, Y. (2019). Tunable thermal and electricity generation enabled by spectrally selective absorption nanoparticles for photovoltaic/thermal applications. *Applied Energy, 236*(October 2018), 117–126. https://doi.org/10.1016/j.apenergy.2018.11.085

Micheli, L., Fernandez, E. F., Caballero, J. A., Smestad, G. P., Nofuentes, G., Mallick, T. K., & Almonacid, F. (2019). Correlating photovoltaic soiling losses to waveband and single-value transmittance measurements. *Energy, 180*, 376–386.

Weather Spark. (n.d.). https://weatherspark.com/y/104950/Average-Weather-in-Dhahran-Saudi-Arabia-Year-Round

Chapter 7
Reflective Analysis of a Volunteer-Installed Off-grid PV System for a Remote Peruvian Community Centre

Shashi Persaud

7.1 Introduction

Electricity is essential for human development and the modern way of life. Its availability, affordability and utilization correlate with societal development and quality of living (Shuwen et al., 2013). For people living in developed societies, the transformative benefits of this energy source often go unnoticed, but for the millions of people residing in off-grid communities without access to affordable electricity, the effects are very visible and life-altering—societal development is stymied, people are deprived of basic amenities, and children's educational opportunities are severely restricted (Jensen et al., 2019). A recent report by the five SDG7 custodian agencies estimated that in 2021 some 675 million people did not have access to electricity (IEA, IEA, IRENA, UNSD, World Bank, WHO, 2023). SDG7 is the seventh of the 17 Sustainable Development Goals (SDGs) adopted by the United Nations in 2015 as the global roadmap for eliminating poverty, protecting the planet and ensuring a peaceful and prosperous world by 2030. SDG7 seeks to "Ensure access to affordable, reliable, sustainable and modern energy for all" (United Nations General assembly, 2015).

Light Up the World (LUTW) is a Calgary-based, Canadian non-profit organization with a mission to "provide renewable energy technologies and high efficiency lighting to communities around the world that do not have access to appropriate and affordable energy solutions". Their mission is grounded in the belief that "access to energy changes lives, and an investment in renewable energy system for those without access to electricity is an investment in their health, education and economic development" (Light Up the World, n.d.). LUTW works with partners and volunteers to implement sustainable solutions for remote off-grid and fringe communities.

S. Persaud (✉)
MacPhail School of Energy, Southern Alberta Institute of Technology, Calgary, AB, Canada
e-mail: Shashi.persaud@sait.ca

A. Sayigh (ed.), *Transition Towards a Carbon Free Future*, Innovative Renewable Energy, https://doi.org/10.1007/978-3-031-61660-0_7

These projects allow LUTW, its partners and volunteers to pursue their shared objective of reducing global energy inequality by providing renewable energy solutions to a disadvantaged fringe community, one project at a time. LUTW activities, over its 28 years of operation, is estimated to have benefited more a million families in 55 countries. These efforts, combined with those of governments, international organizations and other like-minded charities and non-governmental organizations, have been instrumental in reducing the number of people without access to electricity from 1.1 billion in 2010 to 675 billon people in 2021 (IEA, IEA, IRENA, UNSD, World Bank, WHO, 2023). LUTW currently concentrates its overseas projects to Peru where there are many Amazonian and Andean communities still without access to affordable and sustainable electricity (Light Up the World, n.d.).

To date, LUTW has facilitated three (2017, 2018 and 2019), Peru-based field projects for students of the Electrical Engineering Technology (EET) program at the Southern Alberta Institute of Technology (SAIT) in Calgary. These projects have allowed student volunteers the opportunity to experience first-hand, some of the impacts that the lack of electricity can have on the lives of people living in off-grid communities and how simple photovoltaic systems can play a significant role in reducing those inequalities. The scale of the volunteer projects undertaken by LUTW is such that, over a 10-day period, participants get a wide spectrum experience, from basic preparatory training in off-grid PV system technology and design through to the installation and commissioning of an off-grid PV system for an in-need, remote community. This field venture allows for the development of technical skills and cognitive abilities that extend beyond the reach of the EET program curriculum. Very importantly, under LUTW stewardship, volunteers get an introduction to humanitarian engineering and a safe and structured immersive experience that allows for the development of social skills that cannot be obtained in the classroom (Feron & Cordero, 2018; Baaoum, 2018).

7.2 Methodology

This presentation relates to the 2018 LUTW-facilitated field trip to install an off-grid PV system for the community centre at Roma Tambo, a remote pastoral community in the high Andes region of Ancash Province, Peru. Prior to the PV system installation, volunteers were involved in LUTW-led preparatory activities. These included class-room-type training activities to learn about the theory of off-grid PV system design, practical laboratory-type activities to familiarize themselves with PV system components and design principles, wiring practices, and site visits to observe previous systems and to provide on-the-spot maintenance. Volunteers, however, were not involved in the actual design and sizing of the system to be installed or in the selection of components. This was done by LUTW staff prior to the arrival of the group.

One of the expected learning outcomes of the field trip was that it would allow student volunteers to develop a level of proficiency in off-grid PV system design that would strengthen their technical skill sets and enhance their career prospects. The field trip was instrumental in providing students with the initial training and experience necessary for developing competency in off-grid PV system design. However, transforming field experiences into knowledge requires time and the opportunity for learners to critically reflect upon their concrete experiences and to develop the theoretical framework necessary for interpreting, learning and progressing from those experiences (Stern & Powell, 2020). These phases of the experiential learning cycle are most effective when they take place within a structured setting where the reflective activities are matched to learning objectives (Stirling et al., 2016).

To realize the desired cognitive learning outcome in off-grid PV system design, an instructor facilitated, post-trip reflective analysis of the installed PV system design was undertaken. This activity comprised of a series of focused questions and discussions requiring students to reflection on key decisions taken in the various stages of the PV system design process. In each case, the underlying theoretical framework was first discussed and then used as the basis for examining the rationale and validity of those design decisions.

7.3 Results

This section begins with a review of the installed off-grid PV system in Sect. 7.3.1. The data tables and diagrams presented here were developed collectively through reflection, field notes, photographs, and memories. The reflective questions and discussion activities targeting the key stages in the off-grid PV design process aimed are then presented in Sect. 7.3.2.

7.3.1 Technical Project Specifications

Figure 7.1 shows a plan view (not to scale) of the Roma Tambo community centre. The facility comprises of three structures: (i) the main building located in the centre of the photo, (ii) a church located to the left, and (iii) a combined nursery school/ dispensary building. The main hall serves a meeting place and communal facility and is primarily utilized on weekends and in the evenings. The church is primarily used on Sundays and for religious occasions. The third build serves as a nursery school classroom and dispensary. The nursery school operates 5 days per week over the school year.

The layout of the PV system is also shown in Fig. 7.1. The north-facing PV array is in the open area to the right of the centre. Next to the array is the control room housing the charge controller, inverter, AC and DC distribution connector bars, and the battery. From the control centre, the AC and DC loads in the buildings are

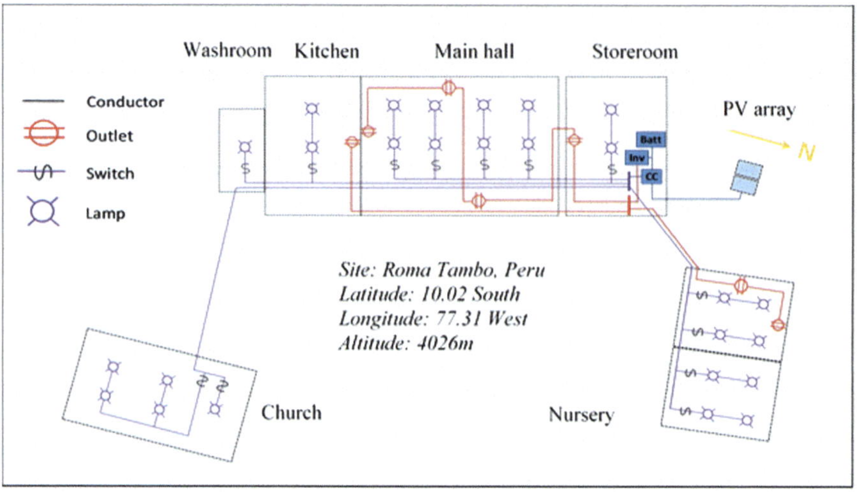

Fig. 7.1 The Roma Tambo community centre off-grid PV system layout

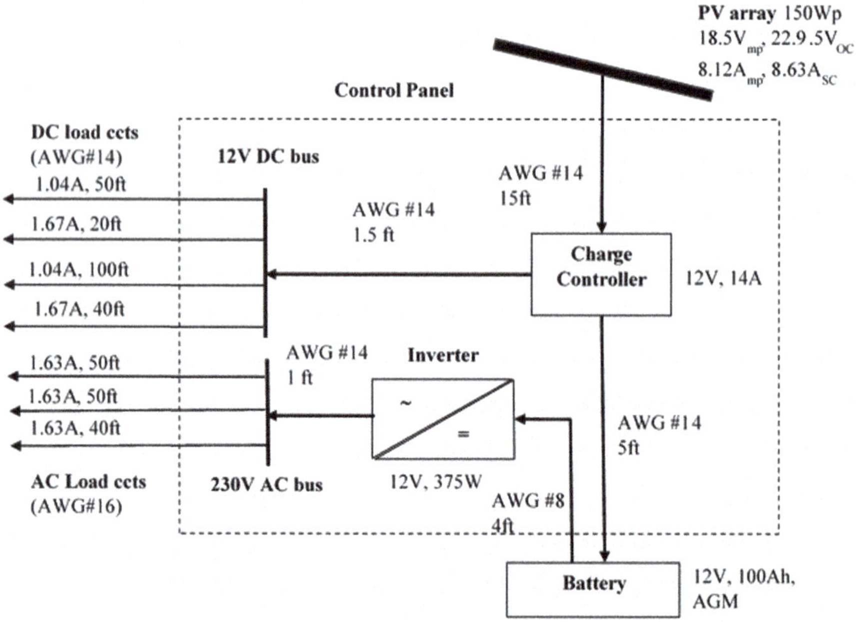

Fig. 7.2 Technical specifications of the installed off-grid PV system

supplied from four radial DC circuits and three AC circuits. The technical specifications of the installation are shown in the schematic of Fig. 7.2.

Based on a prior needs analysis conducted by LUTW, the daily ac and dc load requirements of the facility were estimated. The identified AC loads were a blender

Table 7.1 AC loads

AC loads	Number	P_{rated} (W)	P_{peak} (W)	P_{surge} (W)	Time (h/wk)	Energy (Wh/wk)	Energy (Wh/d)
Blender	1	300	300	600	1	300	42.86
Phones	10	5	50	50	4	200	28.57
Radio	1	25	25	25	10	250	35.71
Total AC		P_{AC}	*375*	*675*	W_{AC}	*750*	*107.14*

Table 7.2 DC loads

DC loads	Number	P_{rated} (W)	P_{peak} (W)	Time (h/wk)	Energy (Wh/wk)	Energy (Wh/d)
Nursery	4	2.5	10	8	80	11.43
Dispensary	4	2.5	10	2	20	2.86
Storeroom	2	2.5	5	2	10	1.43
Main Hall	8	2.5	20	25	500	71.53
Kitchen	4	2.5	10	20	200	28.57
Washroom	1	2.5	2.5	14	35	5.00
Church	5	2.5	12.5	6	75	10.71
Inverter tare	1	1	1	98	98	14.00
Total DC		P_{DC}	*81*	W_{DC}	*1018*	*145.43*

for preparing meals, cellphone chargers for the community and a radio. The DC loads were all low-voltage LED lamps for the seven spaces. Details of the AC and DC loads are presented in Tables 7.1 and 7.2, respectively. Based on the weekly energy usage of the individual loads, it was estimated that the PV system would need to be sized to serve an AC load of 107.14 Watt-hour per day (Wh/d) and a DC load of 145.5 Wh/d, plus estimated system losses.

7.3.2 *Reflective Questions and Discussions*

Question 1
The PV array was positioned so that it is North facing with a tilt angle of 10 degrees to the horizontal. Are these the correct settings for the array orientation and tilt angle?

Discussion 1
In the southern hemisphere, the overhead sun path is offset to the North by the latitude angle. To maximize the available insolation, the PV array should be North facing. A common practice is to set the array tilt angle equal to the site latitude. Assuming uniform atmospheric conditions, this inclination will maximize annual electricity production as it maximizes production in two seasons, spring and fall. In

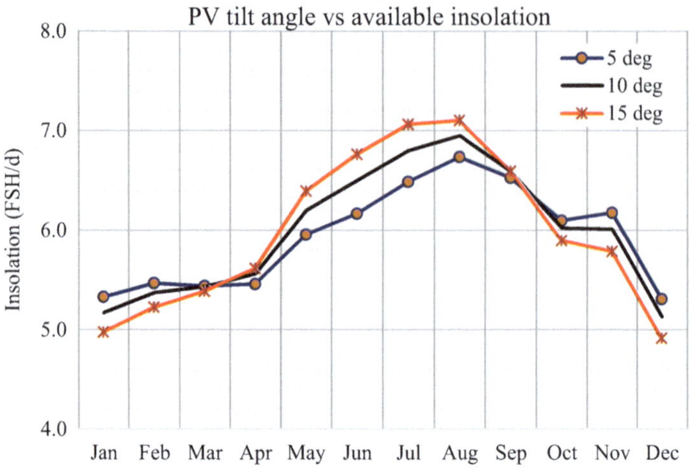

Fig. 7.3 Effect of PV tilt angle on available solar insolation

this case the 10° tilt angle matches the site latitude. It is recognized, however, that in the case of an off-grid PV system that needs to maximize the array output in the month with lowest insolation, the array tilt should be adjusted to maximize production for that month.

Figure 7.3 shows the effect of varying the tilt angle by ±5 degrees about the nominal on the available daily insolation at the Roma Tambo site. These monthly average insolation values were obtained using the NREL PV Watts software (National Renewable Energy Laboratory, n.d.) (PVWatts.nrel.gov). It is noted that lowering the tilt angle by 5 degrees increases the available daily insolation by approximately 0.2 FSH/d during the worst-case months of December, January and February corresponding to the Southern summer or cool season. The lower tilt angle, however, does reduce the array ability to shed precipitation and self-clean, a practical consideration that may offset the minimal increase in available insolation.

Question 2
In LUTW training sessions and in the sizing of the PV array, a daily insolation value of 3.5 full sun hours per day (FSH/d) was recommended for the region. How does this value compare to the available insolation? Is this a good design assumption?

Discussion 2
Referring to Fig. 7.3, it is noted that the lowest average daily insolation for a 10° array tilt is available during the months of December and January and approximately equal to 5.15 FSH/d. This value, however, is the long-term average at the site, and the 3.5 FSH/d value used by LUTW to size the PV array accounts for likely worst-case variation about this average. Assuming that the long-term data is normally distributed with five standard deviations about the mean, the likelihood of the daily insolation falling below 3.5 FSH/d, or −1.6 standard deviations, is only 5.5%. Therefore, the value of 3.5 FSH/d used by LUTW will ensure that the PV system

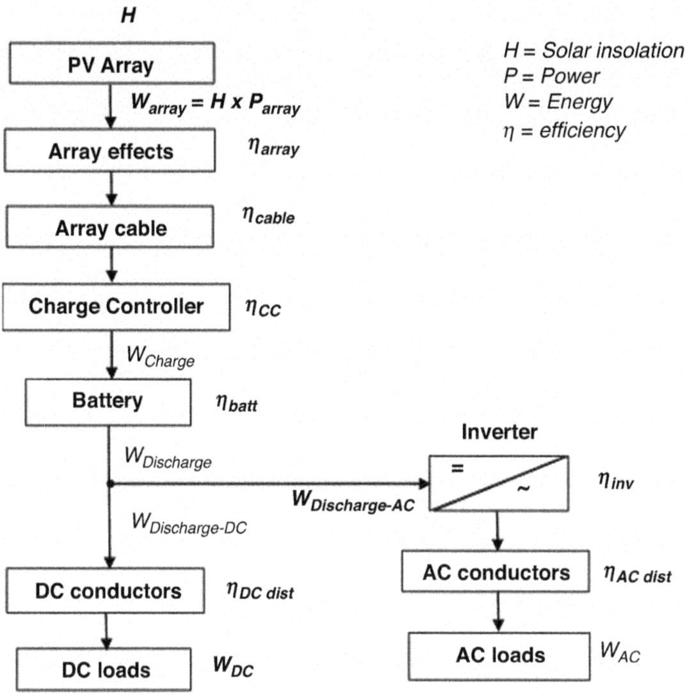

Fig. 7.4 Energy flow diagram of an off-grid PV system

can almost always generate sufficient energy. In those rare cases where the available insolation falls below 3.5 FSH/d, the battery will experience a deeper discharge than the design daily minimum, but the overall impact on battery live will be negligible.

Question 3

A 150Wp PV array was used to in the Roma Tambo installation. Is this array correctly sized?

Discussion 3

Figure 7.4 shows the energy flow diagram for a typical off-grid PV system serving both direct current (dc) and alternating current (ac) loads. The design sizing process starts with assessment of the average daily ac and dc loads to be served the system. These loads are the sum of all the essential individual loads that will be served by the distribution system and usually represent a pared down version of the client's wish list after non-essential loads are eliminated under advice. Of particular importance to the design process are (i) the daily energy requirements of the ac and dc loads, W_{AC} and W_{DC}, (2) the total power rating of the ac and dc loads, P_{AC} and P_{DC}, and (3) the ac surge load, $P_{AC\text{-}surge}$. These load calculations are conveniently done using tables or a spreadsheet, as shown in Tables 7.1 and 7.2. With the loads estimated, it is now possible to proceed with the sizing of the main system components: array, battery, charge controller and cables.

Table 7.3 Assumed component efficiencies and calculated performance ratios

	Array effects	Array cable	Charge controller	Battery	DC wires	Inverter	AC wires	PR_{DC}	PR_{AC}
% efficiency	88	99	98	80	98	87	99	69	59

The sum of the energy served and the losses in the two parallel paths to the battery represent the daily energy discharge (drain) on the battery. Dividing by its round trip, charge/discharge efficiency gives the energy required to re-charge the battery. Dividing further by the combined efficiency of the charge controller, the main array cable and the array to cater for losses in the energy train from the array to the battery gives an equation for calculating the energy output, W_{Array}, of the PV array. This is shown in Eq. (7.1).

$$W_{\text{Array}} = \frac{1}{\eta_{\text{Array}}\eta_{\text{Array cable}}\eta_{\text{Batt}}}\left(\frac{W_{\text{DC}}}{\eta_{\text{DC dist.}}} + \frac{W_{\text{AC}}}{\eta_{\text{AC dist.}}\eta_{\text{Inv}}}\right) = \frac{W_{\text{DC}}}{\text{PR}_{\text{DC}}} + \frac{W_{\text{AC}}}{\text{PR}_{\text{AC}}} \tag{7.1}$$

Here the array efficiency is ratio of actual to rated energy production of the PV array after losses due to shading, soiling, module mismatch, low irradiance, temperature effects, etc., which are taken into consideration. PR_{DC} and PR_{AC} are the performance ratios of the DC only and AC only energy paths, respectively.

If the daily design insolation (H) on the array is specified in available full-sun-hours (FSH), then allowing for a safety factor (SF), the power rating of the array at standard test conditions (STC) is calculated as shown in Eq. (7.2):

$$P_{\text{Array}-\min} > \text{SF}\frac{W_{\text{Array}}}{H} = \frac{\text{SF}}{H}\left(\frac{W_{\text{DC}}}{\text{PR}_{\text{DC}}} + \frac{W_{\text{AC}}}{\text{PR}_{\text{AC}}}\right) \tag{7.2}$$

As noted earlier, the solar insolation available to the PV array varies seasonally, and the PV array needs to be sized so that it is capable of serving the daily load during the time of year with the minimum daily average insolation. Also, since daily insolation values are usually available in the form of long-term monthly averages, an allowance should be made for the likely variances from the daily average minimum. A lower value of design insolation will result in a larger PV array and overall under-utilization, while a higher value of insolation will reduce the size of the PV array but will result in a deeper battery discharge.

Table 7.3 shows typical values for array losses and component efficiencies used by LUTW in their array sizing calculations. Using these field-realistic values, the DC and AC performance ratios are determined to be 0.69 and 0.59, respectively. From Eq. (7.2), the required array peak power rating is calculated as follows:

$$\text{Array power} > \left(\frac{\text{SF}}{H}\right)\left(\frac{W_{\text{dc}}}{\text{PR}_{\text{dc}}} + \frac{W_{\text{ac}}}{\text{PR}_{\text{ac}}}\right) = \left(\frac{1.25}{3.5}\right)\left(\frac{145.43\text{Wh/d}}{0.69} + \frac{107.14\text{Wh/d}}{0.59}\right)$$

$$> 140.13\text{Wp}$$

The 150W_p rating of the installed modules is only slightly larger than the calculated 140.13W_p. This represents a near perfect scenario given that the modules are sourced from a limited stock inventory.

Question 4

A 100 Ah, 12 V, AGM deep-cycle lead acid battery was used in this installation. Is this a good battery choice for this application? How will this battery perform?

Discussion 4

AGM (absorbent glass mat) lead acid batteries are a good choice where cost, availability and maintenance are the dominant considerations. AGM deep-cycle batteries are low-cost, maintenance-free, valve regulated lead acid (VRLA) batteries that are capable of powering loads over long discharge periods. They are the preferred lead acid batteries for off-grid PV applications. Compared to the more expensive lithium batteries, AGM batteries have shorter life and lower round-trip efficiency and cannot sustain the same level of deep discharge. However, they represent a proven, safe, reliable and familiar technology. Replacements are easily sourced and recycling options are widely available. Additionally, AGM lead acid batteries can operate at lower temperatures and can be matched to a wider range of charge controllers (Morningstar, 2022; Svarc, 2020).

The battery capacity was sized for 1 day of autonomous operation and 30% depth of discharge (DoD). A single day of autonomy represents the smallest required battery size, while 30% DoD is the level of discharge that represents a good trade-off between a longer battery cycle life and unused capacity. In general, lead acid batteries should not be discharged below 50%. For example, the Ritar RA12-100 battery, a lower-cost option in the LUTW inventory, will have a 1250 cycles life (or 3.4 years) for a 30% maximum DoD and only 450 cycles with 50% DoD. In both cases, however, the capacity retention of the battery will drop to 60% over the second half of the battery life (HengYang Ritar Power Co. Ltd, n.d.). Temperature also affects the available battery capacity (BC), and for a 16 °C average summer site temperature, a 5% in battery capacity from its 25 °C rating is expected (HengYang Ritar Power Co. Ltd, n.d.). Table 7.4 shows high, low and average seasonal temperatures for the site (Weather Spark, n.d.).

Assuming a single autonomous day, 30% DoD and a temperature derating factor (TDF) of 0.95, the required minimum battery capacity is estimated to be

Table 7.4 Seasonal site temperatures (Weather Spark, n.d.)

Season	Duration	High	Low	Avg.
Warm	4 Aug–17 Oct	18 °C	4 °C	17 °C
Cool	27 Dec–30 Mar	17 °C	1 °C	16 °C

1196.45 Wh. This calculation is shown in Eq. (7.3). This value is approximately equal to the 1200 Wh rating of the 12 V AGM battery used in this installation.

$$\mathrm{BC} > \mathrm{AD}\left(\frac{\frac{W_{dc}}{\eta_{\mathrm{DC\ dist}}}+\frac{W_{ac}}{\eta_{\mathrm{Inv}}\times\eta_{\mathrm{DC\ dist}}}}{\eta_{\mathrm{battery}}\times\mathrm{DD}\times\mathrm{TDF}}\right)=1d\left(\frac{\frac{145.43\mathrm{Wh/d}}{0.98}+\frac{107.14\mathrm{Wh/d}}{0.87\times0.99}}{0.8\times0.3\times0.95}\right)$$
$$=1196.45\mathrm{Wh} \tag{7.3}$$

With this battery capacity and a daily battery drain of 272.79 Wh/d, the maximum daily DoD with zero insolation is determined to be 28.4%. This calculation is shown in Eq. (7.4).

$$\mathrm{DoD}_{\mathrm{max}}=\frac{W_{\mathrm{Discharge}}}{\mathrm{BC}}=\frac{1}{\mathrm{BC}}=\frac{1}{1200Wh}\left(\frac{\frac{145.43\mathrm{Wh/d}}{0.98}+\frac{107.14\mathrm{Wh/d}}{0.87\ \mathrm{x}\ 0.99}}{0.8}\right)=0.284 \tag{7.4}$$

In most cases, the actual daily DoD will be much lower than 28.4% since some load utilization will occur during sunlight hours. This lower DoD would allow for a battery lifespan (projected) of approximately 3.5 years. In the very unlikely case of 2 consecutive days of zero solar insolation, the battery will discharge by 56.8% from its initial state of charge. It must be emphasized that a battery life of 3.5 years can only be realized with careful use which requires that the community be trained to pay careful attention to the state of battery charge and curtail their energy use when required.

Question 5

A 14 A, PWM charge controller with a cut-out voltage of 11.5 V and a maximum voltage of 30 V was used in this installation. What is the function of this system component? Is PWM a suitable technology for this application? Is the CC correctly sized to suit the array SC current and operating voltage range? How should the PV modules be connected?

Discussion 5

The charge controller is an essential component for optimizing battery life. By regulating the current flow to the battery, the charge controller provides optimal charging of the battery for given irradiance and load conditions. Very importantly, it prevents damage to the battery from over- and under-charging and discharging. It also blocks reverse current flow from the battery to the PV array.

Charge controllers can be broadly classified into two groups based on the technique used to control charging current, namely array current switching or array voltage control. Current switching controllers include series, shunt and pulse width modulation (PWM) controllers, with PWM controllers being the current technology of choice. Voltage control charge controllers utilize maximum power point tracking (MPPT) and can extract a significantly higher amount of energy from the PV array than PWM converters. They are also more efficient and can operate with a larger range of array voltages than PWM controllers, but due to greater

Table 7.5 PV module specifications (Yingli Solar, n.d.)

Technology	Polycrystalline	
Model	YL JS 75	
Pmp	75	W
Voc	22.9	V
Vmp	18.5	V
Imp	4.07	A
Isc	4.31	A
Coefficient V_{oc}	−0.08473	V/°C
NOCT	46	°C

technological complexity, they cost more and have a shorter lifespan. In general, PWM charge controllers are best suited to small off-grid systems where efficiency is not of utmost importance and the battery is maintained in a high state of charge (Zientara, 2022). A PWM controller is therefore a suitable technology choice for the RTCC project where cost, reliability, lifespan and replacement are more important than optimizing the PV array energy yield.

When choosing a PWM charge controller, it is important that its current rating is 125% greater than the PV array maximum short-circuit current. The same requirement applies to the main cable from the PV array to charge controller. For these calculations, the maximum short-circuit current is 125% of its rated short-circuit current. The additional 25% caters for conditions of higher than full sun irradiance (NFPA-70, 2023) (Table 7.5).

The array voltage varies with the PV module cell temperature (T_{cell}), and the charge controller must be able to operate safely over the range of possible voltages. Equation (7.5) shows the relationship between array voltage and cell temperature. T_{STC} is the Standard Test Condition temperature of 25 °C.

$$V_{\text{array}}(T_{\text{cell}}) = V_{\text{array}}(T_{\text{STC}}) + \alpha(T_{\text{cell}} - T_{\text{STC}}) \tag{7.5}$$

The array maximum voltage is its open-circuit voltage (V_{OC}) at minimum module cell temperature, while its minimum voltage is its maximum power voltage (V_{mp}) at maximum cell temperature. The PV module cell temperature is estimated from the ambient temperature (T_{ambient}), the module normal operating cell temperature (NOCT) and the irradiance G (in W/m^2) using Eq. (7.6).

$$T_{\text{Cell}} = T_{\text{ambient}} + \left(\frac{G}{800}\right)(T_{\text{NOCT}} - 20) \tag{7.6}$$

Using the minimum and maximum summer temperatures for the site and the NOCT value provided on PV module nameplate and allowing for a ±5 °C worst case temperature variation, the minimum and maximum cell temperatures are found to be −4 °C and 63.63 °C, respectively. It is noted here that the values of G used in the minimum and maximum cell temperature calculations are 0 W/m^2 and 1250 W/m^2, respectively.

Table 7.6 Charge controller specifications (Phocos, n.d.)

Type	PWM	
Model	Phocos CA14	
I_{max}	14	A
V_{max}	30	V
$V_{cut\text{-}out}$	11.5	V
$V_{reconnect}$	12.5	V
Recommended wire size	10	mm^2
Intended use	VRLA batteries only	
Max elevation	4000	m

Using the voltage extrapolation equation of Eq. (7.5), the corresponding maximum and minimum array voltages are found to be 15.23 V and 25.36 V, respectively. These values lie well within the 11.5 V to 30 V operating voltage range of the selected charge controller, indicating that the charge controller will not have high or low operational issues.

Question 6
Was the selected inverter a good choice for the specified ac loads?

Discussion 6
The first requirement of the selected inverter is that it must be capable of converting the nominal 12 V DC input, available at the battery, to a sinusoidal 230 V AC output at the correct frequency. A sine wave inverter is required for this application since the ac loads are sensitive electronic loads (cellphone chargers and radio) and motor load (blender) that require a sinusoidal voltage waveform for efficient operation. Referring to the inverter specification shown in Table 7.6, we can see that the selected inverter is a sine wave inverter with a nominal input voltage of 12 V DC and two possible AC output voltages—120 V AC and 230 V AC.

The second requirement relates to the inverter continuous power rating. To avoid overloading the inverter must be sized so that its continuous power rating is greater than the sum of the continuous power ratings of all the AC loads. Referring to Table 7.1 we see that the continuous load power rating of 375 W exactly matches the inverter continuous power rating. This implies that the inverter is capable of simultaneously powering all of the loads but without any spare capacity. However, given that the largest ac load is the blender with very short run times, the inverter capacity is considered sufficient. Without the blender, the invert will be lightly loaded, and sizing up would further lower overall utilization and efficiency while increasing cost.

The final requirement is that the inverter must have sufficient short-term overload capacity that it can supply the starting current required by the loads, i.e. the peak power rating of the inverter must be greater than the total surge power rating of the connected loads. From the information provided in Table 7.7, we see that inverter peak power rating of 700 W is greater than the 675 W surge rating of the AC loads in Table 7.1. Very importantly also, the inverter also features overload cut out and low battery audible alarm.

Table 7.7 Inverter specifications (Victron Energy, n.d.)

Type	Sine wave	
Model	Phoenix 12 V/375	
$V_{in\text{-}min}$	9.2	V
$V_{in\text{-}max}$	17.2	V
P_{rated}	375	W
P_{max}	700	W
V_{out}	120/230	V
Frequency	50/60	Hz
$P_{no\ load}$	5.6	W
$P_{no\ load\ eco\ mode}$	0.9	W

Question 7

Comment on the sizing of the conductors with respect to their ampacities and full-load voltage drops? Are the conductors correctly sized?

Discussion 7

Conductors can be a significant cost component of an off-grid PV system. Since conductor costs depend on both conductor length and ampacity, an efficient design will feature a layout plan that minimizes wire lengths and sizes. For each circuit, the conductors selected must meet the dual requirements of safety and performance, i.e. the conductors must have sufficient ampacity so that they can safely carry the maximum circuit current without excessive full-load voltage drop. For the conductors directly connected to the charge controller, i.e. those connecting the controller to the array, dc bus and the battery, the minimum current rating will be 156.25% of the PV array STC short-circuit current. For the other conductors, the minimum ampacity is 125% is its full-load current. A full voltage drop of 2% is often used as a guide in sizing DC conductor, but a higher value may be acceptable in cases where performance is not significantly affected.

Table 7.8 shows the calculated ampacities and full-load drop of the various PV system conductors. It is noted that in all cases the ampacity of the conductors utilized exceeds the required ampacity. However, in two cases, the percent full-load voltage drop exceeds 3% and requires investigation. The first case occurs along the main cable from the PV array to charge controller. At 0.62 V, this voltage drop is not large enough to prevent the battery from charging, but it may be sufficient to restrict the charging rate when it is most needed. A #12 AWG or #10 AWG would have been a better array cable choice. If this cable was sized up to an AWG #10 to match the charge controller manufacturer's recommendation, it would reduce the full-load voltage drop to an acceptable 1.32% or 0.24 V. The second case of high voltage drop occurs along the 100 ft. conductor run from the dc bus to the church. Sizing up to AWG #12 would halve the voltage drop, but given the length of this conductor run, this would result in a significant cost increase. Since this circuit only serves dc LED lamps that can operate at much lower than the nominal 12 V, a size up is not required for this conductor.

Table 7.8 Conductor data

Circuit	Length	Required ampacity	Conductor properties (altE, n.d.)			Nominal voltage	FL voltage drop	
	ft	A	AWG	A	Ω/ 1000 ft	V	V	%
Array-CC	15	13.48	14	15	2.525	18.5	**0.62**	**3.33**
CC-battery	5	13.48	14	15	2.525	12	0.21	1.71
CC-DC bus	1.5	13.48	14	15	2.525	12	0.05	0.38
Battery-INV	4	44.93	8	55	0.6282	12	0.04	1.50
DC load cct1	50	1.30	14	15	2.525	12	0.27	2.22
DC load cct2	20	2.50	14	15	2.525	12	0.17	1.39
DC load cct3	100	1.30	14	15	2.525	12	**0.53**	**4.38**
DC load cct4	40	2.50	14	15	2.525	12	0.33	2.28
Inv-AC bus	4	6.11	14	15	2.525	230	0.01	0.01
AC load cct1	50	2.04	16	10	4.015	230	0.62	0.28
AC load cct2	50	2.04	16	10	4.015	230	0.62	0.28
AC load cct3	40	2.04	16	10	4.015	230	0.50	0.23

7.4 Conclusions

Reflection is an effective educational technique for helping students to make sense of and develop deeper insights from their field experiences. This paper has shown, by way of a case study of a 2018 LUTW-EET volunteer field trip to install an off-grid PV system for a remote Peruvian community centre, how a carefully designed post trip reflective activity could be developed to strengthen students' understanding of off-grid PV system design principles. The reflective exercise involved recreating the project specifications and design and then critically analysing the key design stages and decisions. Pre-planning, information gathering and a structured post trip facilitation were therefore essential to the success of the activity. These requirements added a layer of complexity to the field venture, but overall reflective activity was positively received and significantly enhanced the overall student experience.

References

altE. (n.d.). Wire Sizing Chart for 12V, 24V, and 48V DC Systems. https://www.altestore.com/diy-solar-resources/wire-sizing-tool-for-12-24-and-48-volt-dc-systems/

Baaoum, M. (2018). Humanizing engineering education: A comprehensive model for fostering humanitarian engineering education. *International Journal of Modern Education Studies, 2*(1), 01–23.

Feron S. & Cordero, R. R. (2018). Is Peru prepared for large-scale sustainable rural electrification? *Sustainability, 10*(5), 1683.

HengYang Ritar Power Co. Ltd. (n.d.). RA12-100 specification sheet. https://www.ritarpower.com/uploads/ueditor/spec/RA12-100.pdf

IEA, IEA, IRENA, UNSD, World Bank, WHO. (2023). Tracking SDG 7: The Energy Progress Report 2023. Washington, DC: World Bank. © World Bank. License: Creative Commons Attribution—NonCommercial 3.0 IGO (CC BY-NC 3.0 IGO). IRENA IEA, IRENA, UNSD, World Bank, WHO.

Jensen, M., Hopps, E., & Brian R. (2019). Assessing the impact of off-grid solar electrification in rural Peru: Replicability, sustainability and socioeconomics. *International Journal for Service Learning in Engineering, Humanitarian Engineering and Social Entrepreneurship, 14*(3), Special Issue, 61–79.

Light Up the World. (n.d.). Powering opportunity with solar energy, lutw.org

Morningstar. (2022, April 29). Best battery Chemistry for off-grid solar applications. https://www.morningstarcorp.com/news/best-battery-chemistry-for-off-grid-solar-applications/

National Renewable Energy Laboratory. (n.d.). PV Watts calculator, version 8.2.1., https://pvwatts.nrel.gov/

NFPA-70. (2023). National Electric Code Article 690—Solar Photovoltaic (PV) Systems, 2023 Edition, Retrieved online from link.nfpa.org

Phocos. (n.d.). Phocos CA user manual. http://www.solar-bazaar.com/menuis/Phocos-User-Manual-CA.pdf

Shuwen, N., Yanqin, J., Wendie, W., Renfei, H., Lili, H., & Yan, L. (2013). Electricity consumption and human development level: A comparative analysis based on panel data for 50 countries. *Electrical Power and Energy Systems, 53*, 338–347.

Stern, M. J., & Powell, R. B. (2020). Field trips and the experiential learning cycle, 2020. *Journal of Interpretation Research, 25*(1), 47–51.

Stirling, A., Kerr, G., Banwell, J., MacPherson, E., & Heron, A. (2016). A practical Guide for Work-integrated Learning. Effective Practices to Enhance the Educational Quality of Structured Work Experiences Offered through Colleges and Universities. Higher Education Quality Council of Ontario, 2016.

Svarc, J. (2020). Lead-acid vs Lithium-ion Batteries. *Clean Energy Reviews*. https://www.cleanenergyreviews.info/blog/simpliphi-pylontech-narada-bae-lead-acid-battery

United Nations General assembly. (2015, October). Resolution adopted by the General Assembly on 25 September 2015, A/RES/70/1.

Victron Energy. (n.d.). Phoenix inverters: 250VA – 1200VA, 230V and 120V, 50Hz or 60Hz, victronenergy.com.

Weather Spark. (n.d.). Climate and average weather year-rRound in Catac Peru. https://weatherspark.com/y/20517/Average-Weather-in-Catac-Peru-Year-Round

Yingli Solar. (n.d.). JS 75 PV module, Information extracted from the JS 75 PV module nameplate.

Zientara, B. (2022, July 14). What is a solar charge controller and why are they important? *Solar Reviews*. https://www.solarreviews.com/blog/what-is-a-solar-charge-controller.

Chapter 8
Renewable Energy as a Factor of the Sustainable Development of Society

Sergey M. Karabanov, Irina M. Kuvakova, and Yuri A. Merkulov

8.1 Introduction

The modern world is going through a difficult period of revision of the fundamental bases that determined the life of previous generations. The 2015 United Nations Summit proposed for consideration a document with 17 goals for the sustainable development of civilization, which were approved by the leaders of 193 countries. The concept of sustainable development implies a transition from the global development of economic systems (minimizing costs and maximizing profits) to harmonization of the natural, spiritual, and social environments (Kuvakova, 2019). The reasons for creating the concept were the aggravation of problems of globalization and competition of transnational corporations, the differentiation of industrial potential and financial capital, the scientific and technological revolution in IT industry, climate change, and social inequality.

Sustainable development is a type of development in which current generations meet their needs without depriving future generations of the ability to meet their own needs.

The concept of sustainable development suggests the unity of three factors: the economic platform, the social program, and the environmental component.

Analyzing this concept, it is necessary to pay attention to the harmonious relationship of all factors that determine the sustainable development:

1. Sustainable economic development (growth of gross national product, savings and investments, reduction of inflation and unemployment, etc.).

S. M. Karabanov (✉) · Y. A. Merkulov
Ryazan State Radio Engineering University, Ryazan, Russia

I. M. Kuvakova
Academy of Labor and Social Relations, Moscow, Russia

A. Sayigh (ed.), *Transition Towards a Carbon Free Future*, Innovative Renewable Energy, https://doi.org/10.1007/978-3-031-61660-0_8

2. Sustainable growth of living standards (average income, level of education, unemployment, inflation, socio-economic differentiation, etc.).
3. Sustainable demographic trends (population size and composition, total birth and mortality rates, marriages and divorces, etc.).
4. Environmental safety: effective environmental development programs help to create effective sustainable development programs, which provide the conditions for the business development.

In this regard, in the framework of the Paris Climate Agreement of December 12, 2015, the main attention was paid to the threat to the environment due to the emission of greenhouse gases and other harmful substances (Paris Agreement, n.d.). It has been established that the main share of emissions falls on the spheres of energy and human activity.

Modern power industry is represented by two groups of sources: traditional (nonrenewable energy based on mineral raw materials) and renewable energy (solar, wind, water, etc.).

The sustainable development of the world civilization is largely connected with the sphere of energy and the population education.

The purpose of the study is to develop the proposals for improving education in the field of renewable energy to ensure its development, based on key features of specific territories.

8.2 Renewable Energy Development

Renewable energy sources include the following:

- Solar energy.
- Wind energy.
- Tidal energy.
- Wave power of oceans, seas, and rivers.
- Geothermal energy.
- Low-potential thermal energy of the earth, air, and water.
- Biomass (plants, production, and consumption wastes).
- Biogas.

Based on the development analysis of renewable energy, the total global power supply of PV stations could grow nearly sixfold over the next 10 years, reaching 2840 GW by 2030 and rising to 8519 GW in 2050 (Karabanov et al., 2011). This implies a total installed capacity in 2050 that is almost 15 times higher than in 2020. The technical potential of worldwide renewable energy is illustrated in Table 8.1 (Elistratov, 2013).

Table 8.2 shows indicators of renewable energy sources for 2020 and 2021 (Renewable Capacity Statistics, 2022; Statistical Review of World Energy, 2022).

Photovoltaics continues to dominate with other renewable energy sources.

Table 8.1 Technological capacity of renewable energy in the world

	Resources, TWh/year·10^3	
Energy type	Theoretical	Technological
Solar energy		
At the top of the atmosphere	183,000	–
At the ground surface	75,913	5708
At the land surface	26,370	2283
At the ocean surface	49,543	3425
Wind energy	1982	21
Geothermal energy (up to 10 km):		
Pouring springs	34	0,4
Hydrothermal vents	1256	137
Petrogeothermal resources	34,247	2853
Global ocean energy		
From salinity gradient	39,954	399
Heat energy	11,5	0,6
Currents energy	8	0,14
Tidal energy	3	0,8
Wave energy	3,4	0,13
At the land	41	4,6
At the global ocean	22	1,7
Organic waste	2,3	1,4
Hydraulic energy		
First-order stream energy	3,7	1,7
Minor stream energy	1,7	0,85
Total	151482,6	9130,32

8.3 Renewable Energy in the Russian Federation

In terms of natural potential and insolation resource, a part of the Russian Federation territory corresponds to those countries where photovoltaics is developing at a torrid pace (Germany, Italy, and Spain). The southern territories of the Russian Federation, the regions of Transbaikal, and the Far East have a great potential for photovoltaics, which can provide the Russian Federation with a leading position in the global renewable energy industry, and, at the same time, solve the current problems of supplying energy to hard-to-reach and isolated territories. Therefore, there are no restrictions on the use of renewable energy, including photovoltaics (Karabanov et al., 2020). However, the share of renewable energy in the total energy balance of the Republic of Belarus and the Russian Federation currently does not exceed 0.5%, while this figure in some countries (Germany, Denmark) already exceeds 20%.

The theoretical potential of photovoltaics in the Russian Federation is more than 2300 billion tons of standard fuel (tsf), and the economic potential is 12.5 million tons of standard fuel. The potential of photovoltaics supplying the territory of the

Table 8.2 Indicators of renewable energy sources for 2020 and 2021

	2020	2021
Investment		
Total annual investment, bln, US$	342,7	365,9
Installed energy capacity		
Installed renewable energy capacity together with hydroelectric stations, GW	2807,3	3063,9
Installed renewable energy capacity without hydroelectric stations, GW	1596,2	1833,9
Installed capacity of hydroelectric stations, GW	1211,1	1230
Installed capacity of PV power stations, GW	716,8	849,5
Installed capacity of wind power stations, GW	731,8	824,9
Installed capacity of biofuel power plants, GW	133	143,4
Installed capacity of geothermal power stations, GW	14,1	15,6
Installed capacity of midget power plant power stations, GW	0,5	0,5
Share of renewable energy together with hydroelectric stations, %	36,6	38,3
Share of renewable energy without hydroelectric stations, %	19,2	20,4
Capacity addition of renewable energy, GW	266	257
Countries with renewable energy of more than 10 GW (without hydroelectric stations)	20	22
Countries with renewable energy of more than 1 GW (without hydroelectric stations)	57	58

Russian Federation for 3 days exceeds the energy of the entire annual electricity production in the country (Karabanov & Bezrukikh Jr., 2016). Thus, the natural potential makes it possible to develop photovoltaics in the Russian Federation and the Republic of Belarus.

The photovoltaics market in the Russian Federation is developing, showing annual growth. Now the share of renewable energy in the unified energy system of the Russian Federation does not exceed 0.5%. And the installed capacity of PV power plants is 1354.88 MW.

The need to develop photovoltaics is determined by two main reasons (Karabanov et al., 2016):

- Limitation of traditional natural energy resources. Under current conditions and production levels, the Russian Federation is provided with oil reserves for 30–50 years and gas for 50–100 years by various estimates.
- The need of combating climate change.

The quantitative assessment of sustainable development can be differentiated according to various components. For example, sustainable development in the field of ecology can be evaluated by the renewable energy development. To assess the sustainable development, it is necessary that the growth in energy use is to be provided by the use of renewable energy (ΔE_{up}) and the reduction of use of traditional energy types for power generation (ΔE_d). So, the renewable energy volume is $E_{\mathrm{R}} \geq (\Delta E_{\mathrm{up}} + \Delta E_d)$.

The value of E_R should meet the targets for reducing emissions of greenhouse gases and harmful substances, which are defined by the United Nations in the Paris climate agreement.

The index of environmental compatibility of a territory (I_{ect}) may be another indicator of sustainable development. It is calculated as the ratio of the capacity of renewable energy, RE, generating stations (TWh) to the capacity of traditional energy, TE (TWh):

$$I_{ect} = \frac{RE}{TE}$$

The index of environmental compatibility of a territory (I_{ect}) takes into account the gradual replacement of the capacities of traditional energy by renewable energy.

The ecological situation of the territories will be stable with a coefficient equal to 1.

8.4 Trends in the Development of Education in the Field of Renewable Energy

The analysis of current trends (Kuvakova & Melnikov, 2022) in the field of education showed that the most significant of them are as follows:

- Application of modern information and communication technologies.
- Wide use of remote technologies (Karabanov et al., 2022).
- Individualization of the learning process.
- Interdisciplinary studies.
- Active teaching methods.

The population survey for the purpose of knowledge about renewable energy showed that only 20% of respondents have a superficial understanding of this issue. The rest of the respondents have little idea about it.

Modern civilization is characterized by dynamic changes on a global scale in the means of intelligence formation, intellectual interactions (a man, a machine; a computer, network; artificial intelligence, society), growth of competencies and professionalism (professional skills), long-range and strategic management, and self-organization of talents (Kuvakova, 2016). Currently, new information and communication technologies in the field of education, including online learning and the use of artificial intelligence, are widely used (Karabanov & Bezrukikh Jr.).

The need to expand the influence of the education sector on the future of any society and the world as a whole has been prepared by the conditions of continuous renewal, growth, and humanization of all areas of life activity from each person to all mankind. The possibilities of new approaches to education in the field of renewable energy give a strong impetus to the simultaneous development of all aspects of life, business, and the population of the territory, based on its specific character.

The transition to the effective use of the renewable energy education (REE) makes it possible to solve not only the problems of economic growth and the environment, decent employment, and healthcare but also the achievement of the well-being of each family in every country and the world as a whole.

The goal of REE is to make a person active in promoting the ideas of green energy.

The issues of creating a variety of learning processes and a large-scale approach in the field of REE are critical (Karabanov et al., 2022). We offer the following directions in educational processes:

- School education.
- Professional education: college, university, further vocational training, special training courses in technology and practice.
- Pop media education, including using online education, state and non-state pop media and Internet resources, volunteer movement to promote renewable energy, proactive activities of climate campaigners, democratic procedures of civil society, scientific, educational and expert communities.

Government and non-government support for the promotion and involvement of the population in the development and expansion of the use of renewable energy is extremely important. It includes the measures for stimulating tender R&D, innovation for the use of renewable energy, the creation of special funds, and financial support and co-financing (business-population-government) of renewable energy through the authorities. The effective elements for promotion of renewable energy achievements are international organizations and unions and international scientific and research and practical conferences.

A special phenomenon of our time is distance education. Distance learning is a form of educational process that uses a combination of information, software, hardware, and computer tools and technologies in order to provide students with the opportunity to master a certain amount of required knowledge in a non-contact way. This type of training is used both in obtaining basic education and additional education.

In Russia, serious development of the additional education took place in the second half of the 1990s. The key players are the leading Russian universities with great experience in this area: Lomonosov Moscow State University, Bauman Moscow State Technical University, Ryazan State Radio Engineering University, etc. At the moment, Ryazan State Radio Engineering University has additional education programs including aspects of the renewable energy development.

8.5 Conclusion

The paper proposes a mechanism for promoting the development of renewable energy through the education system, including additional education. The development of renewable energy will not only improve the environmental situation in the world but also achieve sustainable development goals. Knowledge of the basics of renewable energy will expand the possibilities of its application and stimulate the development of education.

References

Elistratov, V. V. (2013). *Renewable energy* (2nd ed.). Amplified.

Karabanov, S. M., & Bezrukikh, P.P. Jr. (2016). The Forecast of the world renewable energy development till 2020, In *Conference Proceedings*, EuroSun 2016, https://doi.org/10.18086/eurosun.2016.01.156, ISBN 978-3-9814659-6-9, pp.1654–1660.

Karabanov, S. M., & Bezrukikh, P.P. Jr.. The Study of the efficiency of Photovoltaics and wind power energy, In *Proceedings of 32nd European Photovoltaic Solar Energy Conference*, pp. 2953–2957, ISBN: 3-936338-41-8, Paper https://doi.org/10.4229/EUPVSEC20162016-7DV.4.7

Karabanov, S. M., Kazakova, N. V., & Kuropov, M. V. (2011). The prospects for Photovoltaics development in Russia, In *Proceedings of the 26th European Photovoltaic Solar Energy Conference*, ISBN 3-936338-27-2, pp. 4637–4639.

Karabanov, S. M., Kirakovskiy, V. V., & Bezrukikh, P.P. Jr. (2016). Research of application of renewable energy sources and energy saving technologies in residence houses construction in the Central part of Russia, In *Conference Proceedings*, EuroSun 2016, https://doi.org/10.18086/eurosun.2016.01.156, ISBN 978-3-9814659-6-9, pp.25–32.

Karabanov, S. M., Kistrina, E. I., Bezrukikh, P. P., Kuvakova, I. M., Suvorov, D. V., & Perepelkin, D.A. (2020). Socio-economic aspects of microgrid application in Russia, In *Proceedings of Electro 2020*, pp. 1–4, https://doi.org/10.1109/ELEKTRO49696.2020.9130299

Karabanov, S. M., Merkulov, Y. A., Kuvakova, I. M., & Kistrina, E. I. (2022). Materials science training programs for students and masters with a specialization in electric engineering. *ELEKTRO (ELEKTRO)*, 1–5. https://doi.org/10.1109/ELEKTRO53996.2022.9803541

Kuvakova, I. M, (2016). *Self-organization as a social phenomenon*, Collection of research papers, Issue 49, Moscow, pp. 435–440.

Kuvakova, I. M. (2019). Labor as creativity, reproduction and professional management are the basic tasks of the modern state in the context of the development of the information society and globalization, In *Proceedings of the 2nd International Research-to-Practice Conference "Kostin reading" (April 19, 2019)*, Academy of Labor and Social Relations, Moscow, pp. 105–109.

Kuvakova, I. M., & Melnikov, S. B. (2022). Innovation in the life of territories (regions). Program for increasing the well-being of the population by supporting tools: from problems to solutions, Trends of economic and management education. In *Proceedings of international forum* (pp. 55–62).

Paris Agreement. https://docs.cntd.ru/document/542655698

Renewable Capacity Statistics. (2022). https://www.irena.org/publications/2022/Apr/Renewable-Capacity-Statistics-2022

Statistical Review of World Energy. (2022). https://www.bp.com/content/dam/bp/business-sites/en/global/corporate/pdfs/energy-economics/statistical-review/bp-stats-review-2022-full-report.pdf

Chapter 9
The Influence of Dust on Photovoltaic Performance: Past, Present, and Future Perspectives

Hussein A. Kazem

9.1 Introduction

Solar energy is a renewable and sustainable source of power that harnesses the energy from the sun to generate electricity (Al-Maamary et al., 2016). One of the key technologies used to capture solar energy is photovoltaic (PV) systems, commonly known as solar panels. PV converts sunlight directly into electricity through the "photovoltaic effect", which occurs when certain materials, typically silicon-based semiconductors, absorb photons and release electrons, creating an electric current (Salam et al., 2013). PV systems have gained popularity due to their clean and environmentally friendly nature. They can be installed on various structures, such as rooftops, solar farms, and even portable devices, enabling decentralized energy generation. Solar energy has numerous benefits, including reducing greenhouse gas emissions, decreasing reliance on fossil fuels, and providing electricity in remote areas with limited access to the power grid (Sopian et al., 2020).

As technology advancements continue to improve the efficiency and affordability of PV systems, solar energy is becoming an increasingly viable option for meeting global energy demands. The widespread adoption of solar energy holds great potential to contribute to a more sustainable and renewable future (Al-Waeli et al., 2017a).

PV performance refers to the efficiency and effectiveness of a PV in converting sunlight into electrical energy. It is a key factor in determining the overall effectiveness and economic viability of solar power generation. The performance of a PV is influenced by various factors. One crucial factor is the efficiency η_{pv} of the solar cells themselves, which determines how effectively they can convert irradiance into electricity. Advances in cell technologies, such as "monocrystalline" m-Si,

H. A. Kazem (✉)
Sohar University, Sohar, Oman
e-mail: h.kazem@su.edu.om

A. Sayigh (ed.), *Transition Towards a Carbon Free Future*, Innovative Renewable Energy, https://doi.org/10.1007/978-3-031-61660-0_9

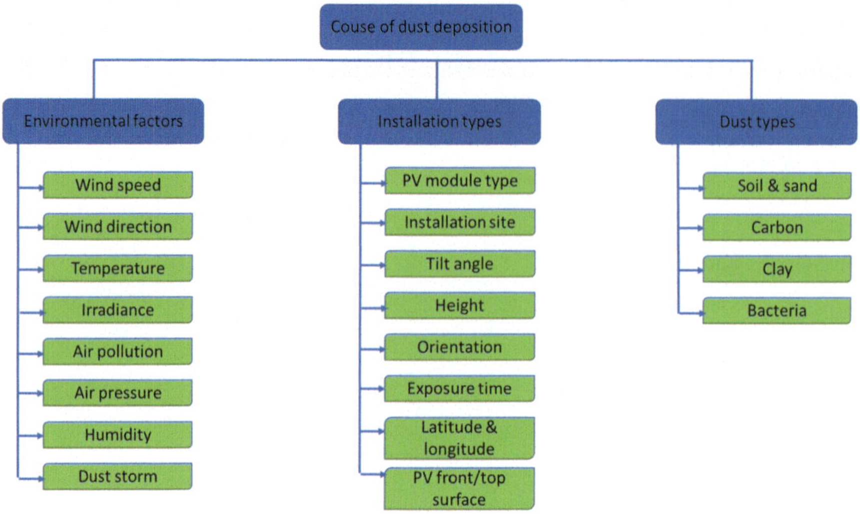

Fig. 9.1 Sources of dust deposition on PV module

"polycrystalline" p-Si, and "thin-film" TF cells, have significantly improved the overall η_{pv} over time. Other factors that affect PV performance include environmental conditions such as "ambient temperature" T_a, shading and dust accumulation (Alnasser et al., 2020). Higher T_a can reduce the efficiency of solar cells, while shading caused by nearby structures, trees, or debris can block irradiance and diminish the system's output. Dust and dirt accumulation on the surface of PVs can also reduce the amount of irradiance reaching the cells, leading to decreased performance (Chaichan & Kazem, 2015). There are various reasons of dust accumulation on PV as shown in Fig. 9.1.

Regular maintenance and cleaning of PV modules are essential to optimize performance. Manual cleaning methods, such as using a brush and mild detergent, or automated techniques like "brushing robots" and "high-pressure water jets" can help remove dust and debris, ensuring maximum light absorption by the cells. Monitoring the PV performance is crucial for identifying any issues or inefficiencies (Al-Waeli et al., 2019). This is typically done through the measurement of electrical parameters like "voltage" V_{pv}, "current" I_{pv}, and "power output" P_{pv}. Continuous monitoring allows for early detection of any malfunctions or performance degradation, enabling timely maintenance and maximizing the system's energy production. Improving PV performance is an ongoing area of research and development. Efforts are focused on increasing η_{pv}, developing innovative cleaning and maintenance techniques, and enhancing system design and integration (Kazem et al., 2020a). Ultimately, optimizing PV performance contributes to maximizing the electricity generation potential of solar energy and promoting its widespread adoption as a clean and sustainable power source (Al-Waeli et al., 2017b).

The accumulation of dust on PV modules has a significant impact on their performance and η_{pv}. Dust particles can settle on the surface of the PVs, reducing the amount of irradiance that reaches the solar cells. This reduction in light absorption directly affects the P_{pv}. Dust acts as a barrier, blocking and scattering the incoming solar irradiance (Kaldellis & Fragos, 2011). This leads to a decrease in the amount of light available for conversion into electricity by the solar cells. As a result, the P_{pv} decreases, reducing its overall performance. The effect of dust on PV behaviour is influenced by various factors. The composition and characteristics of the dust particles, such as their size, shape, and optical properties, play a role in determining the extent of performance degradation. Additionally, environmental conditions like "wind speed", "relative humidity", and local climate can affect the rate and amount of dust deposition on the PV modules (Vedulla et al., 2023; Cordero et al., 2018; Mani & Pillai, 2010).

The performance degradation caused by dust accumulation can be significant, particularly in arid and dusty regions (Kazem et al., 2013). Therefore, regular cleaning and maintenance of PVs are essential to mitigate the negative impact of dust. Manual cleaning methods, such as using soft brushes or sponges, along with mild detergents or water, can help remove the dust particles from the surface (Cordero et al., 2018). Automated cleaning systems, "including robotic" or "water-based cleaning techniques", are also used to efficiently clean large-scale PV installations. Efforts are ongoing to develop advanced dust mitigation strategies and improve the design of PV modules to minimize the impact of dust accumulation. Researchers are exploring the use of self-cleaning coatings, surface treatments, and advanced materials that repel or reduce the adherence of dust particles (Jordan & Kurtz, 2013). Additionally, modelling and simulation techniques are employed to better understand the dust effect and optimize the design and performance of PV systems under dusty conditions (Ghazi et al., 2014). By addressing the challenges posed by dust accumulation and developing effective mitigation strategies, the performance and efficiency of PV systems can be enhanced. This will ensure maximum power generation and contribute to the broader adoption of solar energy as a clean and sustainable source of electricity.

The accumulation of dust on PV modules is a common issue that can significantly impact their performance and efficiency. To maintain optimal power output, regular cleaning of the PV modules is necessary to remove the dust particles that obstruct sunlight absorption (Kazem & Chaichan, 2019). Various cleaning methods are employed to effectively clean the dust accumulated on PV modules. Manual cleaning methods involve using soft brushes, sponges, or cloths to physically remove the dust particles from the surface of the modules. Mild detergents or water can be used in conjunction with these cleaning tools to enhance the cleaning process (Alagoz & Apak, 2020).

Automated cleaning systems are also utilized to clean large-scale PV installations efficiently. These systems often involve the use of robotic mechanisms equipped with brushes or wipers that move along the surface of the modules, removing the dust particles. High-pressure water jets or sprayers can also be employed to dislodge and wash away the dust. In addition to the traditional cleaning methods, there are advanced self-cleaning technologies being developed (Yousif et al., 2019). These technologies include the application of special coatings or surface treatments that repel dust or have a self-cleaning effect. These coatings can prevent the adherence of dust particles or enable them to be easily washed away by rain or other environmental factors. The choice of cleaning method depends on various factors such as the size of the PV installation, accessibility of the modules, the level of dust accumulation, and the specific requirements of the site. It is important to consider the manufacturer's recommendations and guidelines when selecting a cleaning method to ensure that it does not damage the PV modules or compromise their performance (Al-Waeli et al., 2019).

Regular and proper cleaning of PV modules is essential to maintain their efficiency and maximize power generation. Dust accumulation can significantly reduce the amount of sunlight reaching the solar cells, leading to decreased power output. By implementing appropriate cleaning methods, PV system owners can ensure that their modules operate at their optimum performance levels, resulting in improved energy production and overall system reliability.

This research aims to provide a contemporary and comprehensive analysis of the impact of dust on PV modules, considering present and forthcoming conditions. It includes the proposal of a dust impact model, which poses some challenges, alongside the incorporation of the latest developments in the field.

9.2 Dust Effects on Glass and Transparent Materials

The impact of dust on the surface of PV glass and other transparent materials is a significant concern in the field of solar energy. Dust accumulation on these surfaces can have detrimental effects on the performance and efficiency of PVs (Alnasser et al., 2020). It can reduce the amount of light transmitted through the glass, leading to decreased power output as shown in Fig. 9.2. Moreover, dust particles can cause physical damage, such as scratching and etching, which further degrade the optical properties of the materials. Understanding the effects of dust on PV glass and transparent materials is crucial for developing effective cleaning and mitigation strategies to ensure optimal performance and longevity of PV (Kazem & Chaichan, 2016a). In addition to reducing the amount of transmitted light and causing physical damage, dust accumulation on PV glass surfaces and transparent materials can also lead to other negative effects (Kazem et al., 2020b):

- Heat accumulation: Dust acts as an insulating layer on the surface of PV modules, trapping heat and reducing their ability to dissipate heat efficiently. This can lead to an increase in operating temperatures, which negatively impacts the performance and lifespan of the modules (Darwish et al., 2016).
- Moisture retention: Dust particles can hold moisture, creating a favourable environment for the growth of microorganisms such as algae and fungi. This can further deteriorate the optical properties of the glass surface and contribute to corrosion and degradation of the PV modules (Ahmed Darwish et al., 2017).

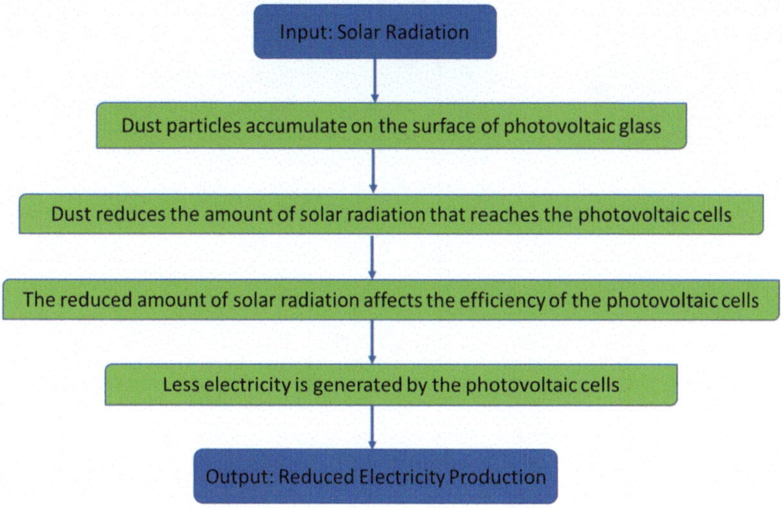

Fig. 9.2 Dust accumulation effect on PV module glass

- Soiling rate variations: The rate at which dust accumulates on PV glass surfaces can vary depending on factors such as location, climate, and environmental conditions. Understanding the rate of soiling and the composition of the dust particles is essential for developing effective cleaning and maintenance strategies (Sayigh, 1979).
- Economic implications: The presence of dust on PV modules reduces their overall energy production, resulting in lower electricity generation and financial losses. Regular cleaning and maintenance of the modules are necessary to minimize these losses and maximize their energy output (Al-Badra et al., 2020).

To address these challenges, ongoing research focuses on developing improved cleaning techniques, such as "automated robotic systems" and "self-cleaning coatings", to efficiently remove dust and maintain optimal PV performance. Additionally, advanced modelling and simulation tools are being developed to better understand the dynamics of dust accumulation and its effects on the optical properties of PV glass surfaces and transparent materials.

Dust accumulation on the surface of PV modules can have both positive and negative effects on their performance, including the "short circuit current" (Isc) and "open circuit voltage" (Voc) as shown in Fig. 9.3 (El-Shobokshy & Hussein, 1993). The effect breakdown:

Short-circuit current (Isc):

- Negative effect: Dust accumulation can reduce the short-circuit current of PV modules. When dust settles on the surface of the modules, it creates a barrier that blocks sunlight from reaching the solar cells (Alnasser et al., 2020). This reduces the amount of incident light available for conversion into electrical current, resulting in a decrease in Isc.

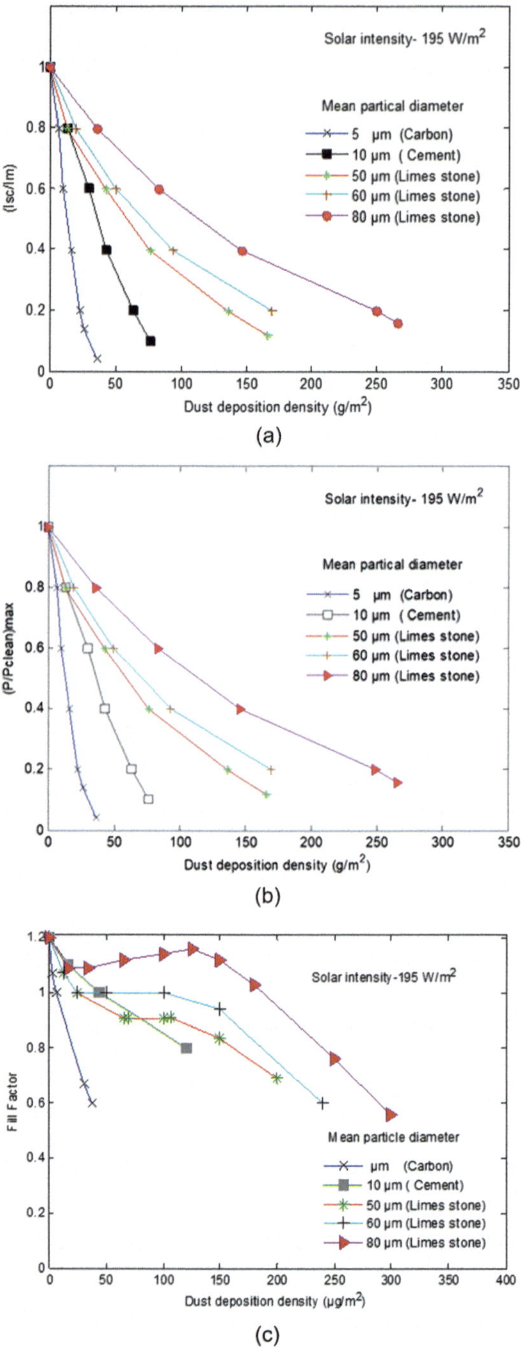

Fig. 9.3 PV module performance degradation due to the dust deposition: (**a**) short-circuit current, (**b**) output power, (**c**) fill factor (El-Shobokshy & Hussein, 1993)

- Positive effect: In certain cases, a thin layer of dust can increase the Isc. This occurs when the dust layer diffuses the incoming sunlight, causing it to spread and reach a larger surface area of the solar cells. This diffusion effect can enhance the current generation, leading to a temporary increase in Isc.

Open-circuit voltage (Voc):

- Negative effect: Dust accumulation can also reduce the open-circuit voltage of PV modules. When dust particles cover the surface of the solar cells, they act as a partial shading mechanism. This shading effect can create local variations in the illuminated area, resulting in non-uniform voltage distribution across the module (Darwish et al., 2018). As a consequence, the overall Voc may decrease due to the presence of dust.
- Positive effect: Similar to Isc, in some cases, a thin layer of dust can increase the open-circuit voltage. This can occur when the dust layer diffuses and scatters the incident light, leading to a more uniform distribution of photons on the solar cell surface. As a result, the overall Voc may experience a temporary increase.

It's important to note that while a thin layer of dust may have temporary positive effects on Isc and Voc, excessive or long-term accumulation of dust can have more significant negative impacts on the overall performance of PV modules (Chaichan & Kazem, 2020). It is generally recommended to keep solar panels clean and free from dust and debris to maximize their efficiency and power output (Darwish et al., 2015). Regular cleaning or employing self-cleaning mechanisms can help mitigate the negative effects of dust accumulation on PV modules.

In addition, wind can have both positive and negative effects on the dust accumulated on PV modules. Wind can influence the dust on PV modules (Hussain et al., 2017):

Dust Redistribution:

- Positive effect: Wind can help in redistributing and removing loose dust particles from the surface of PV modules. When strong winds blow over the modules, they can dislodge and carry away some of the accumulated dust. This can result in a partial cleaning effect, improving the performance of the modules by reducing the shading and blocking of sunlight.
- Negative effect: On the other hand, if the wind is not strong enough to remove the dust completely, it may cause the dust to redistribute and settle in different areas of the module. This can create new shading patterns or uneven dust distribution, potentially impacting the module's performance negatively.

Abrasion and Scratching:

- Negative effect: Wind-blown dust can act as abrasive particles when they come into contact with the surface of PV modules. Over time, the abrasive action of dust carried by wind can lead to scratches and microdamage on the module's protective coatings, such as the "anti-reflective coating". These scratches can reduce the light transmission and overall efficiency of the modules.

Maintenance Requirements:

- Cleaning frequency: Wind conditions can influence the cleaning frequency required for PV modules. In areas with high wind speeds and dust levels, modules may require more frequent cleaning to ensure optimal performance. Windy conditions can lead to faster accumulation of dust, necessitating regular cleaning to maintain efficiency (Kazem & Chaichan, 2016b).

To minimize the negative effects of wind-blown dust on PV modules, regular cleaning and maintenance are essential. Cleaning methods such as rinsing with water, using soft brushes, or employing automated cleaning systems can help remove accumulated dust and maintain the efficiency of the modules (Verma et al., 2011). Additionally, considering the installation of protective measures like windbreakers or barriers around the PV array can reduce the amount of dust carried by wind, reducing the maintenance needs (Alagoz & Apak, 2020).

9.3 Dust Effects on Mirrors

Dust accumulation on mirrors used in solar systems, such as "concentrated solar power" (CSP) or "concentrated photovoltaics" (CPV), can have several effects on their performance (see Fig. 9.4a):

- Dust settling on the surface of mirrors reduces their reflectivity. The presence of dust particles creates a layer that blocks or scatters incoming sunlight, reducing the amount of light that can be concentrated or reflected onto the receiver or PV cells (Niknia et al., 2012). This decreases the overall efficiency and power output of the solar system.
- Dust accumulation on mirrors can lead to shading and the formation of hot spots. When dust settles unevenly on the mirror surface, it can create localized shading, blocking sunlight from reaching certain areas (Zhao et al., 2021). This results in non-uniform illumination on the receiver or PV cells, leading to reduced performance and potential hot spots that can damage the system.
- Dust accumulation on mirrors increases the maintenance requirements of solar systems. Regular cleaning is necessary to remove the accumulated dust and restore the reflectivity of the mirrors (Biryukov, 1998). Failure to clean the mirrors regularly can lead to a progressive reduction in system efficiency over time.
- The rate at which dust accumulates on mirrors depends on several factors, including the local environment, climate, and dust composition. Some regions may experience higher dust deposition rates due to arid or dusty conditions, while others may have lower rates (Bolles et al., 2019). Additionally, the composition of the dust, such as its particle size, shape, and chemical composition, can influence its adhesive properties and the ease of removal during cleaning.

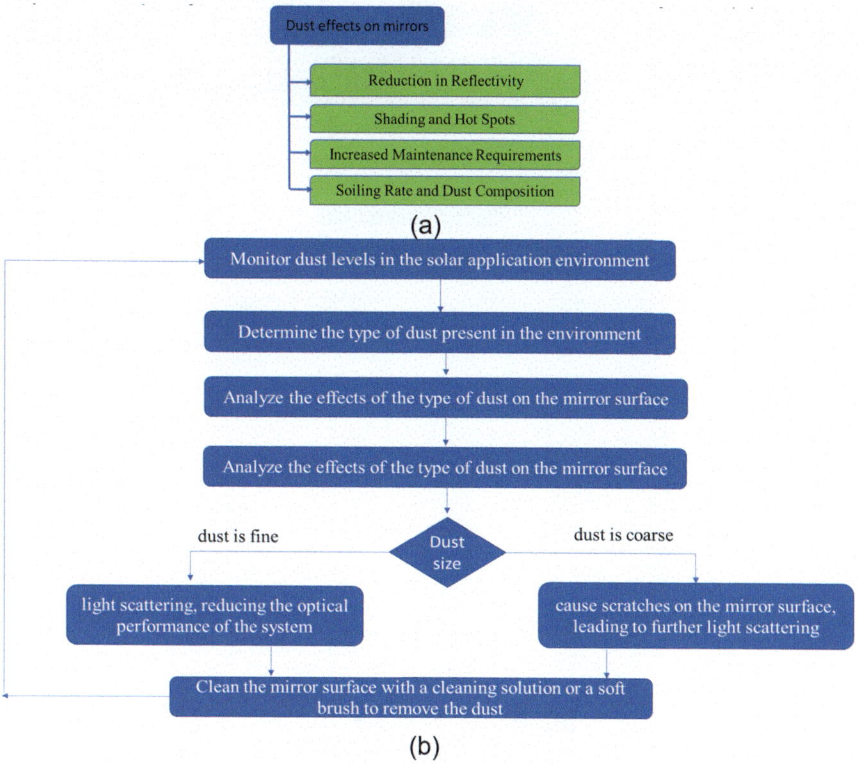

Fig. 9.4 Impact of dust on mirrors: (**a**) effects, (**b**) analysis

To mitigate the negative effects of dust on mirrors in solar systems, regular cleaning and maintenance are crucial. Cleaning methods can vary depending on the system and mirror type but often involve water rinsing, brushing, or the use of specialized cleaning equipment. Implementing preventive measures such as dust barriers, air blowers, or automated cleaning systems can help reduce the frequency of manual cleaning and improve system performance. Additionally, selecting mirror coatings with anti-soiling properties or using hydrophobic coatings can help minimize the adhesion of dust particles and enhance the self-cleaning ability of the mirrors. The steps of analysing dust effect on mirrors are illustrated in Fig. 9.4b.

9.4 Dust Particle Physics and Chemistry

The physical makeup of dust particles can greatly influence the outcome of photovoltaic cells. Particles such as size, shape, electric charge, and chemical composition can reduce the solar radiation that is both taken in and reflected by the PV modules. Dust can also absorb, disperse, and mirror the light of the sun, all of which diminish

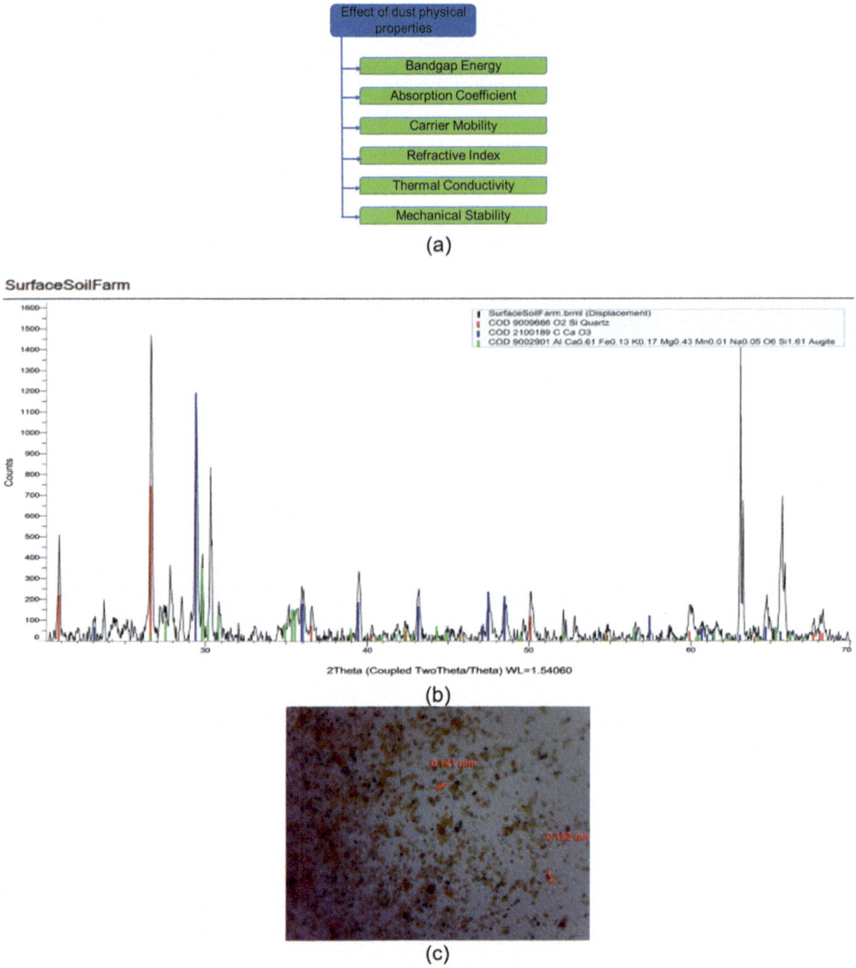

Fig. 9.5 (**a**) Effect of dust physical properties, (**b**) XRD analysis, (**c**) microscopic image

the amount of light reaching the module (Kazem et al., 2020a). Several physical properties can significantly impact the PV performance (Sarver et al., 2013). Some key physical properties and their effects are illustrated in Fig. 9.5a:

- The bandgap energy of the semiconductor material used in PV devices determines the range of light wavelengths that can be absorbed and converted into electrical energy. A suitable bandgap is crucial to match the solar spectrum and maximize photon absorption. Materials with wider bandgaps, such as silicon, are better suited for solar cells that operate under direct sunlight, while narrower bandgap materials, like gallium arsenide, are more suitable for concentrated solar applications.

- The absorption coefficient represents how strongly a material absorbs photons at a given wavelength. Materials with higher absorption coefficients can absorb more photons and convert them into electrical energy more efficiently. Hence, PV materials with higher absorption coefficients are desirable for improved performance.
- Carrier mobility refers to the ability of charge carriers "electrons and holes" to move through a material. Higher carrier mobility facilitates faster transport of charges, reducing the recombination losses and enhancing the overall electrical conductivity of the PV device. High mobility materials, such as crystalline silicon, allow for efficient charge extraction and better device performance.
- The refractive index of a material affects the amount of light that can be transmitted through it or reflected at its surface. For PV devices, a low refractive index helps minimize reflection losses and increase the amount of light absorbed by the active layers. Antireflection coatings are often used to reduce reflection losses by matching refractive indices and optimizing light transmission.
- PV devices generate heat during operation, and efficient dissipation of this heat is crucial to maintain their performance and reliability. Materials with high thermal conductivity can effectively dissipate heat, preventing excessive temperature increases that can degrade device efficiency and lifespan. Good thermal management techniques, such as heat sinks or cooling systems, are necessary to maintain optimal operating temperatures.
- PV devices are subjected to various mechanical stresses, including temperature variations, wind loads, and vibrations. Mechanical stability and durability of materials used in PV modules are essential to ensure long-term performance. Materials with high mechanical strength and resistance to degradation under environmental conditions help maintain the structural integrity of the modules and prevent performance degradation.

Optimizing these physical properties through appropriate materials selection, device design, and manufacturing processes is crucial for improving the efficiency, reliability, and overall performance of PVs. Researchers and engineers continually explore and develop new materials and techniques to enhance these physical properties and advance the field of solar energy conversion.

Reference (Kazem & Chaichan, 2016b) investigated how the physical qualities of dust, like size, shape, weight, and pollutant type, influence the efficiency of PV systems. Several tests were done to examine the dust characteristics, such as "microscopic examination" and "X-ray diffraction" (XRD) as seen in Fig. 9.5b, c.

The size of dust particles plays a significant role in determining the power output of a PV system. When the particles are small, they scatter less light, allowing lighter to be absorbed by the solar cells. This leads to improved system performance. Conversely, larger dust particles can obstruct sunlight and reduce the efficiency of the PV system. Moreover, if the dust particles are both large and have sharp edges, they can cause a higher degree of light scattering, resulting in a decrease in the amount of light reaching the PV modules. Additionally, rough and irregularly shaped dust particles can create shadows on the cell surface, further diminishing the amount of light that reaches the solar cells.

9.5 Dust Mitigation and Cleaning

The act of controlling dust to lower the quantity of dust in the atmosphere is known as dust mitigation. This can be accomplished through several techniques, such as air filtration systems, dust suppression systems, and water-based dust suppression (as depicted in Fig. 9.6). To mitigate the impact of dust accumulation on PV systems, several measures can be taken:

- Regular Cleaning: Regular cleaning of PV modules is essential to remove dust and maintain optimal performance (Maghami et al., 2016). Cleaning can be done using water, soft brushes, or specialized cleaning equipment. The frequency of cleaning depends on factors such as local dust levels, environmental conditions, and system requirements.
- Automated Cleaning Systems: Employing automated cleaning systems can reduce manual labour and ensure consistent cleaning of PV modules (Mondal & Bansal, 2015). These systems may use brushes, wipers, or sprayers to remove dust from the surface of the modules. Some advanced systems utilize "robotic devices" or "self-cleaning coatings" to enhance cleaning efficiency.
- Anti-Soiling Coatings: Applying "anti-soiling coatings" to the surface of PV modules can help repel dust particles and reduce their adhesion (Younis & Onsa, 2022). These coatings can be hydrophobic or self-cleaning, making it easier for dust to be washed away by rain or cleaning processes.
- Dust Barriers: Installing dust barriers or windbreakers around PV arrays can reduce the amount of dust carried by wind and limit its deposition on the modules. These barriers can be physical structures or vegetation strategically placed to shield the modules from windborne dust (Varga & Wiesner, 2021).
- Tilted Mounting: Mounting PV modules at an angle or with a tilt can encourage natural cleaning through rainwater runoff. When rainwater flows across the

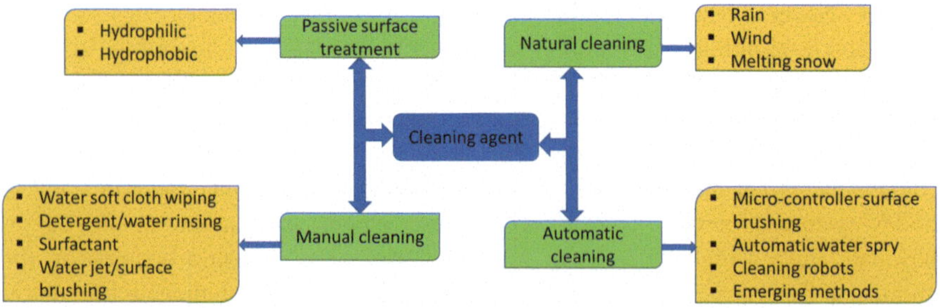

Fig. 9.6 Dust cleaning methods

Fig. 9.7 Dust cleaning: (a) manual, (b) automatic, (c) water, (d) vacuum

surface of tilted modules, it can carry away dust particles, promoting self-cleaning to some extent (Ghosh, 2020).

- Environmental Controls: Implementing environmental controls, such as dust suppression systems or vegetation management, can minimize dust generation in the surrounding area. For example, using water sprays or dust control chemicals in dusty environments can help reduce dust particles in the air and subsequently lower dust deposition on PV modules.
- Monitoring and Maintenance: Regular monitoring of PV system performance, including checking for dust accumulation and any decrease in power output, allows for timely maintenance (Gostein et al., 2014). Prompt detection and cleaning of dust buildup can help prevent significant efficiency losses and prolong the lifespan of the system.

It is important to consider the specific environmental conditions and requirements of the PV system when choosing and implementing dust mitigation strategies. Evaluating the cost-effectiveness and potential benefits of different approaches will help determine the most suitable mitigation measures for a given installation.

The depicted process in Fig. 9.6 demonstrates the removal of dust and debris from the surface of PV modules. This dust, which causes increased light reflection from the module, results in a reduction of the amount of light reaching the surface and subsequently decreases the electricity generated. Manual dust cleaning involves using a brush and mild detergent, while automated methods, as shown in Fig. 9.7, include brushing robots and high-pressure water jets (Mani & Pillai, 2010; Micheli et al., 2021; Al Shehri et al., 2016; Elnozahy et al., 2015; Zorrilla-Casanova et al., 2011).

9.6 Dust Modelling

Modelling the effect of dust on PV modules involves considering various factors and parameters to simulate and quantify the impact on module performance. The main key aspects to consider when modelling the dust effect are as follows:

- Dust Deposition and Distribution: A dust deposition model can be developed based on environmental conditions, dust generation sources, and local dust characteristics. This model estimates the rate and distribution of dust accumulation on the module surface over time. Factors such as wind speed, particle size distribution, and deposition mechanisms need to be taken into account.
- Optical Properties: Modelling the optical properties of dust layers is crucial to assess the impact on light absorption and reflection. The optical model considers factors such as refractive index, absorption coefficient, and scattering properties of dust particles. The interaction between incident sunlight and the dust layer is simulated to determine the amount of light transmitted, absorbed, or reflected.
- Transmittance and Reflection Losses: The presence of dust affects the transmittance and reflection of light within the PV module. Modelling these losses involves considering the absorption and scattering of light within the dust layer and at the interface between the dust layer and the module surface. Optical simulations can provide insights into the reduction in light reaching the solar cells.
- Electrical Performance: Dust accumulation can impact the electrical performance of PV modules. Modelling the effects on parameters such as Isc, Voc, FF, and P_{pv} helps quantify the reduction in module efficiency. Electrical models consider factors like shading losses, reduced light absorption, and altered I–V characteristics due to dust.
- Cleaning Strategies: Modelling the effectiveness of different cleaning strategies is crucial for optimizing maintenance practices. Simulating the impact of various cleaning techniques, intervals, and efficiency factors can help determine the most suitable cleaning approach for minimizing dust-related losses.
- Validation and Calibration: It is important to validate and calibrate the dust model using real-world measurements and data. Experimental data on dust accumulation rates, module performance under different dust levels, and cleaning efficiency can be used to refine and validate the model's accuracy.

Various software tools and numerical models, including "optical simulations", "computational fluid dynamics" (CFD), and "electrical circuit models", can be utilized for dust effect modelling. These models provide valuable insights into the performance degradation of PV modules and aid in optimizing maintenance strategies and system design to mitigate the negative effects of dust. Dust modelling is a technique used to analyse the physical, chemical, and optical properties of dust particles. It can be used to simulate and analyse the dust found in various settings, like planetary atmospheres, interstellar space, galaxies, and laboratory experiments (Hammad et al., 2018; Kazem et al., 2020c). Dust modelling is utilized in many different areas, like climate modelling, astrophysics, astronomy, and planetary science (Parajuli & Kim, 2019). Investigating the characteristics of dust can help scientists better understand the development and form of PV systems. Numerous scientists have simulated the influence of dust on PV modules and measured particular performance metrics, like energy output (Cabanillas & Munguía, 2011), efficiency (Mastekbayeva & Kumar, 2000), power (Massi Pavan et al., 2011), fill factor (Kaldellis & Kokala, 2010; Kaldellis et al., 2010), transmittance ratio (Conceição et al., 2018), etc. Researchers can gain an understanding of the development of dust by modelling its characteristics. However, most suggested models are case studies with basic assumptions and limited accuracy. There is a pressing need for a general model for the effect of dust on photovoltaic systems. Very little research has been done on an overall model. An example of such a model is the one proposed by (Kazem et al., 2022), which is illustrated in Fig. 9.8a.

The proposed model is illustrated in Eqs. (9.1) and (9.2).

$$I_{\mathrm{L}} = I_{\mathrm{G}} - I_{\mathrm{o}}\left\{e^{\left[\frac{q\,(V+I_{\mathrm{L}}R_{\mathrm{S}})}{n\times k\times T}\right]} - 1\right\} - \frac{V + I_{\mathrm{L}}R_{\mathrm{s}}}{R_{\mathrm{p}}} \tag{9.1}$$

$$I_{\mathrm{G}} = I_{\mathrm{PV}}(\,1 - m(a_1k_1 + a_2k_2 + a_3k_3 + \ldots + a_nk_n)) \tag{9.2}$$

where I_G is the "PV photo-generated current considering dust impact" (in A), m is the "dust accumulated mass density ($m = \Delta m/Ac$ in kg/m^2)", a (a_1, a_2, and a_3) is the "dust element coefficients (in m^2.kg^{-1})", k (k_1, k_2, and k_3) is the "dust elements parameter (unitless)", and n is the "number of parameters".

In Fig. 9.8b, the k parameter is established by utilizing XRD to discover the dust components and rates at the site. As an example, if the XRD for the dust factors uncovers numerous parameters and the predominant elements are SiO2, CaO, and Fe2O3, then their separate rates are found to be 41%, 23%, and 19%, respectively, which represent k1, k2, and k3. The dust element coefficients are portrayed as "the effect of the assigned dust element on PV current misfortunes". The model is valid for homogeneous or semi-homogeneous dispersion of dust particles on the PV module. Figure 9.8c exhibits the approval of the proposed model utilizing three diverse dust blends. The I–V characteristic results of the lab estimations and the proposed model (Eqs. 9.1 and 9.2) have been compared to the perfect one-diode model, which shows the predominance of the proposed model.

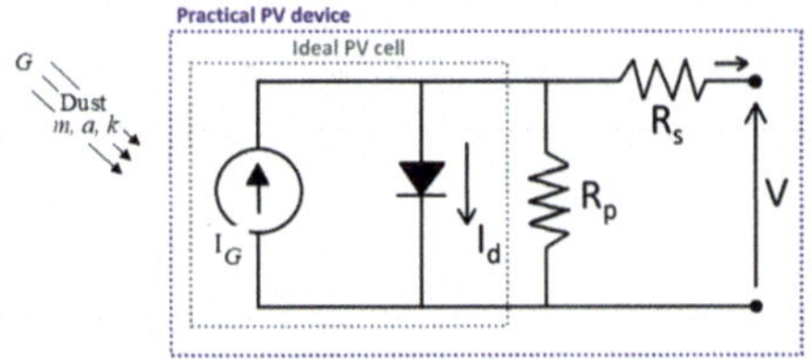

(a)

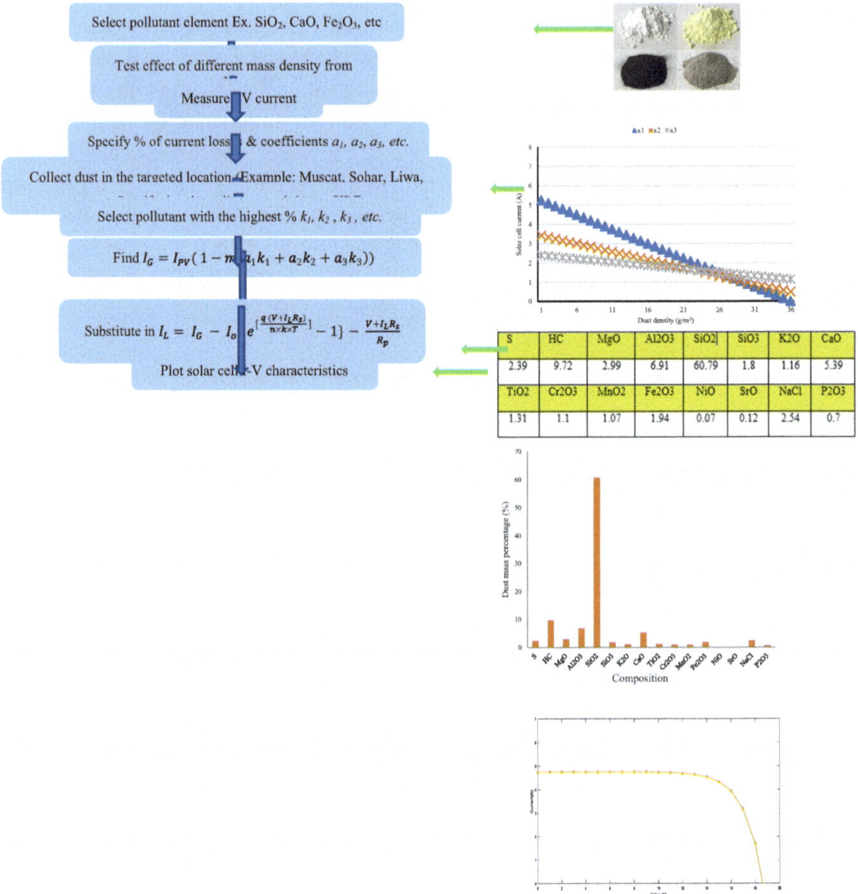

(b)

Fig. 9.8 (**a**) Improved "one-diode model", (**b**) flowchart, (**c**) validation results (Kazem et al., 2022)

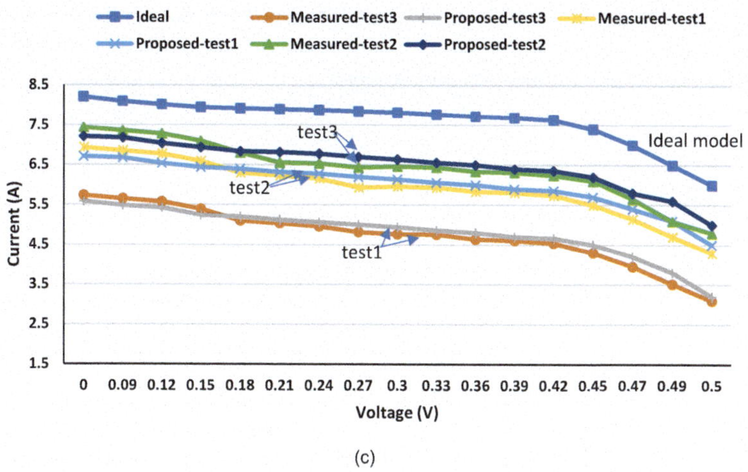

Fig. 9.8 (continued)

9.7 New Prospects and Beyond

The attractiveness of PV systems in desert regions stems from their cost-effectiveness and the availability of inexpensive land for installation. Additionally, their simple installation process and minimal infrastructure requirements make them appealing to individuals in remote areas or those seeking to generate their own electricity. Furthermore, advancements in technology have contributed to a reduction in the cost of PVs, further enhancing their accessibility. As a result, the installation of PV systems in arid regions is expected to continue growing as more people recognize the advantages of solar energy and as technology continues to improve. However, it should be noted that the harsh desert conditions, including high temperatures and dust accumulation, can pose challenges and potentially damage these systems. This study focuses on addressing these issues to provide a promising outlook for PV systems in desert regions. Several points are highlighted to indicate new prospects and advancements in this area:

- There is still a need to address the existing research gap regarding the development of a comprehensive dust model applicable in various scenarios.
- Standardized tests should be established to evaluate the impact of dust on PV modules. These tests should consider factors like the duration of dust accumulation sampling, frequency of cleaning, uniformity of dust distribution, measurement periods, and sampling times.
- Furthermore, the proposed models should incorporate parameters beyond just mass, such as size, shape, composition, weight, uniformity of distribution, and system orientation.
- Environmental factors including temperature, humidity, and wind speed should also be taken into account, as they directly influence dust behaviour.

- Integrating the generalized dust model into PV system design software would be beneficial not only for terrestrial applications but also for potential extraterrestrial projects like Mars colonization and beyond.

9.8 Conclusions

Dust significantly affects the electrical performance of photovoltaic modules by reducing the amount of light available for electricity generation, resulting in decreased efficiency. Moreover, dust can lead to corrosion, electrical shorts, and premature module failure. Additionally, dust accumulation hampers cooling, leading to higher operating temperatures and further performance decline. Therefore, it is crucial to minimize dust levels on PV modules to ensure optimal performance. This study reevaluates the impact of dust and highlights new areas of research requiring further investigation.

- Dust particles pose risks such as etching, discoloration, and scratching of glass surfaces. They can embed themselves into the glass, increasing vulnerability to etching and scratching. Additionally, dust can trap moisture, leading to water spots and subsequent glass discoloration. The complex task of modelling dust while considering these factors requires further research.
- Dust particles vary in size and shape, complicating the study of their effects. These particles reduce PV system performance by absorbing some of the light energy and limiting its availability to reach the PV modules. Consequently, the electricity generation of the cells is diminished. To better comprehend these effects, more research and standardization are needed to explore particle size and shape.
- Generalized models for simulating the effects of dust on PV modules must incorporate the physical and chemical properties of dust particles and their interactions with the environment and other objects. Such models are crucial for understanding the nature of dust in the environment and its impact on physical and chemical processes. Additionally, these generalized models will aid in determining the optimal PV system size, leading to positive effects on technical and economic aspects.

Acknowledgement The research leading to these results has received Research Project Grant Funding from the Research Council of the Sultanate of Oman, Research Grant Agreement No. ORG SU EI 11 010.

Declaration of Interest Statement We the authors declare no affiliations with or involvement in any organization or entity with any financial interest (such as honoraria; educational grants; participation in speakers' bureaus; membership, employment, consultancies, stock ownership, or other equity interest; and expert testimony or patent licensing arrangements), or non-financial interest (such as personal or professional relationships, affiliations, knowledge or beliefs) in the subject matter or materials discussed in this manuscript.

Conflict of Interests The authors declare that there is no conflict of interests regarding the publication of this brief note.

References

Ahmed Darwish, Z., Sopian, K., Kazem, H. A., Alghoul, M. A., & Alawadhi, H. (2017). Investigating the effect of Titanium Dioxide (TiO2) pollution on the performance of the mono-crystalline solar module. In *E3S Web of Conferences*, vol. 23. https://doi.org/10.1051/e3sconf/20172301005

Al Shehri, A., Parrott, B., Carrasco, P., Al Saiari, H., & Taie, I. (2016). Impact of dust deposition and brush-based dry cleaning on glass transmittance for PV modules applications. *Solar Energy, 135*, 317–324. https://doi.org/10.1016/j.solener.2016.06.005

Alagoz, S., & Apak, Y. (2020). Removal of spoiling materials from solar panel surfaces by applying surface acoustic waves. *Journal of Cleaner Production, 253*, 119992. https://doi.org/10.1016/j.jclepro.2020.119992

Al-Badra, M. Z., Abd-Elhady, M. S., & Kandil, H. A. (2020). A novel technique for cleaning PV panels using antistatic coating with a mechanical vibrator. *Energy Reports, 6*, 1633–1637. https://doi.org/10.1016/j.egyr.2020.06.020

Al-Maamary, H. M. S., Kazem, H. A., & Chaichan, M. T. (2016). Changing the energy profile of the GCC states: A review. *International Journal of Applied Engineering Research, 11*(3).

Alnasser, T. M. A., Mahdy, A. M. J., Abass, K. I., Chaichan, M. T., & Kazem, H. A. (2020). Impact of dust ingredient on photovoltaic performance: An experimental study. *Solar Energy, 195*-(November 2019), 651–659. https://doi.org/10.1016/j.solener.2019.12.008

Al-Waeli, A. H. A., Kazem, H. A., Sopian, K., & Chaichan, M. T. (2017a). Techno-economical assessment of grid connected PV/T using nanoparticles and water as base-fluid systems in Malaysia. *International Journal of Sustainable Energy, 37*(6), 558–575. https://doi.org/10.1080/14786451.2017.1323900

Al-Waeli, A. H. A., Sopian, K., Kazem, H. A., & Chaichan, M. T. (2017b). Photovoltaic/Thermal (PV/T) systems: Status and future prospects. *Renewable and Sustainable Energy Reviews, 77*. https://doi.org/10.1016/j.rser.2017.03.126

Al-Waeli, A. H. A., Kazem, H. A., Chaichan, M. T., & Kopian, K. (2019). *Photovoltaic / Thermal (PV/T) systems: Principles, design, and applications* (1st ed.). Springer Nature.

Biryukov, S. (1998). An experimental study of the dry deposition. *Journal of Aerosol Science, 29*(1–2), 129–139.

Bolles, K., Sweeney, M., & Forman, S. (2019). Meteorological catalysts of dust events and particle source dynamics of affected soils during the 1930s Dust Bowl Drought, Southern High Plains, USA. *Anthropocene*, 100216. https://doi.org/10.1016/j.ancene.2019.100216

Cabanillas, R. E., & Munguía, H. (2011). Dust accumulation effect on efficiency of Si photovoltaic modules. *Journal of Renewable and Sustainable Energy, 3*(4). https://doi.org/10.1063/1.3622609

Chaichan, M. T. M. T., & Kazem, H. A. H. A. H. A. (2015). Using aluminium powder with PCM (paraffin wax) to enhance single slope solar water distiller productivity in Baghdad - Iraq winter weathers. *International Journal of Renewable Energy Resources, 5*(1), 251–257. https://doi.org/10.1163/22134468-00002006

M. T. Chaichan and H. A. Kazem, "Experimental evaluation of dust composition impact on photovoltaic performance in Iraq," Energy Sources, Part A: Recovery, Utilization, and Environmental Effects, pp. 1–22, 2020, doi: https://doi.org/10.1080/15567036.2020.1746444.

Conceição, R., et al. (2018). Saharan dust transport to Europe and its impact on photovoltaic performance: A case study of soiling in Portugal. *Solar Energy, 160*(September 2017), 94–102. https://doi.org/10.1016/j.solener.2017.11.059

Cordero, R. R., et al. (2018). Effects of soiling on photovoltaic (PV) modules in the Atacama Desert. *Scientific Reports, 8*(1), 1–14. https://doi.org/10.1038/s41598-018-32291-8

Darwish, Z. A., Kazem, H. A., Sopian, K., Al-Goul, M. A. A., & Alawadhi, H. (2015). Effect of dust pollutant type on photovoltaic performance. *Renewable and Sustainable Energy Reviews, 41*, 735–744. https://doi.org/10.1016/j.rser.2014.08.068

Darwish, Z. A., Sopian, K., Alawadhi, H., Kazem, H. A., & Alghoul, M. A. (2016). The impact of calcium carbonate on the photovoltaic performance: An indoor experimental study. *International Journal of Applied Engineering Research, 11*(3), 2091–2097.

Darwish, Z. A., Kazem, H. A., Sopian, K., Alghoul, M. A. A., & Alawadhi, H. (2018). Experimental investigation of dust pollutants and the impact of environmental parameters on PV performance: an experimental study. *Environment, Development and Sustainability, 20*(1), 155–174. https://doi.org/10.1007/s10668-016-9875-7

Elnozahy, A., Rahman, A. K. A., Ali, A. H. H., Abdel-Salam, M., & Ookawara, S. (2015). Performance of a PV module integrated with standalone building in hot arid areas as enhanced by surface cooling and cleaning. *Energy and Buildings, 88*, 100–109. https://doi.org/10.1016/j.enbuild.2014.12.012

El-Shobokshy, M. S., & Hussein, F. M. (1993). Degradation of photovoltaic cell performance due to dust deposition on to its surface. *Renewable Energy, 3*(6–7), 585–590. https://doi.org/10.1016/0960-1481(93)90064-N

Ghazi, S., Sayigh, A., & Ip, K. (2014). Dust effect on flat surfaces - A review paper. *Renewable and Sustainable Energy Reviews, 33*, 742–751. https://doi.org/10.1016/j.rser.2014.02.016

Ghosh, A. (2020). Soiling losses: A barrier for India's energy security dependency from photovoltaic power. *Challenges, 11*(1), 9. https://doi.org/10.3390/challe11010009

Gostein, M., Caron, J. R., & Littmann, B. (2014). Measuring soiling losses at utility-scale PV power plants. In *2014 IEEE 40th Photovolt. Spec. Conf. PVSC 2014*, pp. 885–890. https://doi.org/10.1109/PVSC.2014.6925056

Hammad, B., Al-Abed, M., Al-Ghandoor, A., Al-Sardeah, A., & Al-Bashir, A. (2018). Modeling and analysis of dust and temperature effects on photovoltaic systems' performance and optimal cleaning frequency: Jordan case study. *Renewable and Sustainable Energy Reviews, 82*(July 2017), 2218–2234. https://doi.org/10.1016/j.rser.2017.08.070

Hussain, A., Batra, A., & Pachauri, R. (2017). An experimental study on effect of dust on power loss in solar photovoltaic module. *Renewables: Wind, Water, and Solar, 4*(1), 1–13. https://doi.org/10.1186/s40807-017-0043-y

Jordan, D. C., & Kurtz, S. R. (2013). Photovoltaic degradation rates - An Analytical Review. *Progress in Photovoltaics: Research and Applications, 21*(1), 12–29. https://doi.org/10.1002/pip.1182

Kaldellis, J. K., & Fragos, P. (2011). Ash deposition impact on the energy performance of photovoltaic generators. *Journal of Cleaner Production, 19*(4), 311–317. https://doi.org/10.1016/j.jclepro.2010.11.008

Kaldellis, J. K., & Kokala, A. (2010). Quantifying the decrease of the photovoltaic panels' energy yield due to phenomena of natural air pollution disposal. *Energy, 35*(12), 4862–4869. https://doi.org/10.1016/j.energy.2010.09.002

Kaldellis, J. K., Kokala, A., & Kapsali, M. (2010). Natural air pollution deposition impact on the efficiency of PV panels in urban environment. *Fresenius Environmental Bulletin, 19*(12), 2864–2872.

Kazem, H. A., & Chaichan, M. T. (2016a). Effect of environmental variables on photovoltaic performance-based on experimental studies. *International Journal of Civil, Mechanical and Energy Science, 2*(4), 1–8.

Kazem, H. A., & Chaichan, M. T. (2016b). Experimental analysis of the effect of dust's physical properties on photovoltaic modules in Northern Oman. *Solar Energy, 139*. https://doi.org/10.1016/j.solener.2016.09.019

Kazem, H. A., & Chaichan, M. T. (2019). The effect of dust accumulation and cleaning methods on PV panels' outcomes based on an experimental study of six locations in Northern Oman. *Solar Energy, 187*(May), 30–38. https://doi.org/10.1016/j.solener.2019.05.036

Kazem, H. A., Khatib, T., Sopian, K., Buttinger, F., Elmenreich, W., & Albusaidi, A. S. (2013). Effect of dust deposition on the performance of multi-crystalline photovoltaic modules based on experimental measurements. *International Journal of Renewable Energy Research, 3*(4), 850–853, [Online]. Available: https://pdfs.semanticscholar.org/0fc2/393de88e647da2d65d1e0022df873fb37e81.pdf

Kazem, H. A., Chaichan, M. T., Al-Waeli, A. H. A., & Sopian, K. (2020a). Analysis of the effects of cell temperature on the predictability of the solar photovoltaic power production. *Solar Energy, 153*(8), 540–561. https://doi.org/10.1016/j.solener.2020.06.043

Kazem, H. A., Chaichan, M. T., Al-Waeli, A. H. A., & Sopian, K. (2020b). Evaluation of aging and performance of grid-connected photovoltaic system northern Oman: Seven years' experimental study. *Solar Energy, 207*(April), 1247–1258. https://doi.org/10.1016/j.solener.2020.07.061

Kazem, H. A., Chaichan, M. T., Al-Waeli, A. H. A., & Sopian, K. (2020c). A novel model and experimental validation of dust impact on grid-connected photovoltaic system performance in Northern Oman. *Solar Energy, 206*(May), 564–578. https://doi.org/10.1016/j.solener.2020.06.043

Kazem, H. A., Al-Waeli, A. H. A., Chaichan, M. T., & Sopian, K. (2022). Modeling and experimental validation of dust impact on solar cell performance. *Energy Sources, Part A: Recovery, Utilization, and Environmental Effects, 00*(00), 1–17. https://doi.org/10.1080/15567036.2021.2024922

Maghami, M. R., Hizam, H., Gomes, C., Radzi, M. A., Rezadad, M. I., & Hajighorbani, S. (2016). Power loss due to soiling on solar panel: A review. *Renewable and Sustainable Energy Reviews, 59*, 1307–1316. https://doi.org/10.1016/j.rser.2016.01.044

Mani, M., & Pillai, R. (2010). Impact of dust on solar photovoltaic (PV) performance: Research status, challenges and recommendations. *Renewable and Sustainable Energy Reviews, 14*(9), 3124–3131. https://doi.org/10.1016/j.rser.2010.07.065

Massi Pavan, A., Mellit, A., & De Pieri, D. (2011). The effect of soiling on energy production for large-scale photovoltaic plants. *Solar Energy, 85*(5), 1128–1136. https://doi.org/10.1016/j.solener.2011.03.006

Mastekbayeva, G. A., & Kumar, S. (2000). Effect of dust on the transmittance of low density polyethylene glazing in a tropical climate. *Solar Energy, 68*(2), 135–141. https://doi.org/10.1016/S0038-092X(99)00069-9

Micheli, L., Fernández, E. F., Aguilera, J. T., & Almonacid, F. (2021). Economics of seasonal photovoltaic soiling and cleaning optimization scenarios. *Energy, 215*(105229). https://doi.org/10.1016/j.energy.2020.119018

Mondal, A. K., & Bansal, K. (2015). Structural analysis of solar panel cleaning robotic arm. *Current Science, 108*(6), 1047–1052.

Niknia, I., Yaghoubi, M., & Hessami, R. (2012). A novel experimental method to find dust deposition effect on the performance of parabolic trough solar collectors. *International Journal of Environmental Studies, 69*(2), 233–252. https://doi.org/10.1080/00207233.2012.664810

Parajuli, S. P., & Kim, H. (2019). Dust emission modeling using a new high - Resolution dust source function in WRF - Chem with implications for air quality, pp. 1–25. https://doi.org/10.1029/2019JD030248

Salam, M. A., Aziz, A., Alwaeli, A. H. A., & Kazem, H. A. (2013). Optimal sizing of photovoltaic systems using HOMER for Sohar, Oman. *International Journal of Renewable Energy Resources, 3*(3), 470–475.

Sarver, T., Al-Qaraghuli, A., & Kazmerski, L. L. (2013). A comprehensive review of the impact of dust on the use of solar energy: History, investigations, results, literature, and mitigation approaches. *Renewable and Sustainable Energy Reviews, 22*, 698–733. https://doi.org/10.1016/j.rser.2012.12.065

Sayigh, E. A. A. M. (1979). *Solar energy applications in buildings* (1st ed.). Academic Press.

Sopian, K., Al-Waeli, A. H. A., & Kazem, H. A. (2020). Energy, exergy and efficiency of four photovoltaic thermal collectors with different energy storage material. *Journal Energy Storage, 29*(October 2019), 101245. https://doi.org/10.1016/j.est.2020.101245

Varga, H. F., & Wiesner, M. R. (2021). Effect of dust composition on the reversibility of photovoltaic panel soiling. *Environmental Science & Technology, 55*(3), 1984–1991. https://doi.org/10.1021/acs.est.0c06196

Vedulla, G., Geetha, A., & Senthil, R. (2023). Review of strategies to mitigate dust deposition on solar photovoltaic systems. *Energies, 16*(109), 1–29.

Verma, L. K., et al. (2011). Self-cleaning and antireflective packaging glass for solar modules. *Renewable Energy, 36*(9), 2489–2493. https://doi.org/10.1016/j.renene.2011.02.017

Younis, A., & Onsa, M. (2022). A brief summary of cleaning operations and their effect on the photovoltaic performance in Africa and the Middle East. *Energy Reports, 8*, 2334–2347. https://doi.org/10.1016/j.egyr.2022.01.155

Yousif, J. H. J. H., Kazem, H. A. H. A., Alattar, N. N. N., & Elhassan, I. I. I. I. (2019). A comparison study based on artificial neural network for assessing PV/T solar energy production. *Case Studies in Thermal Engineering, 13*, 100407. https://doi.org/10.1016/j.csite.2019.100407

Zhao, W., Lv, Y., Wei, Z., Yan, W., & Zhou, Q. (2021). Review on dust deposition and cleaning methods for solar PV modules. *Renewable and Sustainable Energy Reviews, 13*(3), 032701.

Zorrilla-Casanova, J., et al. (2011). Analysis of dust losses in photovoltaic modules, pp. 2985–2992. https://doi.org/10.3384/ecp110572985

Chapter 10
TiO_2-CeO_2 Admixed Photoelectrode for Optimization of Hydrogen Production in Regard to Photoelectrochemical Solar Cell

Shivansh Tripathi, Mridula Tripathi, and Priyanka Chawla

10.1 Introduction

There is a growing demand for energy in order for the globe to expand sustainably, but conventional energy sources are getting more difficult to find. Due to the exponential rise in energy consumption and the acceleration of population expansion worldwide, the twentieth century might be seen as a turning point in the development of civilizations. So, the practical approach to satisfy the continuous demand for energy is to rely on renewable energy sources, sun, wind, water, hydrogen, and other types of energy (Islam et al., 2022; Pooja et al., 2022; Goltsov & Veziroglu, 2001; Chawla et al., 2022; Omer, 2008; Song, 2006). The most reliable source of energy is solar since it has no impact on the environment and emits no greenhouse emissions at all (Dincer & Joshi, 2013; Tilley, 2019). In this regard, using photoelectrochemical solar cells (PECs) to convert solar energy into hydrogen fuel is becoming increasingly important (Ta et al., 2022; Hosseini & Wahid, 2020). This photoelectrochemical solar cell brings up a new perspective and may provide a solution to the environmental problem. The photoelectrochemical (PEC) cell is a very attractive but challenging alternative for capturing solar energy for the manufacture of clean hydrogen fuels today due to its simplicity of installation and on-site storage capabilities (Feng et al., 2022) . A significant improvement is needed to make the PEC cell successful since the conversion rate of the solar energy it receives is insufficient to meet the present energy demand. The PEC's photoanode works as the primary component determining its efficiency, while many other variables also

S. Tripathi
Indian Institute of Space, Science and Technology, Thiruvananthapuram, Kerala, India

M. Tripathi · P. Chawla (✉)
Department of Chemistry, CMP Degree College, University of Allahabad, Prayagraj, UP, India

A. Sayigh (ed.), *Transition Towards a Carbon Free Future*, Innovative Renewable Energy, https://doi.org/10.1007/978-3-031-61660-0_10

affect how well it performs. Diverse semiconductor material types are used for the photoanode component, with an energy band that promotes water oxidation and reduction (Hoffmann et al., 1995; Tripathi et al., 2007; Wang & Wang 2012; Feng et al., 2020).

Among all semiconductors TiO_2 is considered as the suitable material to be used as photoelectrode for PEC cells due to its exceptional stability against photo corrosion on optical excitation in the bandgap. But one of the major problems with TiO_2 as photoelectrode is spectral response so to shift TiO_2's spectral response into the visible or almost visible range; numerous attempts have been tried. It is well known that adding the right oxides to TiO_2 films can enhance their spectrum responsiveness. In the present work, we admixed CeO_2 oxide with TiO_2 as CeO_2 is considered as a suitable material due to its diverse valence states and various crystalline structures; it has been used in various fields such as electrochemical, batteries, functional materials, energy storage devices, and many more. In order to produce hydrogen using the SC-SEP photoelectrochemical solar cell, the surface morphological, structural, and photoelectrochemical characterisation of the naked TiO_2 as well as the TiO_2 overlaid with CeO_2 thin film admixtures has been studied (Tilley, 2019). The analysis of the best photoelectrode area for a semiconductor septum photo electrochemical solar cell using ns-TiO_2-CeO_2 admixed/Ti is the subject of this paper. The goal of the current work was to create an electrode with a high effective surface area, which would improve the PEC activity and quantum yield.

To investigate the impact of electrode area on the output power in a chemical fuel (i.e. H_2) produced by SC SEP PEC cell, the photoelectrochemical response of ns TiO_2-CeO_2 photoelectrodes for five different electrode areas has been tested. This was done to identify the electrode area that would produce hydrogen and produce electricity most efficiently. The photoelectrochemical cell having ns TiO_2- CeO_2 admixed/Ti photoanode of several geometric areas like 0.5, 1.0, 1.5, 2.0, and 2.5 cm^2 was fabricated and characterized.

10.2 Materials and Methods

The following substances were utilized as precursors: sodium hydroxide, titanium isopropoxide ($Ti[OCH(CH_3)_2]_4$), cerium oxide (an analytical grade form Sigma Aldrich), analytical grade liquid ammonia, and sulphuric acid (Merck Pvt. Ltd.). From deionized water, solutions were made.

The approach used to create the ns-TiO_2-CeO_2 admixed photoelectrode is as follows. First, TiO_2 was created using the sol-gel process, and then, using the spin coating method, a film of the thick TiO_2 solution was applied to the titanium sheets (with a thickness of around 0.33 mm). The titanium sheets were first polished with 1000 grit emery paper before being ultrasonically cleaned in acetone for 30 minutes at a frequency of 40 Hz. The sheets were integrated into the design of the photo-resist spinner's spinning substrate holder. The titanium sheets were coated with one drop of a viscous TiO_2 solution using a spin coating process that lasted 30 seconds at 3000 rpm. For obtaining TiO_2-CeO_2 photoanode, cerium oxide was deposited on the above-mentioned ns-TiO_2 coated photoanode by chemical vapour deposition technique. The electrode was then heated to 500 °C and annealed in oxygen.

For optimization study titanium sheets (~0.33 mm thickness) with ns TiO_2-CeO_2 of different areas like 0.5, 1.0, 1.5, 2.0, and 2.5 cm^2 were taken and mechanically polished with emery papers of various grades. For 15 minutes, the films were cleaned by ultrasonic cleaning utilizing acetone. One drop of sol was applied to the substrate over a vacuum-interlocked spin holder at a speed of 3000 rpm to ensure uniform layer deposition. The obtained films were air dried in oven for 15 min at 80 °C and then fired at 450 °C for 30 min. To improve film thickness, this process was repeated four to five times. Finally, samples were annealed at 550 °C for 4 hours in an argon environment. Following their acquisition, electrodes underwent photoelectrochemical analysis.

10.3 Results and Discussion

Figure 10.1 depicts the ns TiO_2-CeO_2 admixed semiconductor's XRD pattern over a titanium sheet. When TiO_2 is mixed with CeO_2, no new composite material is formed, and the XRD displays strong peaks that support the material's crystallinity and are consistent with the observed TiO_2-CeO_2 (Watanable et al., 2009).

Figure 10.2 illustrates the prepared admixed ns TiO_2-CeO_2 under a scanning electron microscope. The fine-grained structure indicative of a nanocrystalline-like matrix is readily visible in the figure. Scherer's equation was used to calculate the

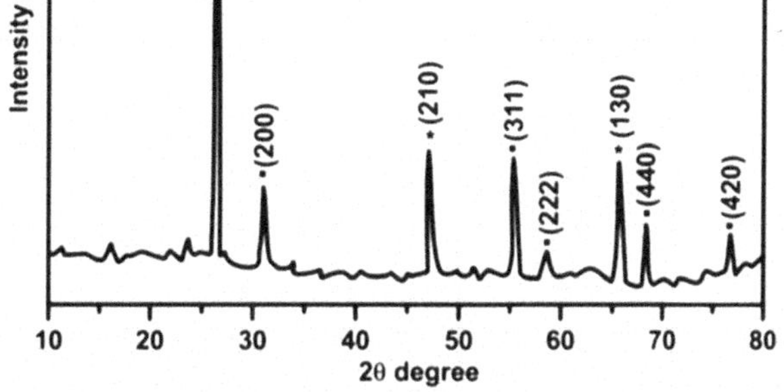

Fig. 10.1 XRD pattern of the ns TiO_2-CeO_2 admixed semiconductor over the titanium sheet

Fig. 10.2 The scanning electron microscope of the prepared admixed ns-TiO_2-CeO_2

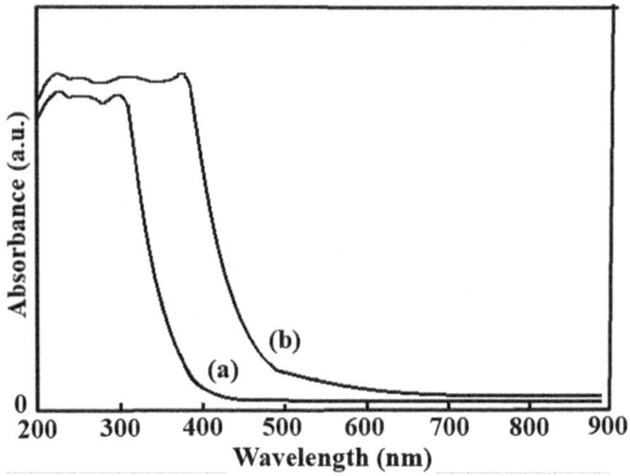

Fig. 10.3 Shows the UV-vis absorption curves of the ns-TiO_2 and ns-TiO_2-CeO_2 nanomaterials

average grain size, which was discovered to range from 50 to 90 nm (Zhang et al., 2003).

The UV-vis absorption curves of ns-TiO_2 and ns-TiO_2-CeO_2 nanomaterials are shown in Fig. 10.3. The presence of CeO_2 clearly improves the ranges of absorption.

For various geometric configurations of photoelectrodes, the SC-SEP-PC cell's I–V characteristic (SCE/1 M NaOH/ns-TiO_2/CeO_2 admixed/Ti/H_2SO_4 ± K_2SO_4/PtCE, PtWE) was measured (Fig. 10.4).

The graph shows that photocurrent density rapidly falls with increasing electrode area before saturating. It can be attributed to a rise in surface state, which is mostly

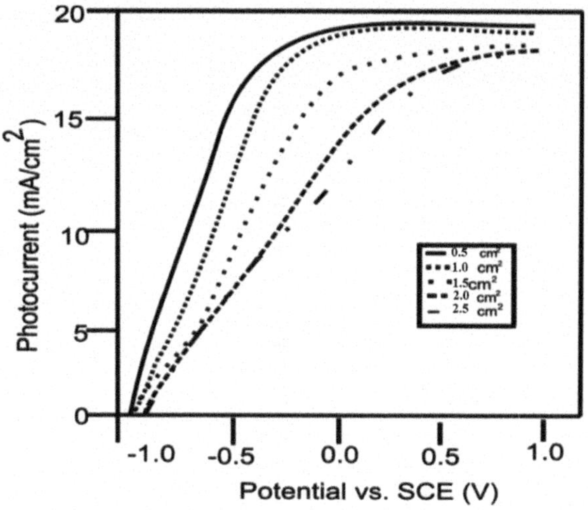

Fig. 10.4 Shows the variation in photocurrent as a function of electrode potential versus SCE for five different photoelectrodes which have been extracted and plotted against electrode of different areas

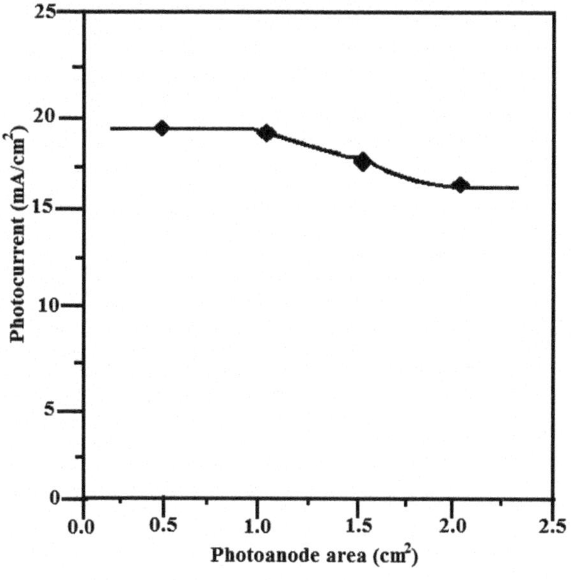

Fig. 10.5 Variation of photocurrent density with photoanode area

caused by surface imperfections and grain boundaries and serves as a charge carrier's recombination centre.

The photocurrent values have been taken and presented in Fig. 10.5 for five different photoelectrode areas. Corresponding to Fig. 10.4, it has been demonstrated that as electrode area increases, the photocurrent density rapidly drops before being saturated.

Figure 10.6 depicts the relationship between hydrogen production rate (HPR) and photoanode area. As can be seen, the maximum photocurrent density and hydrogen

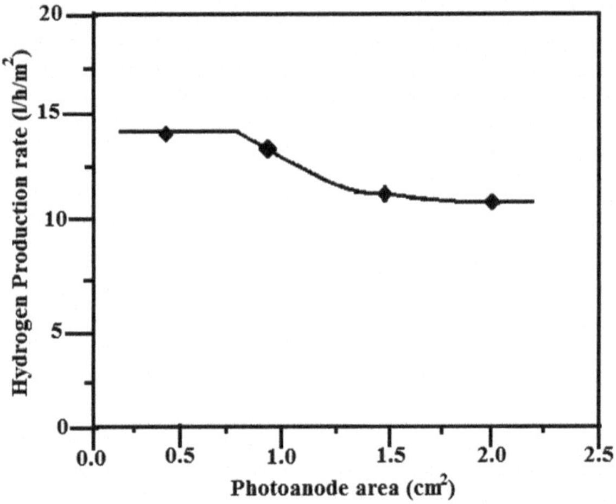

Fig. 10.6 Variation of hydrogen production rate with photoanode area

production rate are correlated with the smallest electrode area and begin to decline with increasing electrode area. Beyond a given area, in this case 1.5 cm^2, the observed decline in HPR is slow. Hence from Figs. 10.5 and 10.6, it can be seen that the modular PEC cells would need enough photoelectrode area to modularize the complexity and increase the pace at which hydrogen is produced. Therefore, choosing an electrode area between the smallest and the one above where the rate of hydrogen production and the decline in photocurrent density almost saturate would be appropriate based on the findings of the current study. The current analysis determined that 1.5 cm^2 was the ideal area for producing hydrogen.

The island of CeO_2 created over the surface of TiO_2 is considered to catalyse the anode reaction, perhaps with the help of the colloidal photochemical system. This is evident via SEM. It is also possible to speculate that the surface modification with thin CeO_2 islands with high conductivity and a broad absorption range (from 200 to 500 nm) is what causes the higher photoactivity brought on by increased gas evolution kinetics. Therefore, it might be argued that ns TiO_2 covered in surface clusters of CeO_2 is a new and improved photoelectrode for the SC-SEP, PEC cell's production of hydrogen.

10.4 Conclusion

Therefore, it can be concluded that the ns TiO_2-CeO_2 represents an enhanced photoelectrode and that it improves hydrogen generation through photosynthesis. To investigate the impact of electrode area on the output power in a chemical fuel (i.e. H_2) generating SC-SEP, PEC cell, the photoelectrochemical response of an ns-TiO_2-CeO_2 photoelectrode for five different electrode areas has been examined. Based on the findings of the current investigations, it has been identified that 1.5 cm^2

is the ideal ns-TiO_2-CeO_2 photoelectrode area for sufficient photoconversion efficiency and hydrogen generation rate.

Acknowledgement The authors are indebted to Padmashri Professor (Late) O.N. Srivastava of the BHU in Varanasi, India's Hydrogen Energy Center and Nano Science & Technology Center, for his enlightening discussions and equipment facilities. One of the authors (Dr. Priyanka Chawla) is also thankful to CSIR-India for the financial support in the form of research associate.

References

Chawla, P., Pandey, A., Pooja, K., & Tripathi, M. (2022). Low cost and efficient counter electrode for solid state dye sensitized solar cells. *Journal of Applied Electrochemistry*, 1–10.

Dincer, I., & Joshi, A. S. (2013). *Solar based hydrogen production systems*. Springer.

Feng, J., Huang, H., Yan, S., Luo, W., Yu, T., Li, Z., & Zou, Z. (2020). Non-oxide semiconductors for artificial photosynthesis: progress on photoelectrochemical water splitting and carbon dioxide reduction. *Nano Today, 30*, 100830.

Feng, H., Liu, D., Zhang, Y., Shi, X., Esan, O. C., Li, Q., Chen, R., & An, L. (2022). Advances and challenges in photoelectrochemical redox batteries for solar energy conversion and storage. *Advanced Energy Materials*, 2200469.

Goltsov, V. A., & Veziroglu, T. N. (2001). From hydrogen economy to hydrogen civilization. *International Journal of Hydrogen Energy, 26*, 909–915.

Hoffmann, M. R., Martin, S. T., Choi, W., & Bahnemann, D. W. (1995). Environmental applications of semiconductor photocatalysis. *Chemical Reviews, 95*, 69–96.

Hosseini, S. E., & Wahid, M. A. (2020). Hydrogen from solar energy, a clean energy carrier from a sustainable source of energy. *International Journal of Energy Research, 44*(6), 4110–4131.

Islam, M. M., Ali, M. I., Ceh, B., Singh, S., Khan, M. K., & Dagar, V. (2022). Renewable and non-renewable energy consumption driven sustainable development in ASEAN countries: do financial development and institutional quality matter? *Environmental Science and Pollution Research, 29*(23), 34231–34247.

Omer, A. M. (2008). Energy, environment and sustainable development. *Renewable and Sustainable Energy Reviews, 12*(9), 2265–2300.

Pooja, K., Pandey, A., Awasthi, K., Tripathi, M., & Chawla, P. (2022). Development of polymer electrolyte based on graphite/MWNTs fillers for sustainable dye sensitized solar cells. *Chemical Papers, 79*, 7323.

Song, C. (2006). Global challenges and strategies for control, conversion and utilization of CO_2 for sustainable development involving energy, catalysis, adsorption and chemical processing. *Catalysis Today, 115*(1–4), 2–32.

Ta, X. M. C., Daiyan, R., Nguyen, T. K. A., Amal, R., TranPhu, T., & Tricoli, A. (2022). Alternatives to water photooxidation for photoelectrochemical solar energy conversion and Green H_2 production. *Advanced Energy Materials, 12*(42), 2201358.

Tilley, S. D. (2019). Recent advances and emerging trends in photoelectrochemical solar energy conversion. *Advanced Energy Materials, 9*(2), 1802877.

Tripathi, M., Pandey, K., & Kumar, S. D. (2007). Surface modification of semiconductor photoelectrode for improved solar cell performance. *Solar Energy Solar Materials, 91*, 1663.

Wang, L., & Wang, W. Z. (2012). Graphene/ semiconductor nanocomposites: preparation and application for photocatalysis. *International Journal of Hydrogen Energy, 37*, 304.

Watanabe, S., Ma, X., & Song, C. (2009). Characterization of structural and surface properties of nanocrystalline TiO_2- CeO_2 mixed oxides by XRD, XPS, TPR, and TPD. *The Journal of Physical Chemistry C, 113*(32), 14249–14257.

Zhang, Z., Zhou, F., & Lavernia, E. J. (2003). On the analysis of grain size in bulk nanocrystalline materials via X-ray diffraction. *Metallurgical and Materials Transactions A, 34*(6), 1349–1355.

Chapter 11
Optimal Heat Recovery and Electricity Production from a Proton Exchange Membrane Fuel Cell's Operation

J. Siecker and K. Kusakana

11.1 Introduction

The global fossil fuel resources are declining rapidly, which urges the need for alternative energy resources. Renewable energy resources (RER), such as solar photovoltaic (PV) arrays, wind turbines, and hydrokinetic turbines, to name a few, may be regarded as credible alternative energy resources to replace/supplement the current declining fossil fuel resources (Mohamed et al., 2012). In contrast, the global electricity demand is increasing rapidly, as well as the CO2 emissions, accordingly. Therefore, with the declining fossil fuel resources and the increasing energy demand, highlights the need to make more use of RER to relieve the burden on any energy grid and effectively reducing the energy cost of the consumer.

Hydrogen may be seen as a possible alternative fuel which can be produced from renewable energy, as mentioned, and a promising contender in the energy storage domain. Furthermore, hydrogen is regarded as a clean fuel, as the product of the chemical reaction is water and heat as the by-product (Nowotny et al., 2016). A proton exchange membrane fuel cell (PEMFC) harnesses the energy contained within the stored H2 and utilizes it to generate clean electrical energy (Ural & Gencoglu, 2013). The generated heat influences the performance of the PEMFC, which means that its performance can be enhanced by increasing the operating temperature until a certain extent (Zhang et al., 2012). However, this enhanced performance comes at a price, which cannot be ignored, causing a reduction in the lifespan of the membrane itself (Chandesris et al., 2015). Various authors have attempted to improve the performance of a PEMFC by effectively controlling the operating temperature, thereof. Also, several authors have attempted to perform the optimal power dispatch of a hybrid power system based on PEMFC technology and

J. Siecker · K. Kusakana (✉)
Central University of Technology, Bloemfontein, South Africa
e-mail: kkusakana@cut.ac.za

A. Sayigh (ed.), *Transition Towards a Carbon Free Future*, Innovative Renewable Energy, https://doi.org/10.1007/978-3-031-61660-0_11

the electrical grid. However, a comprehensive optimal control model which, firstly, contains the optimal power dispatch of a hybrid power system based on fuel cell technology and the electrical grid and, secondly, combining it with the optimal control of a PEMFC by controlling the operating temperature, thereof, is lacking.

The authors proposed a mathematical model of a hybrid power system which integrates solar photovoltaic modules, electrolyser, fuel cell and the electrical grid, supplying electrical energy to a heat pump water heater (HPWH) (Sichilalu et al., 2016). The main aim of the mathematical model is to optimize the energy management of the hybrid system, which is to minimize the operating energy cost of the electrical grid, by taking the timeof-use (TOU) tariff structure into account. The optimal control strategy is evaluated against a thermostatbased controller (baseline), to essentially control/switch the HPWH. The simulation results reveal the HPWH consumed 45.5 kWh of energy when controlled under the optimal control regime, as opposed to 84 kWh when the thermostat-based controller is used.

It can be observed from the literature that a research gap exists, where the main objective of this study is to propose a comprehensive optimal control strategy of a PEMFC, forming part of a hybrid renewable energy micro-grid system connected to the utility grid, which could maximize the heat recovery from the PEMFC by the hot water storage tank (HWST), whilst maximizing the electrical energy supplied from the solar PV modules to the electrical load. Thirdly, the operating energy cost of the utility grid could also be minimized, from a power dispatch perspective. The final additional contribution could be the optimal control model which could also yield a prolonged membrane lifespan of the PEMFC, by controlling its operating temperature, which is one of the highest cost items in this type of system (Chandesris et al., 2015).

11.2 Methodology

11.2.1 Proposed Integrated PEMFC Water Heating System

Figure 11.1 illustrates the schematic of the proposed PEMFC water heating system, for two occupants in a typical two-bedroom residential building. The hydrogen storage tank supplies the PEMFC, which is used to convert H_2 and oxygen (O_2) into electrical energy, H_2O, and thermal energy. The generated electrical energy is used to supply a load, and the thermal energy generated is transferred to a HWST.

The power flow of the proposed PEMFC water heating system used in a renewable energy micro-grid is illustrated in Fig. 11.2. The proton exchange membrane electrolyser (PEMWE) utilizes solar PV energy and the utility grid to generate H_2, which is stored in a hydrogen storage tank. The hydrogen storage tank supplies H_2 to the PEMFC. The PEMFC, solar PV modules, and the utility grid are supplied the load. When H_2 is generated, it is stored in the hydrogen storage tank and is ideally utilized by the PEMFC during the high-cost regions of the TOU tariff structure under

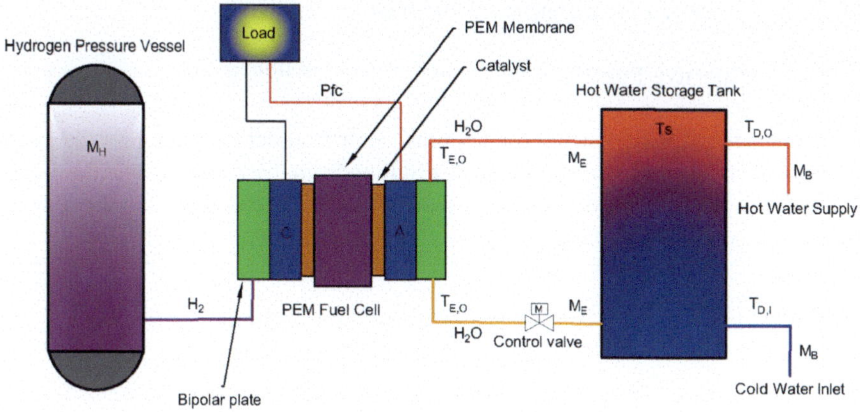

Fig. 11.1 Schematic of the PEMFC water heating system

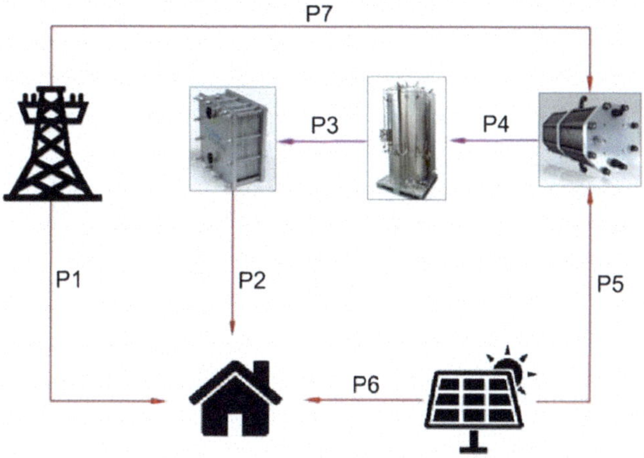

Fig. 11.2 . Power flow diagram of proposed renewable energy micro-grid system

which the utility grid is confined to. The power flows presented in Fig. 11.2 are defined as follows:

P1, electrical power supplied to the load from the utility grid; P2, electrical power supplied from the PEMFC to the load; P3, hydrogen power supplied to PEMFC from hydrogen storage tank; P4, hydrogen power supplied to hydrogen storage tank from PEMWE; P5, electrical power supplied to the PEMWE from PV modules; P6, electrical power supplied to the load from PV modules; P7, electrical power supplied to the PEMWE from the utility grid.

11.2.2 Control Optimization Problem

11.2.2.1 Objective Functions

The first objective, f_1, of the proposed optimal control model is to minimize the daily operational energy cost of the utility grid, resulting from the various power flows, by taking the TOU tariff structure into account, which is illustrated in Eq. (11.1). ρ_c is the electricity pricing tariff, P_1 is the electrical power supplied to the load by the utility grid, P_7 is the electrical power supplied to the PEMWE by the utility grid, and t_s is the sampling time.

$$f_1 = \sum_{k=1}^{N} [\rho_C(P_1(k) + P_7(k))].t_s \tag{11.1}$$

The second objective is to maximize the electrical energy supplied from the solar PV modules to the electrical load, which is depicted in Eq. (11.2).

$$f_2 = \sum_{k=1}^{N} [(P_6(k)].t_s \tag{11.2}$$

The Comfort level of the consumer is addressed, where the difference between the output temperature $T(k)$ of the HWST temperature and the desired temperature $F(k)$ should not be excessive. That is, the value of $(T(k)F(k))^2$ should be minimized. The function f_3 in Eq. (11.3) denotes the discomfort level, which is the third and last objective function to be minimized (Zhang and Xia (2007)):

$$f_3 = \sum_{k=1}^{N} \left[(T(k) + F(k))^2\right].t_s \tag{11.3}$$

The output temperature $T(t)$ is known as the discrete differential temperature equation and is shown in the following equation:

$$T_{k+1} = T_0.\prod_{j=0}^{k}(1 + A_j) + t_s.\sum_{j=0}^{k} B_j.U_j \prod_{i=j+1}^{k}(1 + t_s.A_i) + t_s.\sum_{j=0}^{k} C_j \prod_{i=j+1}^{k} \times (1 + t_s.A_i) \tag{11.4}$$

11.2.2.2 Hydrogen Electrical Power

The electrical power produced by the PEMFC is depicted in Eq. (11.5), where I_F is the current produced by the PEMFC. α_a and α_c are the anode and cathode charge transfer coefficients, respectively. R is the universal gas constant, F is the Faraday constant, J is the current density, J_0 is the exchange current density, t_{mem} is the membrane thickness, σ_{mem} is the membrane conductivity, and J_L is the limiting current density. ρ_{H20} is the partial pressure of water. ρ_{H2} and ρ_{O2} are the partial pressure of hydrogen and oxygen, respectively.

$$P_{FC} = 1.2583.I_F - \frac{I_F.J.t_{mem}}{\sigma_{mem}} + T_F\left[\frac{R.I_F}{2.F}\ln\left(\frac{P_{H_2}.P_{O_2}{}^{0.5}}{P_{H_2O}}\right) - \left(\frac{\alpha_a+\alpha_c}{\alpha_a.\alpha_c}\right).\frac{R.I_F}{2.F}.\ln\left(\frac{J}{J_O}\right) - \frac{R.I_F}{2.F}.\ln\left(\frac{1+\frac{1}{J_L}}{1-\frac{1}{J_L}}\right)\right] \quad (11.5)$$

11.2.2.3 Hydrogen Storage Constraints

The hydrogen storage dynamics is depicted in Eq. (11.6), with its minimum and maximum boundary constraints. M_{H2} (0) is the initial mass within the hydrogen storage tank. $M_{H2}{}^{min}$ is the minimum allowable limit of the hydrogen storage level, and $M_{H2}{}^{max}$ is the maximum allowable limit of the hydrogen storage level.

η_{EL} and η_{FC} are the efficiency of the PEMWE and the PEMFC, respectively. HHV_H is the higher heating value of hydrogen, which is 39.7 kWh/kg (Sichilalu et al. (2016)).

$$M_{H_2}{}^{min}(k) \le M_{H_2}(0) + \eta_{EL}.t_s.\sum_{k=1}^{N}\frac{P_4(k)}{HHV_{H_2}} - \frac{t_s}{\eta_{FC}}.\sum_{k=1}^{N}\frac{P_2(k)}{HHV_H} \le M_{H_2}{}^{max}(k) \quad (11.6)$$

11.3 Case Study Description

In this section, a comprehensive optimal control strategy of a PEMFC, forming part of a hybrid renewable energy micro-grid system connected to the utility grid, is used to maximize the heat recovery from the PEMFC by the HWST, as well as its electricity generation is simulated using the SCIP (solving constrained integer programs) solver in the optimization toolbox of MATLAB (Hohne et al., 2019).

Furthermore, the optimal control strategy is also used to minimize the operating energy cost of the utility grid.

A case study is conducted using meteorological data obtained at the University of Free State, Bloemfontein, South Africa, on the proposed strategy (Sauran, 2017). The heat transferred from the PEMFC to the HWST can be used to assist the main water heating source for a typical two-bedroom residential building. The utility grid, assisted by the PEMFC used in a hybrid renewable energy micro-grid system, is utilized to supply the load of a typical two-bedroom residential building comprising of an air-to-air heat pump and an air-to-water heat pump. The sampling time, ts = 30 minutes, will be implemented to simplify the simulations.

11.3.1 Data Presentation

Table 11.1 shows the parameters of the HWST for a typical two-bedroom residential building, which is typically for two occupants (Sichilalu et al., 2015). As is the surface area of the tank, M is the volume of the tank, C is the specific heat capacity of water, h is the surface, k is the thermal conductivity, and dx is the insulation thickness.

The air-to-air heat pump used to perform space heating to the residential building is rated at 5000 W of electrical power and 14,000 W of thermal power (AC Direct, 2020). Then, the air-to-water heat pump used to perform the space heating requirements to the residential building is rated at 1004 W of electrical power and 3200 W of thermal power (Right Air, 2022).

Table 11.2 shows the parameters of the PEM fuel cell used for electrical energy production (Zhang et al., 2012; Kai et al., 2019; Bareis et al., 2019; Bernt et al., 2020). The exchange current densities at the anode and cathode are identical, due to the anode exchange being the more dominant of the two current densities (Barbir, 2012). α_a and α_c are the anode and cathode charge transfer coefficients, respectively. R is the universal gas constant, F is the Faraday constant, J is the current density, J_0 is the exchange current density, t_{mem} is the membrane thickness, σ_{mem} is the membrane conductivity, and J_L is the limiting current density. ρ_{H20} is the partial pressure of water. ρ_H and ρ_O are the partial pressure of hydrogen and oxygen, respectively.

Table 11.1 Hot water storage tank (HWST) parameters

Parameters	Symbol	Value	Unit
Tank surface area	A_s	1.19	m2
Tank volume	M	50	L
Specific heat capacity	C	4186	J/kgK
Surface heat transfer coefficient	h	6.3	W/m^2K
Thermal conductivity	k	0.055	W/mK
Thickness of insulation	dx	0.07	m

Table 11.2 PEM electrolyser parameters

Parameters	Symbol	Value	Unit
Anode/Cathode transfer coefficient	α_a/α_a	0.5	–
Universal gas constant	R	8.314	J/mol.K
Faraday constant	F	96,485	C/mol
Current density	J	1	A/cm^2
Exchange current density	Jo	0.00000173	A/cm^2
Membrane thickness	Tmem	0.0178	cm
Membrane conductivity	σ_{mem}	0.0001	S/cm
Limiting current density	J_L	2	A/cm^2
Partial pressure of water/hydrogen/oxygen	$\rho_{H2O}/\rho_H/\rho_O$	101,325	Pa

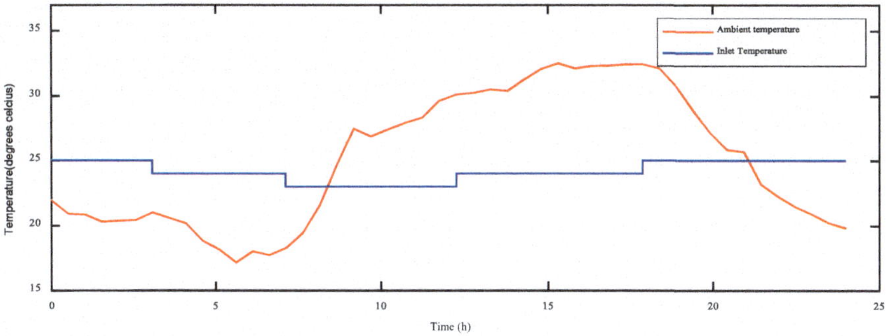

Fig. 11.3 Ambient and inlet water temperature

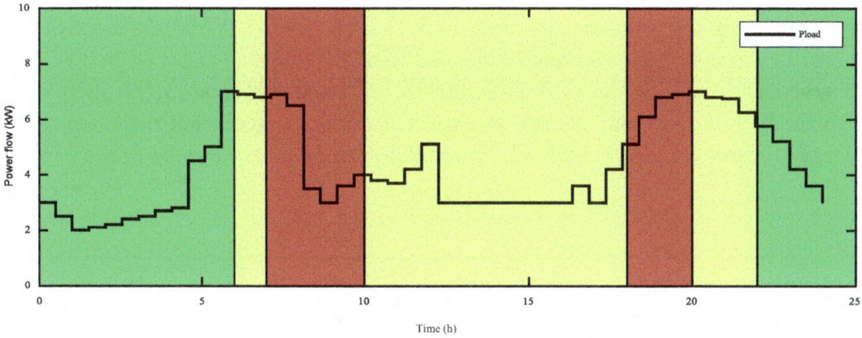

Fig. 11.4 Load profile of the residential building

Figure 11.3 presents the ambient temperature in degrees Celsius during a typical winter day in Bloemfontein, Free State, South Africa, as well as the inlet water temperature. The ambient air temperatures have been retrieved from the radiometric station situated at the University of Free State, forming part of the South African Universities Radiometric Network (SAURAN) (Sauran, 2017).

Figure 11.4 illustrates the load demand of the typical two-bedroom residential building as well as the regions of the TOU tariff structure. The off-peak regions are illustrated by the green areas, the standard regions are illustrated by the yellow areas, and the red regions are illustrated by the red areas.

11.4 Simulation Results and Discussion

The proposed optimal control regime is evaluated by comparing it to a standard system which, in this case, the load is simply only supplied by the utility grid and no other means of supplying the load and energy storage system.

The optimal control strategy of a PEM fuel cell, forming part of a hybrid renewable energy micro-grid system connected to the utility grid, is discussed in Sect. 11.2. This model optimally controls the switching of the PEM fuel cell system operating within the renewable energy micro-grid. It attempts to minimize the operating energy cost of the utility grid, whilst maximizing the electrical energy production and the heat recovery from the PEM fuel cell. Furthermore, the undesired heat which is recovered negatively influences the lifespan of the PEM fuel cell, as mentioned in Sect. 11.1.

11.4.1 Standard Utility Grid System

The cumulative energy consumption of the load which is for a typical two-bedroom residential building is illustrated in Fig. 11.5. From 00 h00 to 05 h00, the slope of the energy consumption is relatively stable, due to no excessive changes to the load are observed, as seen from Fig. 11.5. From 06 h00 to 07 h00, the slope slightly increases, compared to the previous period, as expected due to a significant increase in the energy consumption from the load. From 07 h00 to 17 h00, the slope of the energy

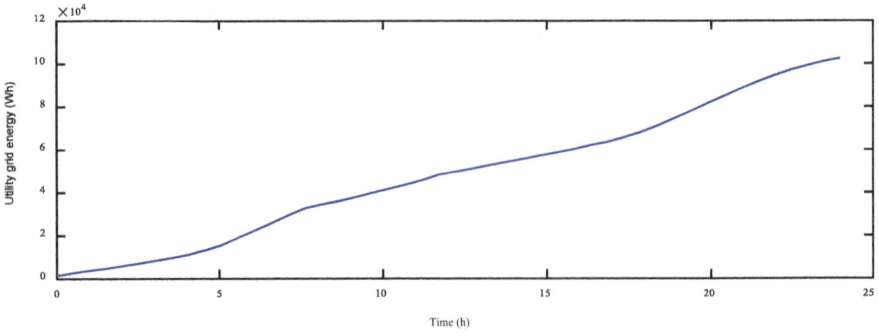

Fig. 11.5 Cumulative Energy consumption of the Utility Grid

consumption slightly decreases because of a reduction to the energy consumption of the load. From 17 h00 to 21 h00, the slope increases once again, due to the increase in the energy consumption of the load. From 23 h00 until midnight, the slope once again decreases slightly, as a result of the reduction in the energy consumption to a maximum total of 102.48 kWh.

11.4.2 Optimal Control of the PEMFC Hybrid Renewable Energy Micro-Grid System Connected to the Utility Grid

The power flows to the load are illustrated in Fig. 11.6. From 00 h00 to 06 h00, it can be observed that the utility grid, P_1, hardly supplies power to the load, whereas the PEMFC, P_2, supplies the vast majority power to the load. The solar PV modules, known as P_6, cannot supply the load, due to the absence of solar irradiance during this period.

From 06 h00 to 07 h00, P_2 is supplying most of the power required by the load, where P_1 does not supply the load. It can be observed that P6 begins to supply power to the load due to this when solar irradiance becomes present. From 07 h00 to 10 h00, P2 decreases until it stops supplying power to the load, the power supplied by P6 increases significantly, and the utility grid, P1, remains dormant during this period which is the first peak pricing region of the TOU tariff structure. The power consumption continues to decrease until zero, which is when the utility grid no longer supplies the load, due to the system is entering the first peak pricing period of the TOU tariff structure and the utility grid remains dormant for the remainder of the peak pricing period. From 10 h00 to 18 h00, the only source of power to the load is the solar PV modules, P6. From 18 h00 to 20 h00, P2 begins to supply the load, as this is the second peak pricing region of the TOU tariff structure; the energy consumption of the load remains unchanged at zero, which is during the second standard pricing period of the TOU tariff structure due to PEMFC is

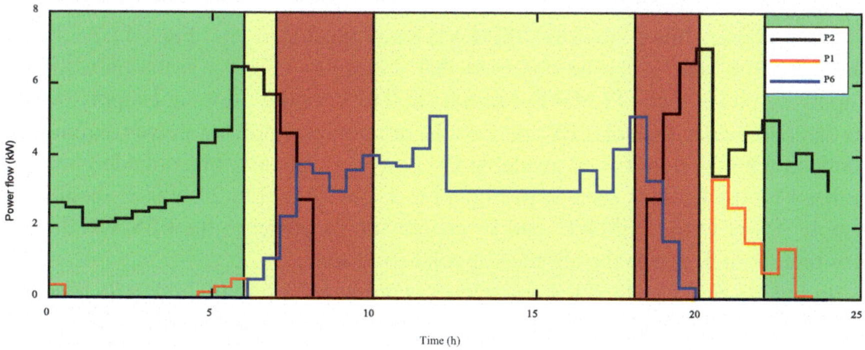

Fig. 11.6 Power flows to the load

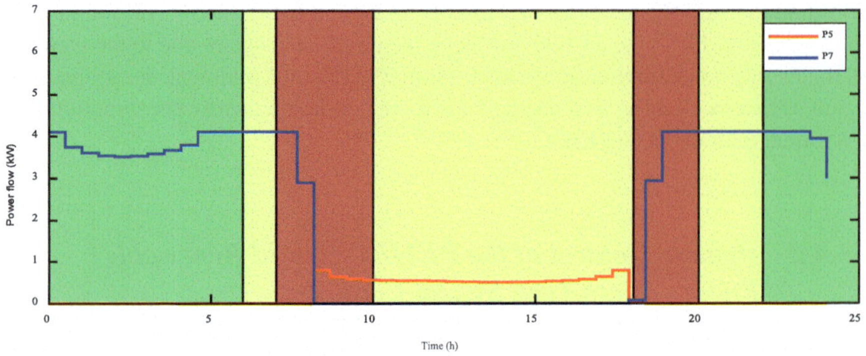

Fig. 11.7 Power flows to the PEMWE

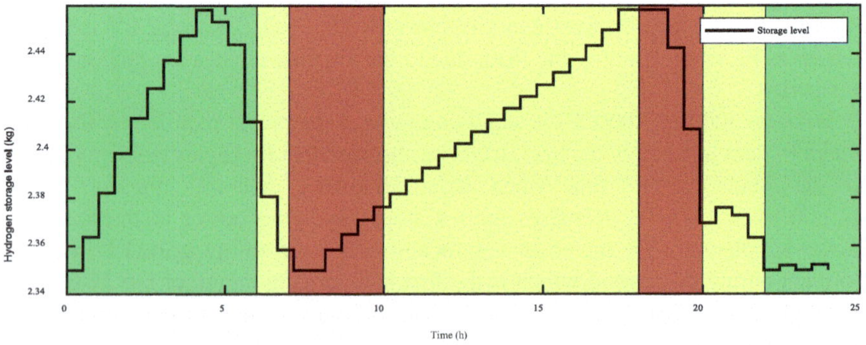

Fig. 11.8 Hydrogen storage tank dynamics

supplying the load during this period, as seen from Fig. 11.7. Furthermore, it can be observed from Fig. 11.6 the solar PV modules P6 are assisting the PEMFC to supply the load with the majority of the available PV power. From 20 h00 to 24 h00, the utility grid begins to supply the load with power, and the PEMFC continues to supply power to the load.

The power flows to the PEMWE are illustrated in Fig. 11.7. From 00 h00 to 08 h00, it can be observed that the utility grid, P_7, is the only source supplying power to the PEMWE. From 08 h00 to 18 h00, P_7 does not supply power to the load, whilst the solar PV modules P_5 are purely supplying power required to the load with the minority of available PV power, as majority of available power is utilized by the load, as seen from Fig. 11.6. From 18 h00 to 24 h00, P_7 begins to supply power to the PEMWE, and P_5 cannot supply power to the PEMWE during this time period, due to the absence of solar irradiance.

In Fig. 11.8, the hydrogen storage level is presented. From 00 h00 to 04 h00, the hydrogen storage level is initially at 2.35 kg and increases until it reaches 2.46 kg, which is due to the PEMFC is consuming less hydrogen than the PEMWE is supplying the hydrogen storage tank. From 04 h00 to 07 h00, the hydrogen storage

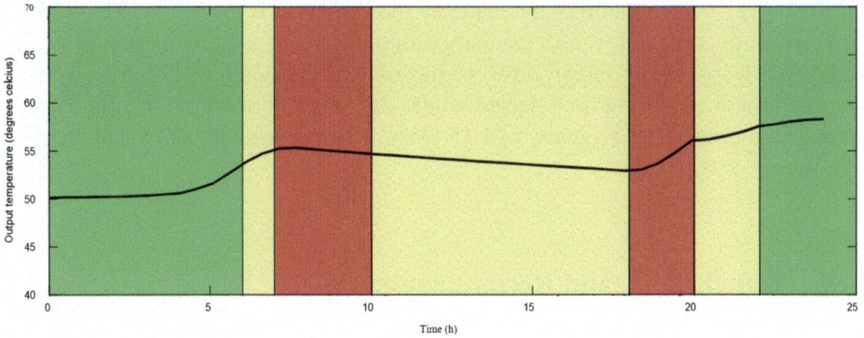

Fig. 11.9 Hot water storage tank temperature

level decreases slightly to 2.35 kg, due to the PEMFC is starting to consume more hydrogen power than the PEMWE is supplying the hydrogen storage tank. From 07 h00 to 17 h30, the hydrogen storage level increases until it reaches 2.46 kg, due to the PEMFC does not supply any power to the load and the PEMWE is supplying hydrogen to the hydrogen storage tank. From 17 h30 to 20 h00, the hydrogen storage level decreases significantly to 2.37 kg, due to the load demands significantly more power from the PEMFC. From 20 h00 to 00 h00, the hydrogen storage level decreases until 2.35 kg, due to the PEMFC is consuming more hydrogen than the PEMWE is supplying the hydrogen storage tank.

From Fig. 11.9, the HWST temperature is initially 50 °C at 00 h00, after which it begins to increase. The HWST temperature increases until 08 h00, due to the heat generated by the PEMFC during electrical energy production. From 08 h00 to 18 h30, the HWST temperature begins to decrease, which is due to the PEMFC not being in operation during this period. From 18 h30 to 00 h00, the HWST temperature begins to increase until it reaches 56.71 °C at the end of the day. Furthermore, based on the research conducted by Chandesris et al. (2015), if the generated heat is recovered from the PEMFC and transferred to the HWST, the time for the membrane to degrade to a thickness of 50% could be prolonged by 0.91 years, by means of the interpolation of the simulation results.

11.4.3 Daily Economic Analysis

The proposed optimal control algorithm economically evaluated against the standard system, which in this case the load, is purely supplied by the utility grid and no other means of supplying the load and energy storage system. The daily operating energy cost of the standard system (baseline) consumed 102.48 kWh, which corresponds to USD 8.6. The optimal control regime yields a daily grid energy consumption of 5.57 kWh and 53.69 kWh, by the load and PEMWE, respectively. Moreover, the optimal control regime resulted in a daily operational energy cost of USD 0.47 and USD 4.5, for the energy consumption of the load and the PEMWE, respectively. The simulation results, in conjunction with the daily economic analysis, of the proposed model yield a daily operating energy cost saving of 42.21% compared to the standard system (baseline), during the selected winter day. Furthermore, the thermal energy recovery amounted to a recovery of 1.15 kWh, which corresponds to a daily energy cost saving of 0.14 USD.

11.5 Conclusions and Recommendations

In this paper, relevant and recent literature are used to evaluate the proposed comprehensive optimal control of the PEMFC water heating system, as well as it is operating within a renewable energy micro-grid, connected to the utility grid. The optimal control strategy is modeled mathematically after which the multiobjective function, together with all the associated constraints, is described in detail, in Sect. 11.2. Accurate data is used in this study to perform the necessary simulations based on the proposed optimal control model. The proposed optimal control regime is evaluated by comparing it to a standard system (baseline). The standard utility grid system (baseline), where the load is purely supplied by the utility grid, produced 102.48 kWh. The simulation results of the proposed optimal control model yield the utility grid supplied the load 5.57 kWh and the PEMWE 53.69 kWh, whilst the PEMFC produced 55.05 kWh to supply the total load demand. The simulation results of the proposed optimal control model also illustrate a maximum thermal energy recovery of 4.15×10^6 J, which is equivalent to 1.15 kWh, is achieved.

Furthermore, by recovering the generated heat from the PEMFC, the time for the membrane to degrade to a thickness of 50% could be prolonged by 0.91 years, by means of the interpolation of the simulation results, obtained from Reference (Chandesris et al., 2015). Therefore, when the membrane goes beyond 50%, the membrane degradation occurs non-linearly.

The main aim of this study is to develop a comprehensive optimal control model of the PEMFC water heating system, as well as it is operating within a renewable energy micro-grid, connected to the utility grid, where further research may be conducted on more advanced control techniques, such as model predictive control (MPC), which could adapt to any changes to its external environment.

References

AC DIRECT. (2020). Samsung Ducted HSP Split 72000 Btu/hr Inverter Air Conditioner. [Online]. Available: Samsung Ducted MSP Split 48000 Btu/hr (3 Phase) Non Inverter Air Conditioner – ACDirect. Accessed 27 April 2021.

Barbir, F. (2012). *PEM fuel cells: Theory and practice*. Academic Press.

Bareis, K., de la Rua, C., Mockl, M., & Hamacher, T. (2019). Life cycle assessment of hydrogen from proton exchange membrane water electrolysis in future energy systems. *Applied Energy, 237*, 862–872.

Bernt, M., Schroter, J., Mockl, M., & Gasteiger, H. A. (2020). Analysis of gas permeation phenomena in a PEM water electrolyzer operated at high pressure and high current density. *Journal of Electrochemical Society, 167.*

Chandesris, M., Medeau, V., Guillet, N., Chelghoum, S., Thoby, D., & Fouda-Onana, F. (2015). Membrane degradation in PEM water electrolyzer: Numerical modeling and experimental evidence of the influence of temperature and current density. *International Journal of Hydrogen Energy, 40*, 1353–1366.

Hohne, P. A., Kusakana, K., & Numbi, B. P. (2019). Optimal energy management and economic analysis of a grid connected hybrid solar water heating system: A case of Bloemfontein, South Africa. *Sustainable Energy Technologies and Assessments., 31*, 273–291.

Kai, J., Saito, R., Terabaru, K., Li, H., Nakajima, H., & Ito, K. (2019). Effect of temperature on the performance of polymer electrolyte membrane water electrolysis: Numerical analysis of electrolysis voltage considering gas/liquid two-phase flow. *Journal of Electrochemical Society, 166*, 246–254.

Mohamed, Y. A., Zeinneldin, H. H., & Salama, M. M. A. (2012). Seamless formation and robust control of distributed generation Micogrids via direct voltage control and optimized dynamic power sharing. *IEEE Transactions on Power Electronics, 27*, 3.

Nowotny, J., Hoshino, T., Dodson, J., Atanacio, A. J., Ionescu, M., Peterson, V., Prince, K. E., Yamawaki, M., Bak, T., Sigmund, W., & Veziroglu, T. N. (2016). Generation of hydrogen fuel using nuclear energy. *International Journal of Hydrogen Energy.*

RIGHT AIR. (2022). Alliance Domestic Hot Water Heat Pump or Geyser 3.2 kW [Online]. Available: Alliance Domestic Hot Water Heat Pump for Geyser 3.2 kW – RightAir. Accessed 02 July 2022.

Sauran.net. (2017) Station Details - Southern African Universities Radiometric Network. [online] Available at http://www.sauran.net/ShowStation.aspx?station=7. Accessed 15 Nov 2017.

Sichilalu, S., Mathaba, T., & Xia, X. (2015). Optimal control of a wind-PV-hybrid powered heat pump water heater. *Applied Energy, 185*, 1173–1184.

Sichilalu, S. M., Tazvinga, H., & Xia, X. (2016). Integrated energy management of grid-tied-PV-fuel cell hybrid system. *Energy Procedia, 103*, 111–116.

Ural, Z., & Gencoglu, M. T. (2013). Design and simulation of a solar hydrogen system for different situations. *International Journal of Hydrogen Energy*, 1–8.

Zhang, J., & Xia, X. (2007). Best switching time of hot water cylinder-switched optimal control approach. In *AFRICON*, pp. 1–7.

Zhang, H., Su, S., Lin, G., & Chen, J. (2012). Efficiency calculation and configuration design of a PEM Electrolyzer system for hydrogen production. *International Journal of Electrochemical Science, 7*, 4143–4157.

Chapter 12
The Impact of Natural Disaster on Renewable Energy: The Role of Foreign Direct Investment and Infrastructure in Asia

Yessi Rahmawati, Rossanto Dwi Handoyo, Deni Kusumawardani, and Widya Sylviana

12.1 Introduction

Natural disasters pose a significant threat to both developing and developed countries worldwide (van Bergeijk & Lazzaroni, 2015; Abbas Khan et al., 2019). As climate change issues continue to rise, there has been an increasing discussion about the relationship between natural disasters and climate change. Consequently, researchers have focused on investigating the impact of natural disasters on various aspects of life, including socio-economic factors (Abbas Khan et al., 2019) and macroeconomic consequences (van Bergeijk & Lazzaroni, 2015). Recovery speed following natural disasters depends on numerous factors, and advanced technologies (Abbas Khan et al., 2019) and net foreign direct inflow (Khan & Anwar, 2022) have emerged as areas of particular interest. Asia, as a continent, is characterized by regions prone to disasters (Asian Development Bank, 2013). Countries such as India, Bangladesh, Nepal, the Philippines, Vietnam, Indonesia, Pakistan, and Afghanistan face multiple hazards, thereby increasing the mortality risk for their citizens (see Fig. 12.1). Consequently, this research aims to examine the impact of natural disasters on selected Asian countries in relation to renewable energy, foreign direct investment, and infrastructure.

The existing literature highlights the concept of natural disasters as inverse creative destruction (Crespo Cuaresma et al., 2008; Monllor & Altay, 2016). This perspective draws from Schumpeter's theory of creative destruction; however, in the case of natural disasters, it is associated with negative effects on economic growth (Qureshi et al., 2019). In other words, researchers argue that natural hazards can

Y. Rahmawati (✉) · R. D. Handoyo · D. Kusumawardani · W. Sylviana
Departement of Economics, Economic and Business Faculty, Airlangga University, Surabaya, Indonesia
e-mail: yessi.rahmawati@feb.unair.ac.id

A. Sayigh (ed.), *Transition Towards a Carbon Free Future*, Innovative Renewable Energy, https://doi.org/10.1007/978-3-031-61660-0_12

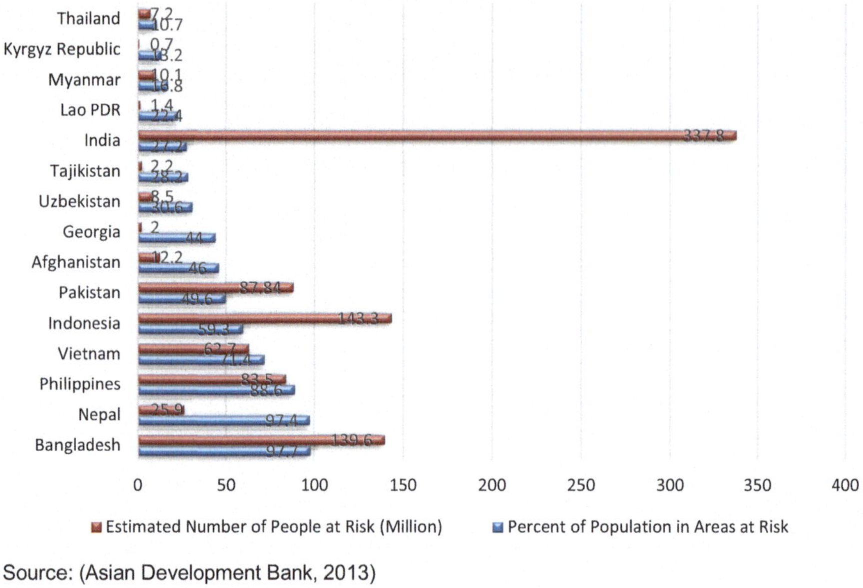

Fig. 12.1 High mortality rates due to natural hazard in Asia. (Source: Asian Development Bank, 2013)

cause significant destruction, even in well-established economies such as Indonesia and India (Barbhuiya & Chatterjee, 2020). Empirical studies further support this notion, demonstrating that natural disasters can increase poverty levels in affected countries (Awasthi, 2019; Bui et al., 2014), exacerbate inequality (Keerthiratne, 2018), and impact national energy consumption (Doytch & Klein, 2018; Lee et al., 2021).

Natural disasters impose immense tangible and intangible costs, causing widespread suffering in countries worldwide. In the context of the relationship between natural disasters and renewable energy, it is important to consider two key aspects. Firstly, natural disasters disrupt the inflows of goods, services, money, and technology transfers between countries. This disruption can diminish the crucial role of foreign direct investment inflows, consequently affecting macroeconomic conditions (Khan & Anwar, 2022). Furthermore, natural disasters can impede the transfer of renewable energy technologies from one country to another. Secondly, as highlighted by Doytch and Klein (2018), natural disasters not only disrupt macroeconomic conditions but also impact energy consumption, encompassing both renewable and non-renewable energy sources. Reduced energy consumption or diminished energy demand can undermine the economic capacity to utilize renewable energy effectively. Moreover, natural disasters can significantly decrease national consumption levels (Arouri et al., 2015). Consequently, natural catastrophes can hinder the development of renewable energy in two ways: by impeding technology transfer and by diminishing energy demand.

The impact of natural disasters on renewable energy can be observed through their disruption of foreign direct investment (FDI). The relationship between renewable energy and FDI has been extensively explored by scholars in both developed and developing countries. For instance, Samour et al. (2022) conducted a study on the United Arab Emirates (UAE) and found that an increase in FDI statistically improves renewable energy in both the short and long term, using the bootstrap autoregressive distributed lag (ARDL) approach. Similar results were also reported by Emre Caglar (2020) regarding the relationship between FDI and renewable energy. In this present paper, the author employs the same methodology, namely, the bootstrap ARDL approach, to examine nine selected countries based on the Climate Change Performance Index (CCPI) report from 2018. The empirical research study demonstrates a short- and long-term positive unidirectional causality between FDI and renewable energy in all selected countries, including Denmark, Finland, France, India, Italy, Morocco, Norway, Portugal, and Sweden. In the context of Asia, there is limited research focusing on the relationship between FDI and renewable energy. However, some studies have investigated this relationship in specific Asian countries. For example, Liu et al. (2016) examined FDI and renewable energy in China, while Kang et al. (2021) and Murshed et al. (2022) explored South Asia. Furthermore, studies conducted by Murshed et al. (2022) and Rej et al. (2022) investigated FDI and renewable energy in Bangladesh and India, respectively. These studies reinforce the previous arguments, providing comprehensive explanations of the crucial role of FDI in promoting renewable energy.

In the realm of renewable energy, the transportation of various types of energy sources requires specific and unique modes of transportation. Bamati and Raoofi (2020) argue that the development of better and tailored transportation modes for each type of renewable energy can enhance the consumption of renewable energy. This, in turn, reduces transportation costs and leads to more affordable prices, thereby increasing consumer willingness to purchase renewable energy at lower prices. In contrast, Liu et al. (2016) present conflicting findings regarding the importance of infrastructure investment in improving renewable energy. The authors collected data from 2000 to 2015 in 25 selected countries and failed to demonstrate the significance of infrastructure investment for renewable energy development. The authors attribute this ineffectiveness to the diverse range of advanced technologies employed in the renewable energy sector, as well as the unique regional characteristics of each country. Given the contrasting results from the existing literature, there is a need to further explore and explain the relationship between infrastructure and renewable energy.

The first objective of this research is to contribute novel empirical evidence by examining the impact of foreign direct investment (FDI) and infrastructure on renewable energy consumption in Asia. To achieve this, the study utilizes the updated technique of the two-step generalized method of moments (GMM) introduced by Roodman (2009). The aim is to provide valuable insights into the relationship between FDI, infrastructure, and renewable energy consumption in the context of natural disasters, with the goal of generating effective mitigation plans for disaster-prone countries, particularly in Asia. Despite the abundant availability of resources and the significant economic potential, the research reveals that the region's renewable energy potential, including wind, biomass, solar, and hydropower, remains underutilized. These resources are dispersed throughout the continent, highlighting the need for further exploration and development to harness their full capacity.

12.2 Methodology

12.2.1 Data

The primary objective of this research is to address the existing gap in empirical research regarding the impact of natural disasters on renewable energy consumption, specifically focusing on the mediating role of foreign direct investment (FDI) and infrastructure. To achieve this, panel data from 27 selected countries in Asia are utilized. The analysis employs the two-step generalized method of moments (TS-GMM) to examine the relationship between these variables using a dataset spanning from 2010 to 2020. The secondary data used in this study is sourced from the World Bank and the Center for Research on the Epidemiology of Disasters (CRED), which provides emerging data statistics. The dependent variable in this research is renewable energy consumption, while the independent variables include lagged renewable energy consumption, the number of affected people due to natural disasters, and measures of infrastructure. Additionally, to enhance the robustness of the findings, gross domestic product per capita (GDPP) is included as a control variable in the model.

The measurement of the impact of natural disasters remains a subject of ongoing debate among researchers. Various approaches have been suggested, including using indicators such as the number of deaths, the number of affected individuals, or the total economic losses incurred (Karim & Noy, 2016). However, it is important to consider that certain types of natural disasters may not necessarily result in fatalities but can still inflict significant suffering on affected populations, as is the case with drought. In order to account for such events, this study adopts the measure of the number of affected people, consistent with previous literature (Hallegatte et al., 2020; Keerthiratne, 2018). The term "affected people" encompasses both deaths and individuals who have been affected by any type of natural disaster, such as floods, droughts, landslides, volcanic eruptions, tsunamis, and others. However, this study does not include technological disasters, such as nuclear accidents or terrorist incidents, to ensure the avoidance of biased results. A descriptive overview of the data utilized in this research is presented in Table 12.3.

Table 12.3 Statistic descriptive

Variables	Obs	Mean	Std dev	Min	Max
Ln affected	250	11.96187	3.273793	0	19.66356
Ln RE	250	2.644881	1.38179	−0.7133499	4.469236
Ln GDPP	250	25.66024	1.973145	22.38406	30.31318
Ln infrastructure	250	11.09061	1.93809	6.327937	15.41767

Source: Data Processing Output

12.2.2 *Two-Step GMM*

The Two-Step GMM is a widely used econometric technique for estimating dynamic panel models (Arellano & Bond, 1991). It addresses potential endogeneity issues and provides consistent and efficient parameter estimates. This approach is particularly suitable for analyzing the impact of natural disasters on renewable energy consumption, taking into account the influence of FDI and infrastructure (Tan & Uprasen, 2022). In the first step of the Two-Step GMM, the model is differenced to eliminate individual-specific effects and create a set of moment conditions (Arellano & Bover, 1995). Differencing the model helps remove unobserved heterogeneity and potential endogeneity biases. This step is crucial for obtaining consistent estimates. The moment conditions used in the Two-Step GMM estimation are based on the orthogonality principle, which states that the average value of the model errors should be zero (Hansen, 1982). The GMM estimator selects parameters that minimize the discrepancy between the observed moments and the theoretical moments implied by the model. These moments are typically expressed as the product of the instrument matrix and the model's residuals (Arellano & Bond, 1991).

In the second step, the model is estimated using the moment conditions derived in the first step. This estimation step utilizes instrumental variables to address endogeneity concerns and produce efficient parameter estimates (Blundell & Bond, 1998). The instruments used in the GMM estimation should satisfy certain criteria, such as being relevant for the variables of interest and exogenous to the error term (Hansen, 1982). The Two-Step GMM estimation accounts for both the contemporaneous and lagged relationships between the variables in the panel data, capturing the dynamics of the system (Arellano & Bond, 1991). By considering the dynamic nature of the model, it provides more accurate estimates compared to other techniques like fixed effects or pooled ordinary least squares.

The validity of the Two-Step GMM estimation can be assessed through diagnostic tests. The Sargan test is commonly used to evaluate the overidentification restrictions, ensuring that the instruments used in the GMM estimation are valid and do not suffer from endogeneity (Sargan, 1958). The Hansen test is employed to examine the validity of the instruments in the overidentified model, providing further assurance of the estimation's reliability (Hansen, 1982). Additionally, serial correlation tests are conducted to detect and address any potential autocorrelation issues (Arellano, 2009).

To enhance the robustness of the findings, the estimated model is compared with alternative models, including the fixed effects model (FEM) and panel least squares (PLS) (Nickell, 1981). The focus is placed on the coefficient of the lagged renewable energy variable to assess the consistency of the results across different model specifications. By employing the Two-Step GMM methodology and conducting rigorous diagnostic tests, the present research aims to provide reliable and robust insights into the relationship between natural disasters, foreign direct investment, infrastructure, and renewable energy consumption in the selected Asian countries.

The panel data model is specified as follows:

$$RE_{it} = \beta_0 + \beta_1\, Lag\_RE_{it} + \beta_2 \text{Affected People}_{it} + \beta_3 \text{FDI}_{it} + \beta_4 \text{Infras}_{it} + \beta_5\, \text{GDPP}_{it} + e_{it}$$

where:

RE_{it}: Renewable energy consumption in country *i* at time *t*
Lag_RE_{it}: Lagged renewable energy consumption in country *i* at time *t*-1
Affected People_{it}: Natural disasters in country *i* at time *t*
FDI_{it}: Foreign direct investment inflows in country *i* at time *t*
Infras_{it}: Infrastructure development in country *i* at time *t*
GDPP_{it}: Gross domestic product per capita in country *i* at time *t*.

12.3 Results

The present research utilizes panel data collected from the World Bank to investigate the impact of natural disasters on renewable energy, with a focus on the mediating effects of foreign direct investment (FDI) and infrastructure. The empirical findings from the panel dynamics analysis, employing first differences and system generalized method of moments (GMM), are presented in Table 12.5. Prior to discussing the main research findings, diagnostic tests were conducted to ensure the validity of the instruments used in the study. The Sargan test, Hansen test, and tests for serial correlation were employed as diagnostic checks, all of which confirm the reliability of the instruments. To enhance the robustness of the results, a comparative analysis was performed using various models, including first differences GMM, system GMM, fixed effects model (FEM), and partial least squares (PLS). In terms of robustness checks, particular attention was given to the coefficient estimation for the lag of renewable energy. Following the recommendation of Roodman (2009), the most suitable model for the present research was determined to be the one-step first difference GMM. Consequently, the empirical results focus on the findings obtained from the one-step first difference GMM estimation.

The impact of natural disasters on renewable energy has garnered increasing attention within the academic community. The objective of this study is to address a gap in the existing literature by examining the influence of natural disasters on

Table 12.5 Panel dynamics empirical result

	(1)	(2)	(3)	(4)	(5)	(6)
Renewable energy	One-step FDGMM	Two-step FDGMM	One-step SYSGMM	Two-step SYSGMM	FEM	PLS
L1.RE	0.1871***	0.1879***	−0.1552***	−0.0901	0.1252**	0.1312**
	(0.0292)	(0.0638)	(0.0639)	(0.2434)	(0.0622)	(0.0606)
Affected people	0.1658***	0.1678***	0.1664***	0.1649***	0.1736***	0.1731***
	(0.0202)	(0.0328)	(0.0178)	(0.0275)	(0.0244)	(0.0237)
FDI	−0.0490	−0.0434	−0.0518	−0.0710	−0.0526	−0.0504
	(0.0498)	(0.0968)	(0.0514)	(0.0766)	(0.0688)	(0.0673)
Infrastructure	0.1700***	0.1605***	0.3222***	0.2665	0.2027***	0.1911**
	(0.0917)	(0.2349)	(0.1204)	(0.2187)	(0.1190)	(0.1105)
GDPP	−0.4962***	−0.4940***	−0.5939**	−0.5241***	−0.5287**	−0.5168***
	(0.0777)	(0.2039)	(0.1100)	(0.2043)	(0.1185)	(0.1111)
Cons			9.120684		13.01593***	12.78008***
			(15.1033)		(1.7586)	(1.6617)
Sargan test	0.000	0.000	0.000	0.000		
Hansen	0.990	0.990	0.980	0.998		
AR1	0.009	0.013	0.013	0.046		
AR2	0.479	0.582	0.755	0.868		

Source: Data Processing Output
Standard errors in parentheses
$^{*} p < 0.1$, $^{**} p < 0.05$, $^{***} p < 0.01$

renewable energy consumption in 27 selected Asian countries. The findings of this research reveal a significant and positive coefficient for the lagged renewable energy variable. This suggests that the level of renewable energy in the previous year has a substantial impact, leading to an approximate 18% increase in renewable energy consumption in the subsequent year. These results are consistent with a previous study conducted by Lee et al. (2021), which investigated 123 countries worldwide. The positive and significant relationship between lagged renewable energy and future renewable energy consumption highlights the importance of sustained efforts and investment in renewable energy development. It underscores the potential for renewable energy to serve as a reliable and growing source of energy in the face of natural disasters and emphasizes the need for policies and strategies that promote its adoption and utilization.

The findings of this study indicate a positive and significant relationship between natural disasters and renewable energy. The increase in the number of affected people is associated with a 16% increase in renewable energy consumption. These findings align with previous research conducted by Khan and Anwar (2022), Lee et al. (2021), and Doytch and Klein (2018). Doytch and Klein (2018) suggest that the positive relationship between natural disasters and renewable energy is due to the destruction of conventional infrastructure, which prompts governments to invest in renewable energy during post-disaster recovery efforts. This shift reduces reliance on conventional energy sources and stimulates the demand for renewable energy. Moreover, higher demand and consumption of renewable energy are often associated with increased productivity (Khan & Anwar, 2022). Lee et al. (2021) support this notion by stating that post-disaster periods are characterized by decreased oil consumption and increased renewable energy consumption. An illustrative example is Japan, which successfully enhanced its renewable energy consumption and infrastructure by 100% following the Fukushima Nuclear Disaster (Esteban & Portugal-Pereira, 2014). Furthermore, a study by Fang et al. (2023) found that natural disasters have an impact on conventional energy efficiency within the same year, resulting in a decrease in conventional energy use and its replacement by renewable energy. It is worth noting that the coefficient values in the present study are relatively high compared to previous literature. However, it should be acknowledged that the majority of Asian countries included in the analysis are middle- to low-income countries, with only a few countries such as Japan, South Korea, and Singapore having high incomes. Overall, these findings highlight the potential of natural disasters to drive the transition toward renewable energy, particularly in countries where infrastructure damage necessitates reconstruction and provides an opportunity for investment in sustainable energy sources.

The third finding of this study highlights a strong positive and significant relationship between infrastructure and renewable energy. This implies that an increase in the quality and quantity of infrastructure can lead to a significant increase in renewable energy consumption. Rydin et al. (2018) emphasize the importance of aligning renewable energy supply with the development of infrastructure such as roads, pipelines, and bridges to facilitate efficient transportation of renewable energy resources. Similarly, Abrell and Rausch (2016) argue that developed countries have been more successful in adopting and expanding renewable energy due to their sustainable infrastructure compared to developing countries. For example, the availability of affordable electric vehicle charging infrastructure in developed countries may not be accessible in developing countries (Alkawsi et al., 2021; Gorbunova & Anisimov, 2020). It is important to note that advanced infrastructure for renewable energy can also have a multiplier effect on other sectors, such as logistics, thereby contributing to the overall economy (Bagheri et al., 2018; Khan et al., 2022). This highlights the interconnected nature of infrastructure development and renewable energy expansion, demonstrating how improvements in infrastructure can create a supportive environment for the growth and integration of renewable energy sources.

The present research includes gross domestic product per capita (GDPP) as the only control variable, and it is found to have a negative and significant impact on renewable energy. This suggests that an increase in GDPP leads to a decrease in renewable energy consumption, indicating that the selected countries in this study, which rely heavily on conventional energy sources, may not be fully prepared for the transition to renewable energy. This finding aligns with the research conducted by Fotourehchi (2017), who found a positive and significant impact of GDPP on renewable energy in the long term but not in the short term. Similar findings were also reported by Halkos and Tzeremes (2014), who observed an M-shaped curve relationship between GDPP and renewable energy consumption in developing countries. Wang et al. (2022) and Sahlian et al. (2021) propose a possible explanation for this phenomenon, suggesting that many middle-income countries lack access to advanced technologies necessary for renewable energy adoption. Consequently, instead of reducing non-renewable energy consumption, these countries prioritize generating more income to afford advanced technologies and pursue environmentally friendly practices.

12.4 Conclusions

The present study investigated the impact of natural disasters on renewable energy consumption, considering the role of Foreign Direct Investment (FDI) and infrastructure in 27 selected Asian countries. The findings can be summarized as follows, first Lagged renewable energy: The previous year's investment and utilization of renewable energy positively and significantly influence its consumption in the subsequent year. Second, both natural disasters and infrastructure have a positive and significant impact on renewable energy consumption. Investing in renewable energy infrastructure is crucial for promoting its utilization. Third, surprisingly, a negative relationship was observed between GDPP and renewable energy consumption. This suggests that middle-income countries face challenges in adopting renewable energy due to technological and infrastructure limitations. Based on these findings, policymakers should consider the following recommendations:

1. Governments should prioritize investments in renewable energy infrastructure to facilitate its adoption and utilization. This includes developing transportation networks, transmission systems, and storage facilities.
2. Promote technological advancements: Middle-income countries should focus on advancing renewable energy technologies to overcome barriers to adoption. This can be achieved through research and development initiatives, international collaborations, and technology transfer programs.
3. Supportive policies and incentives: Governments should implement policies that encourage renewable energy adoption, such as feed-in tariffs, tax incentives, and regulatory frameworks that promote clean energy. This can attract FDI and create a favorable environment for renewable energy investments.
4. Capacity building and knowledge sharing: It is crucial to invest in capacity building programs to enhance the skills and knowledge of local communities, policymakers, and industry professionals in renewable energy technologies and practices. Sharing best practices and experiences among countries can also accelerate the transition to renewable energy.

By implementing these recommendations, countries can unlock the full potential of renewable energy, reduce dependence on fossil fuels, mitigate the impact of natural disasters, and promote sustainable economic growth.

The present research paper has several limitations that should be acknowledged. First, the analysis focuses on a selected set of 27 countries in Asia, which may limit the generalizability of the findings to other regions or countries. The results may be influenced by specific characteristics of the countries included in the study, such as their level of economic development, political stability, or renewable energy policies. Second, the study relies on panel data from a specific time period, ranging from 2010 to 2020. This time frame may not capture long-term trends or account for recent developments in renewable energy technologies and policies. Future research could consider extending the time period to provide a more comprehensive analysis of the relationship between natural disasters, FDI, infrastructure, and renewable energy consumption. Third, while the present research considers FDI and infrastructure as channels through which natural disasters affect renewable energy consumption, it does not explicitly examine the mechanisms or pathways of these effects. Further investigation into the specific mechanisms by which natural disasters influence FDI, infrastructure development, and, subsequently, renewable energy consumption would enhance our understanding of these relationships. Fourth, the present research relies on secondary data sources, such as the World Bank and the Center for Research on the Epidemiology of Disasters (CRED). Although these sources provide valuable information, they are subject to potential data limitations and measurement errors. Future research could consider incorporating primary data collection methods or exploring alternative data sources to validate and strengthen the findings. Lastly, the present research focuses primarily on the quantitative analysis of panel data and, therefore, does not fully capture the qualitative aspects or contextual factors that may influence the relationship between natural disasters and renewable energy consumption. Incorporating qualitative research methods, such as interviews or case studies, could provide deeper insights into the specific challenges and opportunities associated with natural disasters and renewable energy adoption. Acknowledging these limitations is important for interpreting the findings of the present research and for guiding future studies in this field.

References

Abbas Khan, K., Zaman, K., Shoukry, A. M., Sharkawy, A., Gani, S., Sasmoko, Ahmad, J., Khan, A., & Hishan, S. S. (2019). Natural disasters and economic losses: Controlling external migration, energy and environmental resources, water demand, and financial development for global prosperity. *Environmental Science and Pollution Research, 26*(14), 14287–14299. https://doi.org/10.1007/s11356-019-04755-5

Abrell, J., & Rausch, S. (2016). Cross-country electricity trade, renewable energy and European transmission infrastructure policy. *Journal of Environmental Economics and Management, 79*, 87–113. https://doi.org/10.1016/j.jeem.2016.04.001

Alkawsi, G., Baashar, Y., Abbas, U., Alkahtani, A. A. D., & Tiong, S. K. (2021). Review of renewable energy-based charging infrastructure for electric vehicles. *Applied Sciences, 11*(9), 3847. https://doi.org/10.3390/app11093847

Arellano, M. (2009). Practitioners' corner: Computing robust standard errors for within-groups estimators: Practitioners' corner. *Oxford Bulletin of Economics and Statistics, 49*(4), 431–434. https://doi.org/10.1111/j.1468-0084.1987.mp49004006.x

Arellano, M., & Bond, S. (1991). Some tests of specification for panel data: Monte Carlo evidence and an application to employment equations. *The Review of Economic Studies, 58*(2), 277. https://doi.org/10.2307/2297968

Arellano, M., & Bover, O. (1995). Another look at the instrumental variable estimation of error-components models. *Journal of Econometrics, 68*(1), 29–51. https://doi.org/10.1016/0304-4076(94)01642-D

Arouri, M., Nguyen, C., & Youssef, A. B. (2015). Natural disasters, household welfare, and resilience: evidence from rural Vietnam. *World Development, 70*, 59–77. https://doi.org/10.1016/j.worlddev.2014.12.017

Asian Development Bank (Ed.). (2013). *The rise of natural disasters in Asia and the Pacific: Learning from ADB's experience*. Independent Evaluation ADB.

Awasthi, P. R. (2019). A relationship between natural disaster and poverty. *Journal of APF Command and Staff College, 2*(1), 67–69. https://doi.org/10.3126/japfcsc.v2i1.26733

Bagheri, M., Shirzadi, N., Bazdar, E., & Kennedy, C. A. (2018). Optimal planning of hybrid renewable energy infrastructure for urban sustainability: Green Vancouver. *Renewable and Sustainable Energy Reviews, 95*, 254–264. https://doi.org/10.1016/j.rser.2018.07.037

Bamati, N., & Raoofi, A. (2020). Development level and the impact of technological factor on renewable energy production. *Renewable Energy, 151*, 946–955. https://doi.org/10.1016/j.renene.2019.11.098

Barbhuiya, M. R., & Chatterjee, D. (2020). Vulnerability and resilience of the tourism sector in India: Effects of natural disasters and internal conflict. *Tourism Management Perspectives, 33*, 100616. https://doi.org/10.1016/j.tmp.2019.100616

Blundell, R., & Bond, S. (1998). Initial conditions and moment restrictions in dynamic panel data models. *Journal of Econometrics, 87*, 115.

Bui, A. T., Dungey, M., Nguyen, C. V., & Pham, T. P. (2014). The impact of natural disasters on household income, expenditure, poverty and inequality: Evidence from Vietnam. *Applied Economics, 46*(15), 1751–1766. https://doi.org/10.1080/00036846.2014.884706

Crespo Cuaresma, J., Hlouskova, J., & Obersteiner, M. (2008). Natural disasters as creative destruction? Evidence from developing countries. *Economic Inquiry, 46*(2), 214–226. https://doi.org/10.1111/j.1465-7295.2007.00063.x

Doytch, N., & Klein, Y. L. (2018). The impact of natural disasters on energy consumption: An analysis of renewable and nonrenewable energy demand in the residential and industrial sectors. *Environmental Progress & Sustainable Energy, 37*(1), 37–45. https://doi.org/10.1002/ep.12640

Emre Caglar, A. (2020). The importance of renewable energy consumption and FDI inflows in reducing environmental degradation: Bootstrap ARDL bound test in selected 9 countries. *Journal of Cleaner Production, 264*, 121663. https://doi.org/10.1016/j.jclepro.2020.121663

Esteban, M., & Portugal-Pereira, J. (2014). Post-disaster resilience of a 100% renewable energy system in Japan. *Energy, 68*, 756–764. https://doi.org/10.1016/j.energy.2014.02.045

Fang, Z., Chiu, Y., Lin, T.-Y., Chang, T.-H., & Lin, Y.-N. (2023). Analysis of natural disasters and energy efficiency in China. *Environment, Development and Sustainability., 26*, 10999. https://doi.org/10.1007/s10668-023-03182-4

Fotourehchi, Z. (2017). Renewable energy consumption and economic growth: A case study for developing countries. *International Journal of Energy Economics and Policy, 7*(2).

Gorbunova, A. D., & Anisimov, I. A. (2020). Assessment of the use of renewable energy sources for the charging infrastructure of electric vehicles. *Emerging Science Journal, 4*(6), 539–550. https://doi.org/10.28991/esj-2020-01251

Halkos, G. E., & Tzeremes, N. G. (2014). The effect of electricity consumption from renewable sources on countries' economic growth levels: Evidence from advanced, emerging and developing economies. *Renewable and Sustainable Energy Reviews, 39*, 166–173. https://doi.org/10.1016/j.rser.2014.07.082

Hallegatte, S., Vogt-Schilb, A., Rozenberg, J., Bangalore, M., & Beaudet, C. (2020). From poverty to disaster and back: A review of the literature. *Economics of Disasters and Climate Change, 4*(1), 223–247. https://doi.org/10.1007/s41885-020-00060-5

Hansen, L. P. (1982). Large sample properties of generalized method of moments estimators. *Econometrica, 50*(4), 1029. https://doi.org/10.2307/1912775

Kang, X., Khan, F. U., Ullah, R., Arif, M., Rehman, S. U., & Ullah, F. (2021). Does foreign direct investment influence renewable energy consumption? Empirical evidence from South Asian countries. *Energies, 14*(12), 3470. https://doi.org/10.3390/en14123470

Karim, A., & Noy, I. (2016). Poverty and natural disasters—A qualitative survey of the empirical literature. *The Singapore Economic Review, 61*(01), 1640001. https://doi.org/10.1142/S0217590816400014

Keerthiratne, S. (2018). Impact of natural disasters on income inequality in Sri Lanka. *World Development, 14*, 217.

Khan, M. T. I., & Anwar, S. (2022). Natural disasters and foreign exchange reserves: The role of renewable energy and human capital. *Renewable Energy, 192*, 838–848. https://doi.org/10.1016/j.renene.2022.05.022

Khan, M. T. I., Anwar, S., & Batool, Z. (2022). The role of infrastructure, socio-economic development, and food security to mitigate the loss of natural disasters. *Environmental Science and Pollution Research, 29*, 52412. https://doi.org/10.1007/s11356-022-19293-w

Lee, C.-C., Wang, C.-W., Ho, S.-J., & Wu, T.-P. (2021). The impact of natural disaster on energy consumption: International evidence. *Energy Economics, 97*, 105021. https://doi.org/10.1016/j.eneco.2020.105021

Liu, W., Xu, X., Yang, Z., Zhao, J., & Xing, J. (2016). Impacts of FDI renewable energy technology spillover on China's energy industry performance. *Sustainability, 8*(9), 846. https://doi.org/10.3390/su8090846

Monllor, J., & Altay, N. (2016). Discovering opportunities in necessity: The inverse creative destruction effect. *Journal of Small Business and Enterprise Development, 23*(1), 274–291. https://doi.org/10.1108/JSBED-10-2014-0172

Murshed, M., Elheddad, M., Ahmed, R., Bassim, M., & Than, E. T. (2022). Foreign direct investments, renewable electricity output, and ecological footprints: Do financial globalization facilitate renewable energy transition and environmental welfare in Bangladesh? *Asia-Pacific Financial Markets, 29*(1), 33–78. https://doi.org/10.1007/s10690-021-09335-7

Nickell, S. (1981). Biases in dynamic models with fixed effects. *Econometrica: Journal of the Econometric Society, 49*(6), 1417–1426

Qureshi, M. I., Yusoff, R. M., Hishan, S. S., Alam, A. F., Zaman, K., & Rasli, A. M. (2019). Natural disasters and Malaysian economic growth: Policy reforms for disasters management. *Environmental Science and Pollution Research, 26*(15), 15496–15509. https://doi.org/10.1007/s11356-019-04866-z

Rej, S., Bandyopadhyay, A., Das, N., Hossain, M. E., Islam, M. S., Bera, P., & Yeediballi, T. (2022). The asymmetric influence of environmental-related technological innovation on climate change mitigation: What role do FDI and renewable energy play? *Environmental Science and Pollution Research, 30*(6), 14916–14931. https://doi.org/10.1007/s11356-022-23182-7

Roodman, D. (2009). *How to do Xtabond2: an introduction to difference and system GMM in stata* (Vol. 9, p. 86).

Rydin, Y., Natarajan, L., Lee, M., & Lock, S. (2018). Do local economic interests matter when regulating nationally significant infrastructure? The case of renewable energy infrastructure projects. *Local Economy: The Journal of the Local Economy Policy Unit, 33*(3), 269–286. https://doi.org/10.1177/0269094218763000

Sahlian, D. N., Popa, A. F., & Crețu, R. F. (2021). Does the increase in renewable energy influence GDP growth? An EU-28 analysis. *Energies, 14*(16), 4762. https://doi.org/10.3390/en14164762

Samour, A., Baskaya, M. M., & Tursoy, T. (2022). The impact of financial development and FDI on renewable energy in the UAE: A path towards sustainable development. *Sustainability, 14*(3), 1208. https://doi.org/10.3390/su14031208

Sargan, J. D. (1958). The estimation of economic relationships using instrumental variables. *Econometrica, 26*(3), 393. https://doi.org/10.2307/1907619

Tan, Y., & Uprasen, U. (2022). The effect of foreign direct investment on renewable energy consumption subject to the moderating effect of environmental regulation: Evidence from the BRICS countries. *Renewable Energy, 201*, 135–149. https://doi.org/10.1016/j.renene.2022.11.066

van Bergeijk, P. A. G., & Lazzaroni, S. (2015). Macroeconomics of natural disasters: Strengths and weaknesses of meta-analysis versus review of literature: Meta-analysis versus review of literature. *Risk Analysis, 35*(6), 1050–1072. https://doi.org/10.1111/risa.12372

Wang, J., Hassan, M. S., Alharthi, M., Arshed, N., Hanif, I., & Saeed, M. I. (2022). Inspecting non-linear behavior of aggregated and disaggregated renewable and non-renewable energy consumption on GDP per capita in Pakistan. *Energy Strategy Reviews, 39*, 100772. https://doi.org/10.1016/j.esr.2021.100772

Chapter 13
Numerical Modelling of a Sedimentation Tank for Enhancing Sediment Removal in Open-Loop Ground Source Heat Pump System of Type Standing Column Well

Dorian Hernandez, Philippe Pasquier, and François Guibault

13.1 Introduction

13.1.1 SCW-IW System

Ground source heat pump (GSHP) systems have the potential to significantly lower the amount of energy consumption, the power demand, and greenhouse gas emissions (Banks, 2008; Hughes, 2008). In particular, a standing column well (SCW), which is a kind of open-loop system that directly exploits the groundwater as a heat transfer medium (Banks, 2008), can be integrated efficiently in dense urban areas (Laroche et al., 2022; Pasquier et al., 2016). Consisting of a borehole of 75 m to 450 m deep with a 150 mm diameter (Orio et al., 2005), SCWs have demonstrated a real advantage for building energy saving, by being more compact than conventional U-tube closed-loop systems and leading to less drilling cost for the same energy output (Orio et al., 2006). SCWs can be coupled with an injection well (IW) to return a portion of the groundwater to the aquifer during peak power periods. This process, called bleed, boosts the flow in fractures. This leads to a renew of the water present in the SCW with a more suitable temperature and to a better efficiency of the overall GSHP system (Spitler et al., 2002).

D. Hernandez (✉) · P. Pasquier
Polytechnique Montréal, Department of Civil, Geological and Mining Engineering, Montréal, QC, Canada
e-mail: dorian.hernandez@polymtl.ca

F. Guibault
Polytechnique Montréal, Department of Computer and Software Engineering, Montréal, QC, Canada

A. Sayigh (ed.), *Transition Towards a Carbon Free Future*, Innovative Renewable Energy, https://doi.org/10.1007/978-3-031-61660-0_13

13.1.2 Clogging and Maintenance of the Injection Well

The use of an IW during the bleeding process requires special attention. Indeed, open-loop systems must deal with three main disturbances related to the use of groundwater: mineral precipitation, biological growth, and sediment transport (Bakema, 2001). One of the elements most impacted by these processes is the IW (Bakema, 2001). While proposing a monitoring strategy for fouling and clogging of open-loop GSHP systems, Gjengedal et al. (2019) showed the presence of sediments at the bottom of seven IWs that suffered from clogging problems (Gjengedal et al., 2019). These sediments were mainly composed of sand, silt, clay, and ionic precipitates. In the continuity of their work, Stenvik et al. (2022) observed that the coupled effect of the precipitation and the deposition of sediments can explain the reason for the significant clogging of IWs (Stenvik et al., 2022). To postpone the maintenance of the IW of an open-loop GSHP system and to avoid possible clogging of the IW, partly due to the sediment load present in the groundwater, a sedimentation tank can be helpful and is therefore studied in this work.

13.1.3 Design of Settling Tank

Frequently used for wastewater treatment and in water treatment plants, the design of sedimentation tanks is historically based on Hazen's theory (Hazen, 1904), which states that the settling velocity of particles must be greater than the overflow rate inside the tank. This sizing approach, still employed nowadays, uses particle properties and water properties without considering the different hydrodynamics phenomena present inside the tank and largely observed by the community, such as density currents, waterfall phenomena, and recirculation (Augustus et al., 1985; Larsen, 1977). With the development of numerical simulation and enhanced computational power, it is now possible to study those phenomena impacting sedimentation tank efficiency (Hirom & Devi, 2022) and to optimize tank geometry. These phenomena are mainly related to the interior geometry of the tank, such as the position of the inlet and the configuration of baffles. Inlet hydraulic properties and particle properties, such as inlet Reynolds number and Froude number, are also important parameters impacting tank flow and efficiency (Tamayol et al., 2010).

In this study, we are investigating the potential use of a commercially available septic tank to enhance the sedimentation of the particles directed to an IW to postpone maintenance and avoid eventual clogging phenomena. To do this, a numerical model was set up based on the mixture model of Manninen et al. (1996) to simulate the process of flow and sedimentation inside a tank, according to typical SCW operation constraints. The behavior of the tank for two particle diameters is then studied, as well as the impact on the efficiency of different baffles added at the top and bottom of the tank.

13.2 Methodology

13.2.1 Sediment Characterization

A particle characterization has been done to obtain the particle size distribution, density, and volume fraction. To do so, groundwater samples were collected at the end of a pumping test performed on an experimental SCW located in Tétreaultville (Montréal, Canada) where geological context and drilling reports showed that the rock surrounding the well was in majority composed of calcite of density $\rho_p = 2.7$ g/cm^3. The concentration of particles in the mixture ($C_p = 134$ mg/l) was obtained by filtration of the samples with a Buchner and a 0.45 micrometer filter following the ASTM Standard D5907–18. After filtration, a diffraction laser test was performed to obtain the particle size distribution. A Gaussian curve with a mean diameter of $d_{\text{mean}} = 13$ μm was obtained as shown in Fig. 13.1.

13.2.2 Model Equation

To simulate a multiphase flow inside the tank, composed of water and dispersed particles, the $k - \epsilon$ mixture model was applied. This is a two-way coupled dispersed phase model where the fluid and particles are considered as two interpenetrating materials with average properties (Manninen et al., 1996). Considered as a simplification of the Euler-Euler model (Messa et al., 2021), the mixture model uses one set

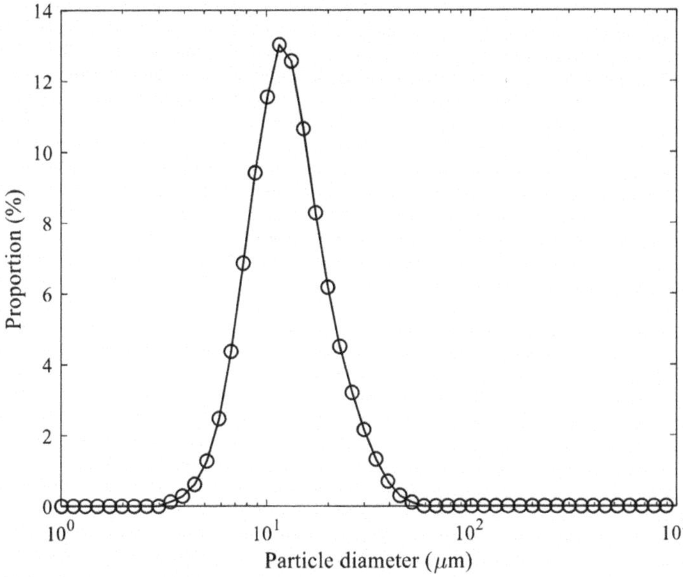

Fig. 13.1 Particle size distribution from laser diffraction analysis

of continuity and momentum equations for the mixture, linked with a constitutive equation. Less computationally expensive than a Euler-Lagrange approach (Hirom & Devi, 2022), it has a real interest for engineering application such as sedimentation due to the direct use of volume fraction and mass flow rate of the sediments (Jamshidi et al., 2019). It can also handle multiple types of dispersed phases, which allow considering sediments of different sizes. The main assumptions related to this model are a constant density ρ_i for each phase i, the same pressure field p for all phases, and a relative velocity $u_{\text{slip},\, i}$ between the continuous and each dispersed phase obtained under a local equilibrium assumption (Manninen et al., 1996).

13.2.2.1 Average Properties and Variables

Considering one continuous phase (water : $i = 1$) and N-1 different spherical dispersed phases ($i = 2, \ldots, N$), the mixture velocity u_m can be written as:

$$u_m = \frac{1}{\rho} \sum_{i=1}^{N} \Phi_i \rho_i u_i$$

where ϕ_i, ρ_i and u_i are, respectively, the volume fraction, the density, and the velocity of phase i.

The density of the mixture is:

$$\rho = \sum_{i=1}^{N} \Phi_i \rho_i$$

In the case of one dispersed phase, a Krieger type mixture viscosity μ_m with a maximum packing $\Phi_{\max} = 0.62$ is used (Krieger & Dougherty, 1959):

$$\mu_m = \mu_1 \left(1 - \frac{\Phi_{i=2}}{\Phi_{\max}}\right)^{2.5\Phi_{\max}}$$

The continuous phase volume fraction Φ_1 is:

$$\Phi_1 = 1 - \sum_{i=2}^{N} \Phi_i$$

and the mean relative velocity of the different dispersed phases with regard to the continuous phase are:

$$u_1 = \frac{1}{\rho} \sum_{i=2}^{N} \Phi_i \rho_i u_{\text{slip},i}$$

13.2.2.2 Continuity and Momentum Equations

The continuity equation of a single dispersed phase can be written as:

$$\frac{\partial}{\partial t}(\rho_i \Phi_i) + \nabla \cdot \left(\rho_i \Phi_i \left(u_m - u_r + u_{\text{slip},i}\right) - \rho_i D_{md} \nabla \Phi_i\right) = 0$$

with D_{md} the turbulent diffusivity defined as:

$$D_{md} = \frac{\mu_t}{\rho \sigma_t}$$

The continuity and momentum equations for the mixture can be written as:

$$\begin{cases} \dfrac{\partial}{\partial t}(\rho) + \nabla \cdot (\rho u_m) = 0 \\ \rho \dfrac{\partial u_m}{\partial t} + \rho (u_m \cdot \nabla) u_m + \nabla \cdot \left(\displaystyle\sum_{i=2}^{N} \Phi_i \rho_i u_{\text{slip},\, i} u_{\text{slip},\, i}^T - \rho\ u_r u_r^T \right) \\ \quad = -\nabla p + \nabla \cdot \tau_m + \rho g \end{cases}$$

where $g = 9.81\ m^2/s$ is the gravity acceleration and τ_m represent the viscous and turbulent $k - \epsilon$ constraint on the mixture:

$$\tau_m = (\mu_m + \mu_t) \left(\nabla\ u_m + (\nabla\ u_m)^T - \frac{2}{3}(\nabla \cdot u_m) I\right) - \frac{2}{3}\rho k I$$

with I the unit tensor.

13.2.2.3 Constitutive Equation and Drag Model

The system is closed by the implementation of a local equilibrium for $i = 2, \ldots, N$:

$$\frac{3}{4}\frac{C_{D,i}}{d_i} + \left|u_{\text{slip},i}\right| u_{\text{slip},i} = -\frac{\rho - \rho_i}{\rho_t}\left(-\frac{\partial u_m}{\partial t} - (u_m \cdot \nabla) u_m + g\right)$$

where $C_{D,\, i}$ is the drag coefficient of the Schiller-Naumann model used to obtain the relative velocity $u_{\text{slip},\, i}$ (Schiller & Naumann, 1933):

$$C_{D,i} = \begin{cases} \dfrac{24}{Re_{p,i}}\left(1 + 0.15\, Re_{p,i}^{0.687}\right) & Re_{p,i} < 1000 \\ 0.44 & Re_{p,i} > 1000 \end{cases}$$

$$Re_{p,i} = \frac{d_i \rho_1 \mid u_{\text{slip},i} \mid}{\mu_m}$$

In the previous equation, $Re_{p,\ i}$ is the Reynolds number which depends on the relative velocity of the particles. An iterative implicit process is thus used with an initial guess of $u_{\text{slip},\ i} = u_{\text{initial},\ i}$ equal to the Stokes terminal velocity given by:

$$u_{\text{initial},i} = \frac{1}{18}\frac{d_i^2 g(\rho_i - \rho_1)}{\mu_1}$$

13.2.2.4 Turbulence Equations

Finally, to consider turbulent effects, the standard $k - \epsilon$ model for the mixture is used (Launder & Spalding, 1973):

$$\begin{cases} \rho\frac{\partial k}{\partial t} + \rho u_m \cdot \nabla k = \nabla \cdot \left(\left(\mu_m + \frac{\mu_t}{\sigma_k}\right)\nabla k\right) + P_k - \rho\epsilon \\ \rho\frac{\partial \epsilon}{\partial t} + \rho u_m \cdot \nabla \epsilon = \nabla \cdot \left(\left(\mu_m + \frac{\mu_t}{\sigma_\epsilon}\right)\nabla \epsilon\right) + C_{\epsilon 1}\frac{\epsilon}{k}P_k - C_{\epsilon 2}\rho\frac{\epsilon^2}{k} \end{cases}$$

with P_k the production term and μ_t the turbulent viscosity:

$$P_k = \mu_t\left(\nabla u_m : \left(\nabla u_m + (\nabla u_m)^T\right) - \frac{2}{3}(\nabla \cdot u_m)^2\right) - \frac{2}{3}\rho k \nabla \cdot u_m$$

$$\mu_t = \rho C_\mu \frac{k^2}{\epsilon}$$

The coefficients used in the previous equations are $C_\mu = 0.09$, $C_{\epsilon 1} = 1.44$, $C_{\epsilon 2} = 1.92$, $\sigma_k = 1.0$, $\sigma_\epsilon = 1.3$ and $\sigma_t = 0.35$. The latter is the particle Schmidt number.

13.2.3 Numerical Model

13.2.3.1 Model Description and Boundary Conditions

A two-dimensional transient model was set up based on the geometry of a commercial sedimentation tank (Fig. 13.2).

Constant mixture velocity $u_{m,\ \text{inlet}}$ and particle volume fraction Φ_{inlet} are imposed at the inlet of the tank.

A constant hydrostatic pressure is imposed at the outlet, while a slip condition on the upper surface is used. No slip conditions are imposed everywhere else. The finite element method was used with the linear approximations for the velocity, pressure,

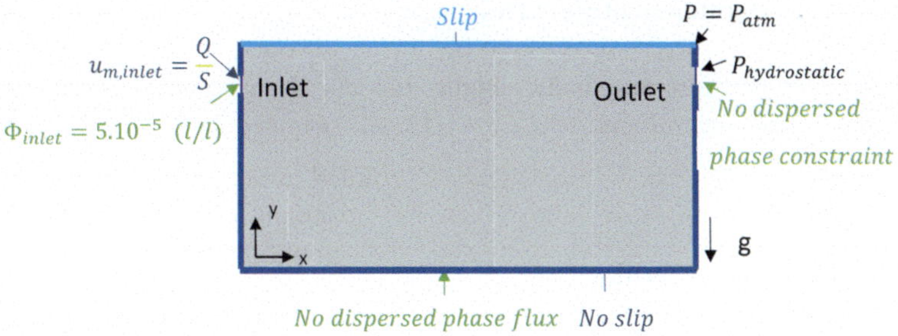

Fig. 13.2 Boundary conditions used to represent a sedimentation tank

and volume fraction. Consistent streamline and crosswind stabilization techniques were also used to ensure numerical convergence.

13.2.3.2 Numerical Algorithm

A fast and robust stationary solution was obtained after two main steps. The first step consists of computing the flow for a given inlet velocity, without any particles. A hydrostatic pressure and a zero velocity profile have been implemented as initial conditions. This calculation was accelerated by applying a coarser mesh with also viscosity ramping. A constant damped Newton Raphson method was applied. The second step calculates the transient solution of the mixture model (fluid and particles) with the fluid flow solution obtained in step 1 as the initial condition. A BDF scheme was used with a first-order interpolation and a constant damped Newton Raphson method at each time step. Convergence is reached when the residual error is less than 10^{-4}. Mesh independence was evaluated by comparing three mesh sizes and examining the stabilization of the variables of interest. In addition, a hydrodynamic validation of the model was carried out based on the work of Shahrokhi et al. (2013).

13.2.4 Investigated Parameters

13.2.4.1 Constant Flow Rate

SCW operation parameters such as flow rate and bleeding rate depend on the energy demand of the building (Beaudry et al., 2022). A typical range of flow rates for SCW systems is between 100 l/min and 1000 l/min. Considering that the bleed rate Q_{bleed} is usually between 5 and 20 percent of the total flow rate (Spitler et al., 2002), this gives a range of $Q_{\text{bleed}} = [5 - 200]$ l/min. The present study focuses on a flow rate $Q_{\text{bleed}} = 20$ l/min.

13.2.4.2 Two Diameters of Particles

Because bigger particles may induce higher risks of clogging, two diameters were investigated. The mean diameter $d_{\mathrm{mean}} = 13\ \mu\mathrm{m}$ obtained by the PSD test and a diameter $d = 33\ \mu\mathrm{m}$.

13.2.4.3 Addition of Baffles

Baffles installed within a sedimentation tank have been widely investigated to improve the hydrodynamic inside the tank and its efficiency (Deldar et al., 2018; Goula et al., 2008; Zhou et al., 1992). In most cases, they are used to control the flow, to dissipate the kinetic energy, and to guide or disperse sediments. In the present study, four different geometries have been investigated:

1. A baseline geometry, which consists of a rectangular tank of length $L = 5$ m and height $H = 2.3$ m
2. The initial geometry with a top-mounted baffle of height $h = 0.75$ m placed between 0.5 m and 4.5 m from the inlet.
3. The initial geometry with a bottom-mounted baffle of height $h = 1$ m placed between 0.5 m and 4.5 m from the inlet.
4. The initial geometry with multiple bottom-mounted baffles of two different heights $h = 0.5$ and 1 m placed between 0.5 m and 4.5 m from the inlet.

13.3 Results

13.3.1 Baseline Geometry Behavior

Figure 13.3 presents the evolution of the mass of sediments injected in the tank and collected by the tank for a duration of 12 hours for the two diameters of particles.

After 12 h of operation, respectively, 190 *g* and 1050 *g* of sediments were collected for the diameters $d = 13\ \mu\mathrm{m}$ and $d = 33\ \mu\mathrm{m}$, which correspond to a 10% and 54% tank efficiencies, respectively. It can also be observed in Fig. 13.4 that the tank efficiency decreases with time as the sedimented mass tends to stabilize. This is due to the filling of the tank with sediments. With time, particles will settle and form a settling bed (high concentration zone) at the bottom of the tank as seen in Fig. 13.4.

The height of this settling bed will increase, while the particles settle, until the moment when no more sedimentation will be possible due to the scouring effect present on the upper part of the settling bed (Patziger et al., 2012). This stationary regime, where no more sediments settle, is reached quickly in the case of the $d = 13\ \mu\mathrm{m}$ particles (after 6 hours of operation) as shown in Fig. 13.3. Comparing the volume fraction repartition (Fig. 13.4) with the shear stress repartition

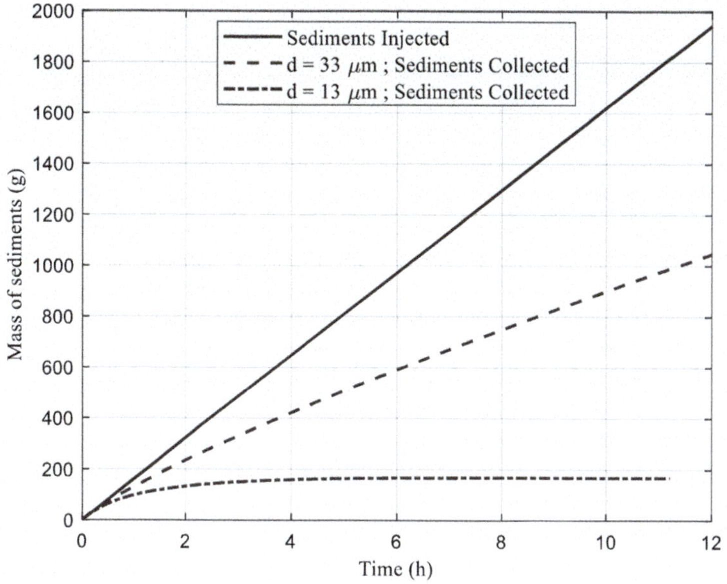

Fig. 13.3 Evolution of the mass of sediment injected and collected

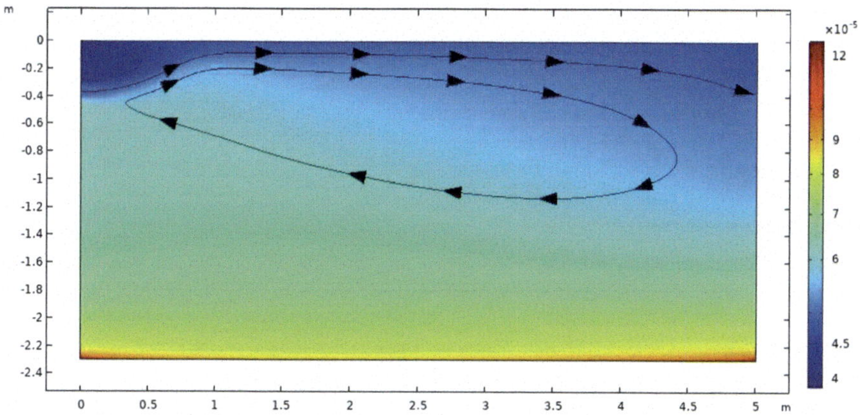

Fig. 13.4 Volume fraction of sediments and flow velocity streamlines at $t = 40,000$ s (Baseline geometry) and $d = 33$ μm

(Fig. 13.5), it is clear that sedimentation takes place in low shear stress places at the bottom of the tank.

Those results show that bigger particles are more likely to settle than smaller particles. Moreover, it demonstrates that for this geometry, inlet hydraulic properties and particle properties, the tank efficiency is highly dependent on the particle size.

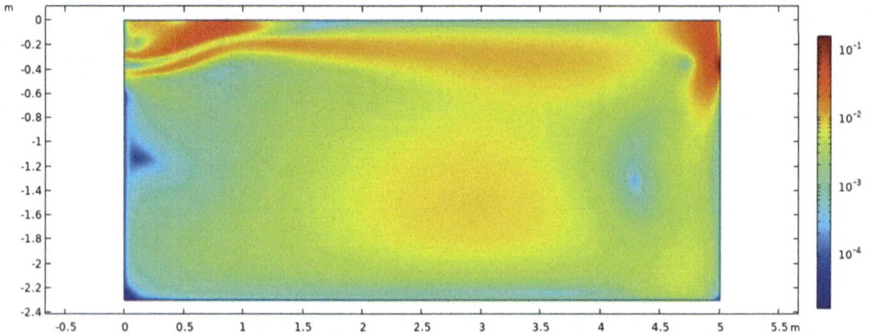

Fig. 13.5 Shear stress (Pa) repartition for the stationary flow step (baseline geometry)

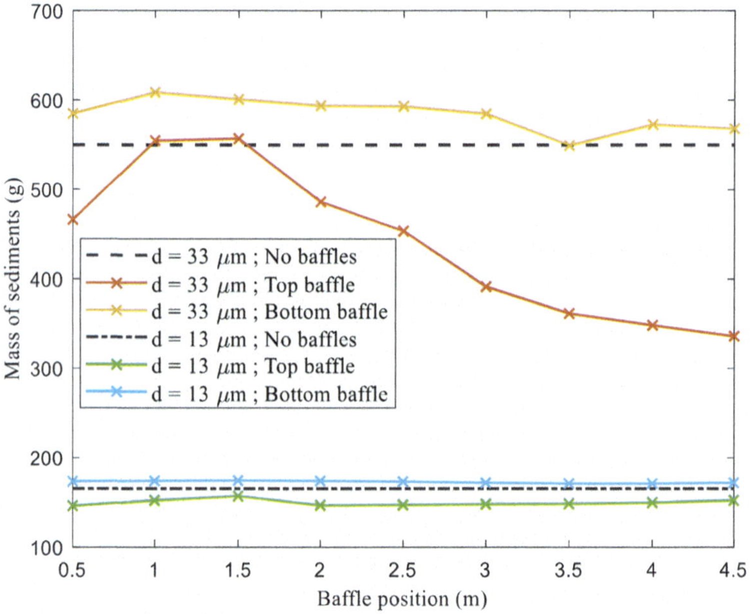

Fig. 13.6 Mass of sediments function of the top and bottom baffle position

13.3.2 *Top and Bottom Baffle Impact*

Figure 13.6 illustrates the mass of sediments collected after 5 h30 (20,000 s) for the different top and bottom baffle configurations (Geometries 2 and 3).

A small benefit of the top-mounted baffle is observed for the diameter $d = 33\ \mu m$ and for a position x between 1 m and 1.5 m. The same behavior has been observed by Deldar et al. (2018) who demonstrated an increase of tank efficiency when adding a top baffle near the top inlet due to the reduction of fluid turbulent energy and fluid flow enhancement (Deldar et al., 2018). However, attention needs to be paid when using this type of baffle because it can also enhance scouring of the particles (Patziger et al., 2012). This may explain the negative impact of the top baffle on the 13 μm diameters (Fig. 13.6).

In the case of the bottom baffle, an increase in the amount of sediments collected is obtained for both particle diameters. When looking at the sediments volume fraction field and the velocity streamlines, it can be seen in Fig. 13.7 that a sedimentation cell appears due to the presence of the bottom baffle.

Bottom mounted baffles seem to reduce the intensity of recirculation zones with a dissipation of the kinetic energy, leading to a reduction of the scouring effect of particles.

13.3.3 *Multiple Bottom Baffle Impact*

As a bottom baffle seems to enhance particle deposition with the creation of a settling area, the impact of multiple bottom mounted baffles has been investigated (geometry 4). Figure 13.8 compares the mass of sediments collected as a function of the spacing and height of bottom baffles for the two particle diameters after 5 h30 of simulated time.

Results show that sedimentation is enhanced when spacing Δx is reduced and the height h of the bottom baffles increases. Figure 13.9 shows the sediment volume fraction field and the velocity streamlines for the best case ($\Delta x = 0.5$ m; $h = 1$ m).

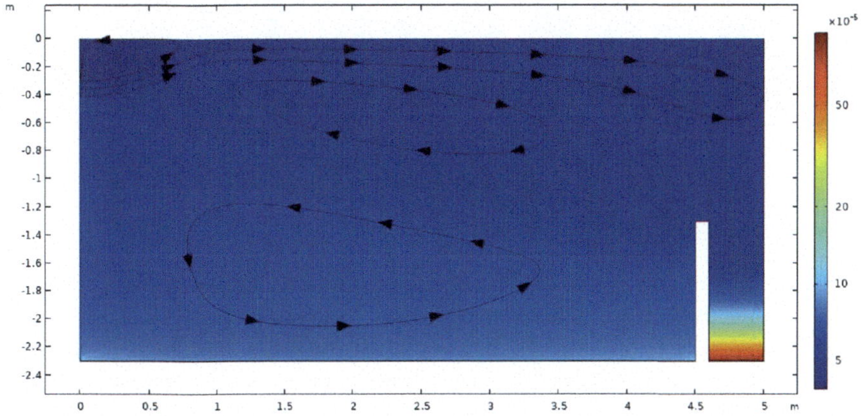

Fig. 13.7 Volume fraction of sediment at $t = 20,000$s (Bottom baffle $x = 4.5$ m ; $h = 1$ m) and $d = 33\ \mu m$

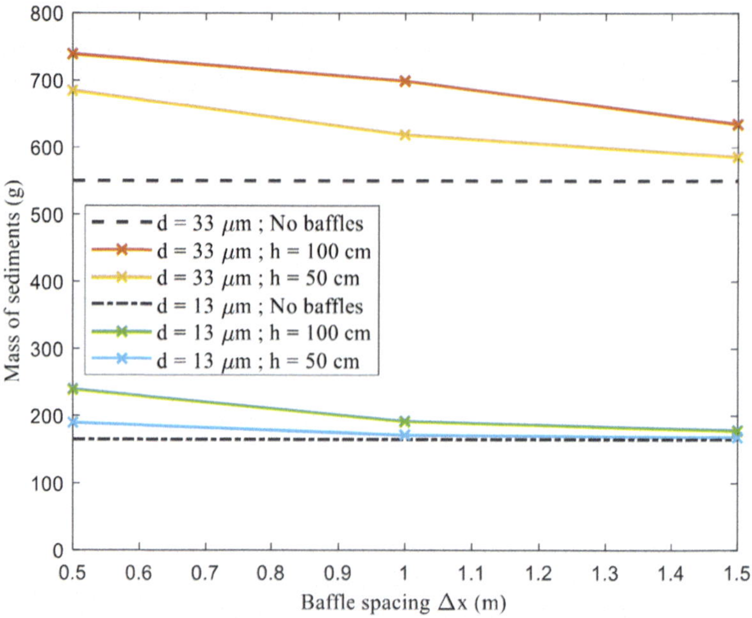

Fig. 13.8 Mass of sediments function of the height and spacing of bottom baffles

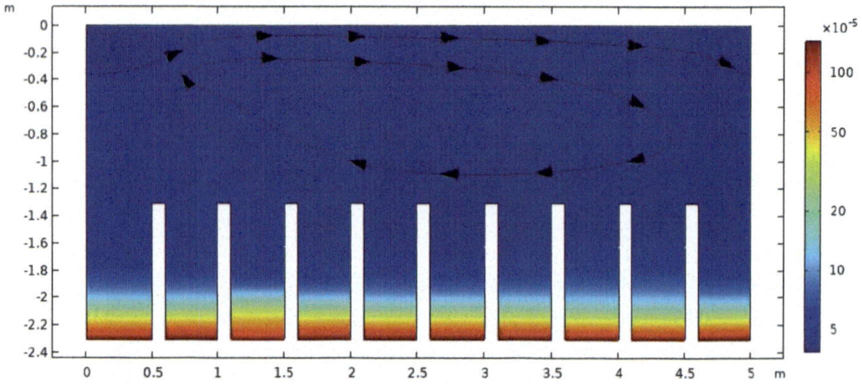

Fig. 13.9 Volume fraction of sediments at $t = 20,000$s (Multiple bottom baffles $\Delta x = 0.5$ m; $h = 1$ m) and $d = 33$ μm

The use of multiple bottom baffles leads to the creation of multiple sedimentation cells, with relatively high particle concentration. Looking at the shear stress repartition for this solution (see Fig. 13.10) and comparing it to the standard geometry, the use of multiple baffles reduces by almost 4 orders of magnitude the bottom shear stress of the tank.

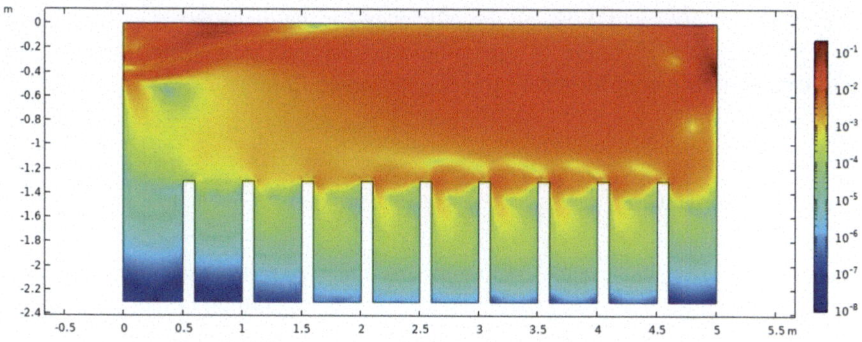

Fig. 13.10 Shear stress (Pa) repartition for the stationary flow step (Multiple bottom baffles $\Delta x = 0.5$ m; $h = 1$ m)

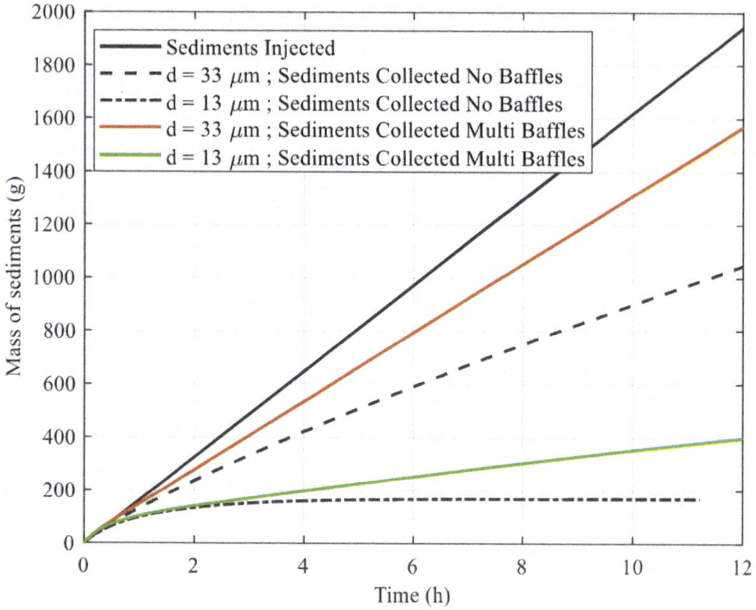

Fig. 13.11 Evolution of the mass of sediment injected and collected (best geometry)

Comparing the evolution of the mass of sediments collected by the best tank geometry (multiple bottom baffles with $\Delta x = 0.5$ m and $h = 1$ m) to the baseline geometry for a duration of 12 hours for the two investigated diameters, a significant improvement of the sedimentation tank behavior is observed, as shown in Fig. 13.11.

Increases of 10% and 37%, on the efficiency for the diameters $d = 13$ μm and $d = 33$ μm, after 12 hours of operation are observed. Moreover, there is still sedimentation occurring for the 13 μm particles after 12 hours.

13.4 Conclusions

Water treatment systems such as sedimentation tanks are passive solutions that have a great potential for improving particle removal in an open-loop GSHP application. This paper has shown a significant dependence of sedimentation tank efficiency on particle properties, such as particle size, and on sedimentation tank geometry. Results show that top inlet baffle may be a good solution if placed between $\frac{x}{L} = [0.2 - 0.3]$. However, it may produce negative scouring effects, which tend to re-suspend particles.

The results also show that the use of bottom baffles seems to reduce the intensity of recirculation in the tank, leading to the creation of sedimentation cells that are favorable to particle settling, while reducing the effects of scouring. Improvements of 10% and 37% on the efficiency have been found for particle diameters $d = 13$ μm and $d = 33$ μm, respectively, with the configuration of multiple bottom baffles with $\Delta x = 0.5$ m and $h = 1$ m. Finally, this work provides an idea of a sedimentation tank geometry sizing in order to reduce the sediment flow going to the injection well in the scope of standing column well operation.

References

Augustus, M., Baler, J. M., Chen, C., Daigger, G. T., Grifth, G. T., & TeKippe, R. J. (1985). *Clarifier design*. Lancaster Press USA.

Bakema, G. (2001). *Well and borehole failures and solutions in underground thermal energy storage*. IF Technology. https://doi.org/10.13140/RG.2.2.10214.98885

Banks, D. (2008). *An introduction to thermogeology: Ground source heating and cooling*. Blackwell Publishing.

Beaudry, G., Pasquier, P., Marcotte, D., & Zarrella, A. (2022). Flow rate control in standing column wells: A flexible solution for reducing the energy use and peak power demand of the built environment. *Applied Energy, 313*, 118774. https://doi.org/10.1016/j.apenergy.2022.118774

Deldar, S., Dehkordi, A. J., & Arani, H. K. (2018). Investigating the effect of flow entrance and existence of baffle on sedimentation efficiency using Discrete Phase Model (DPM). *Transport Phenomena in Nano and Micro Scales, 6*, 29–36. https://doi.org/10.22111/tpnms.2018.23893.1143

Gjengedal, S., Ramstad R. K., Hilmo B. O., & Frengstad B. S. (2019). Fouling and clogging surveillance in open loop GSHP systems. *Bulletin of Engineering Geology and the Environment, 79*(1): 69–82. https://doi.org/10.1007/s10064-019-01556-5

Goula, A. M., Kostoglou, M., Karapantsios, T. D., & Zouboulis, A. I. (2008). A CFD methodology for the design of sedimentation tanks in potable water treatment. Case study: The influence of a feed flow control baffle. *Chemical Engineering Journal, 140*(1 3), 110 121. https://doi.org/10.1016/j.cej.2007.09.022

Hazen, A. (1904). On sedimentation. *Transactions of the American Society of Civil Engineers, 53*(53), 45–71. https://doi.org/10.1061/TACEAT.0001655

Hirom, K., & Devi, T. T. (2022). Application of computational fluid dynamics in sedimentation tank design and its recent developments: A review. *Water, Air, and Soil Pollution, 233*(1), 10.1007/s11270-02105458-9.

Hughes, P. J. (2008). Geothermal (ground-source) heat pumps: Market status, barriers to adoption, and actions to overcome barriers in United States https://doi.org/10.2172/948543

Jamshidi, R., Panagiota, A., & Mazzei, L. (2019). On the closure problem of the effective stress in the Eulerian and mixture modeling approaches for the simulation of liquid-particle suspensions. *Physics of Fluids, 31*(1), 013302. https://doi.org/10.1063/1.5081677

Krieger, I. M., & Dougherty T. J. (1959). A mechanism for non-Newtonian flow in suspensions of rigid spheres.

Laroche, V., Pasquier, P., & Courcelles, B. (2022). Integration of standing column Wells in urban context: A numerical investigation-case in the City of Montreal. *Sustainable Cities and Society, 78*, 103513. https://doi.org/10.1016/j.scs.2021.103513

Larsen, P. (1977). *On the hydraulics of rectangular settling basins: Experimental and theoretical studies*. Department of Water Resources Engineering.

Launder, B. E., & Spalding, D. B. (1973). The numerical computation of turbulent flows. *Computer Methods in Applied Mechanics and Engineering, 3*, 269–289.

Manninen, M., Taivassalo, V., & Kallio, S. (1996). On the mixture model for multiphase flow. *VTT*. Publications.

Messa, G. V., Yang, Q., Adedeji, O. E., Chara, Z., Duarte, C. A. R., & Matoušek, V. (2021). Computational fluid dynamics modelling of liquid-solid slurry flows in pipelines: State-of-the-art and future perspectives. *PRO, 9*. https://doi.org/10.3390/pr9091566

Orio, C., Johnson, C., Rees, S., Chiasson, A., Deng, Z., & Spitler, J. (2005). A survey of standing column well installations in North America. *ASHRAE Transactions, 111*, 109e121.

Orio, C. D., Johnson, C. N., & Poor, K. D. (2006). Geothermal standing column wells: Ten years in a New England school. *ASHRAE Transactions, 112*(2), 57e64.

Pasquier, P., Nguyen, A., Eppner, F., Marcotte, D., & Baudron, P. (2016). Standing column wells. In *Advances in ground-source heat pump systems* (pp. 269–294). Elsevier. https://doi.org/10.1016/B978-0-08-100311-4.00010-8

Patziger, M., Kainz, H., Hunze, M., & Józsa, J. (2012). Influence of secondary settling tank performance on suspended solids mass balance in activated sludge systems. *Water Research, 46*(7), 2415–2424. https://doi.org/10.1016/j.watres.2012.02.007

Schiller, L., & Naumann, A. (1933). *A drag coefficient correlation*. Zeitschrift Verein Deutscher Ingenieure.

Shahrokhi, M., Rostami, F., Md, S. M., & A., & Syafalni. (2013). Numerical modeling of baffle location effects on the flow pattern of primary sedimentation tanks. *Applied Mathematical Modelling, 37*(6), 4486–4496. https://doi.org/10.1016/j.apm.2012.09.060

Spitler, J. D., Rees S. J., Deng Z., Chiasson A., Orio C. D., & Johnson C. (2002). R&D studies applied to standing column well design (No. ASHRAE 1119-RP).

Stenvik, L. A., Gjengedal, S., & R. Ramstad R. K. & Frengstad B. S. (2022). Hydrochemical and biotic control on iron incrustations in groundwater heat pump systems: Case study from a saline, anoxic aquifer in Melhus, Norway. *Geothermics, 100*(mars), 102349. https://doi.org/10.1016/j.geothermics.2022.102349

Tamayol, A., Firoozabadi, B., & Ashjari, M. A. (2010). Hydrodynamics of secondary settling tanks and increasing their performance using baffles. *Journal of Environmental Engineering, 136*(1), 32 39. https://doi.org/10.1061/(asce)ee.1943-7870.0000126

Zhou, S., McCorquodale, J. A., & Vitasovic, Z. (1992). Influences of density on circular clarifiers with baffles. *Journal of Environmental Engineering, 118*(6), 829–847. https://doi.org/10.1061/(asce)0733-9372(1992)118:6(829)

Chapter 14
Comparison of CLOT-Adjusted AHI-8/9 and FY-4A Solar Irradiance Products for Solar PV Power Output Forecasting Using LSTM

I. B. Benitez, J. A. Ibañez, C. D. Lumabad III, J. M. Cañete, and J. A. Principe

14.1 Introduction

Solar photovoltaic (PV) systems have become one of the most in-demand solutions to provide cleaner energy. Aside from the 5declining cost, solar PV systems are also one of the cheapest sources of electricity resulting in widespread installation and increased integration into the electricity grid (Solar PV—Analysis, 2022). Meanwhile, electricity grids are designed to provide a constant and consistent electricity supply to meet the load demand. Integrating variable renewable energy (VRE) technologies, such as solar photovoltaics (PV), into the grid introduces grid management issues. The variable and unreliable nature of solar PV output, being highly correlated with solar irradiance, poses a unique challenge in balancing power supply and demand in the grid (Harrou et al., 2020; Pawar et al., 2020). One way to address this is through solar PV output power (PPV) forecasting for grid planning. On the one hand, having accurate PPV forecasts ensures that sufficient power is available during periods of high demand. On the other hand, inaccurate forecasts lead to a mismatch between predicted and actual solar PV output, resulting in frequency fluctuation, voltage deviation, or, worse, a power outage.

There are various ways to forecast solar PV output, including statistical techniques, machine learning, and hybrid methodologies (Antonanzas et al., 2016). Models derived from machine learning (ML) techniques are widely used in

I. B. Benitez · J. A. Ibañez · C. D. Lumabad III · J. M. Cañete
National Engineering Center, University of the Philippines Diliman, Quezon City, Philippines

J. A. Principe (✉)
Department of Geodetic Engineering, University of the Philippines Diliman, Quezon City, Philippines
e-mail: japrincipe@up.edu.ph

A. Sayigh (ed.), *Transition Towards a Carbon Free Future*, Innovative Renewable Energy, https://doi.org/10.1007/978-3-031-61660-0_14

forecasting PPV and often achieve higher forecast accuracy than statistical methods because of their better patterns compared to the latter (Akhter et al., 2022; Das et al., 2018). Meanwhile, the long short-term memory (LSTM) technique addresses the major shortcoming of traditional recurrent neural networks (RNN): its inability to utilize information it learned from a previous time step into its current step (Abdel-Nasser & Mahmoud, 2019). In the study of (He et al., 2018), the LSTM forecast models under various weather conditions (spring, summer, autumn, and winter) performed better in forecasting PPV than the extreme learning machine (ELM) and artificial neural networks (ANN). Finally, LSTM with Nadam optimizer was reported to perform better than the two well-documented statistical techniques used in forecasting, ARIMA and SARIMAX (Sharma et al., 2022).

The accuracy of PPV forecasts depends on the quality of input data. The availability of geostationary satellite data has led to high-resolution solar irradiance data products often used in PPV forecasting. Other studies also investigated different datasets used in PPV forecasting, including the effects of clouds (Benitez et al., 2022a, b; Bonkaney et al., 2017; Haegele, 2020; Kenny & Fiedler, 2022; Kim et al., 2020; Lorenz et al., 2016; Mahachi & Rix, 2016; Suri et al., 2014; Zhen et al., 2017). Nonetheless, the study of (Sotto et al., 2023) showed the need to adjust such satellite products to account for cloud effects which tend to underestimate solar irradiance values. This study aims to determine whether or not it is better to use adjusted satellite-derived solar irradiance data in solar PV output forecasting. Following (Harrou et al., 2020; Park et al., 2021), this work compares the performance of PPV forecasting via the LSTM algorithm using the Advanced Himawari 8/9 (AHI-8/9) shortwave radiation (SWR) and Fengyun-4A (FY-4A) surface solar irradiance (SSI) as inputs and adjusted using correction factor derived from the cloud optical thickness (CLOT) product of AHI-8/9.

14.2 Methodology

The overall methodology of the study is illustrated in Fig. 14.1. Firstly, the input datasets were processed to ensure that they have the same temporal resolution. Outliers were then removed, and the data gaps were filled. The dataset was also divided into train (80%), test (10%), and validation (10%) before proceeding to LSTM model training and implementation. Finally, the validation forecast accuracies were compared in terms of root mean squared error (RMSE), mean absolute error (MAE), and mean absolute percentage error (MAPE).

14.2.1 Data

This study used solar irradiance and solar PV production data from various sources, summarized in Table 14.1. R stands for the original solar irradiance data, while R'

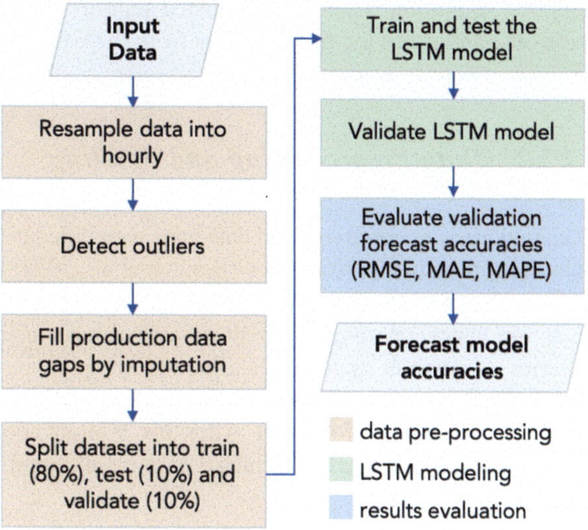

Fig. 14.1 Overview of study methodology

Table 14.1 Summary of input data used in the study

Source	Data	Resolution	
		Spatial	Temporal
SolarEdge	PV production	In situ	5 min
AHI-8/9	CLOT	5 km	10 min
	R_{SWR}	5 km	10 min
FY-4A	R_{SSI}	4 km	15 min
Project SINAG	R'_{SWR}	5 km	10 min
	R'_{SSI}	5 km	15 min

stands for the solar irradiance data adjusted using cloud optical thickness (*CLOT*). For all the data, the acquisition period was 8:00 AM to 3:00 PM from March to May 2021.

The solar PV output dataset was extracted from the SolarEdge database (SolarEdge, 2023) at a solar PV installation in La Union. The extracted data were consolidated into a single comma-separated value (CSV) file and resampled to hourly resolution. The CLOT and R_{SWR} data were downloaded via file transfer protocol (FTP), and the NetCDF data were extracted using Python script with date-time adjusted to UTC + 8 for the Philippine Time. The extracted data were consolidated into a single comma-separated values (CSV) file and then resampled to hourly temporal resolution. The R_{SSI} data were downloaded from the Fengyun-4A database (National Satellite Meteorological Center, 2016) and preprocessed using image-image georeferencing. SSI raster files were then clipped to the Philippine boundary, and the NetCDF data were extracted using Python script with date-time adjusted to UTC + 8 for the Philippine Time. The extracted SSI data were consolidated into a single comma-separated values (CSV) file and then resampled to hourly temporal resolution. The CLOT-adjusted solar irradiance datasets (R'_{SWR} and R'_{SSI}) were obtained using the methodology from the work of (Sotto et al., 2023).

14.2.2 Data Preprocessing and Splitting

Data outliers were removed, and data gaps were filled using an imputation method to enhance forecasting performance (Benitez et al., 2023; Kim et al., 2021; Zhu et al., 2019). Collected datasets were resampled to hourly temporal resolution with missing solar irradiance removed from the whole dataset using pandas and pyresample libraries in Python (pandas.DataFrame.resample—pandas 2.0.3 documentation, 2023). The remaining datasets were divided (80%), test (10%), and validation (10%) sets for the LSTM model. After this, 973 data points were retained, obtained from March to June 2021, of which 779 were allocated for training and 97 each for testing and validation.

14.2.3 LSTM Model

The long short-term memory (LSTM) architecture is a recurrent neural network (RNN) architecture that has emerged as a fundamental architecture for time series data processing and forecasting. Long-term dependencies and temporal patterns are particularly well captured by LSTM, making them ideal for processing sequential data with intrinsic time dependence. LSTM models can successfully preserve and update information over long periods by merging memory cells and gating processes, allowing them to duplicate complicated relationships within the time series data (Hochreiter & Schmidhuber, 1997).

Four LSTM layers were stacked with 64, 32, 16, and 8 units per layer. The complex features and patterns of the data from solar PV output were intended to be captured by this configuration. Two crucial techniques were used during the training process to optimize choosing the best model. Moreover, the training was stopped using the EarlyStopping callback if the model's performance did not improve after a predetermined number of epochs to save time and avoid overfitting. The best-performing model using the validation set at each epoch was automatically saved using the ModelCheckpoint callback.

The hyperparameters used in the LSTM model are listed in Table 14.2. Adam optimizer was used with a learning rate of 0.001. Moreover, the mean squared error (MSE) loss function was used to calculate the difference between the predicted and actual values after training the model for 500 epochs. The four hidden layers that made up the LSTM layers gradually reduced from 64 to 8 LSTM units per layer. The output layer employed a dense layer with a linear activation function, while the LSTM layers used the hyperbolic tangent (tanh).

Table 14.2 Summary of LSTM hyperparameters

Hyperparameter	Value
Optimizer	Adam
Learning rate	0.001
Epochs	500
Loss function	MSE
Hidden layers	
LSTM	4
LSTM Units	64, 32, 16, 8
Activation	tanH
Output layers	
Layer	Dense
Activation	Linear

To determine the optimal configuration of forecasting solar PV output, various hyperparameter combinations of LSTM networks were systematically examined. Lastly, optimal hyperparameters and architecture were identified to improve the efficiency of solar photovoltaic (PV) system planning and grid management.

14.2.4 Data and Model Evaluation

The Euclidean distance (d) (Liberti et al., 2014) was calculated using Eq. 14.1 to determine the distance between the adjusted irradiance datasets.

$$d = \sqrt{\sum_{i=1}^{n} (q_i - p_i)^2} \tag{14.1}$$

where d is the distance of two time series data, q_i and p_i are the values of the original solar irradiance datasets (R_{SWR} and R_{SSI}) and the CLOT-adjusted irradiances (R'_{SWR} and R'_{SSI}), and n is the total number of observations.

The metrics used to assess the accuracy of the forecasting models obtained in this study were based on a study by Alcañiz (2023), which identified the top three error metrics. Based on (Alcañiz et al., 2023), the most commonly used metrics for solar PV output forecasting are the root mean square error (RMSE), mean absolute error (MAE), and mean absolute percentage error (MAPE).

$$\text{RMSE} = \sqrt{\frac{\sum_{i=1}^{n} (y_i - x_i)^2}{n}} \tag{14.2}$$

$$\text{MAE} = \frac{\sum_{i=1}^{n} | y_i - x_i |}{n} \tag{14.3}$$

$$\text{MAPE} = \frac{100\%}{n} \sum_{t=1}^{n} \left| \frac{y_i - x_i}{y_i} \right| \tag{14.4}$$

where y_i is the actual value, x_i is the predicted value, and n is the total number of observations.

14.3 Results and Discussions

This study compared the performance of solar PV output forecasting using long short-term memory (LSTM) with two data inputs: the Advanced Himawari 8/9 (AHI-8/9) shortwave radiation (SWR) and Fengyun-4A (FY-4A) surface solar irradiance (SSI).

Figure 14.2a shows the comparison of the R_{SWR} and R'_{SWR} values for the randomly selected period from the dataset, while Fig. 14.2b shows the comparison of the sample R_{SSI} and R'_{SSI} values. In terms of Euclidean distance, there are 2933.198 units between R_{SWR} and R'_{SWR} and 292.126 units between R_{SSI} and R'_{SSI}. This implies that the CLOT

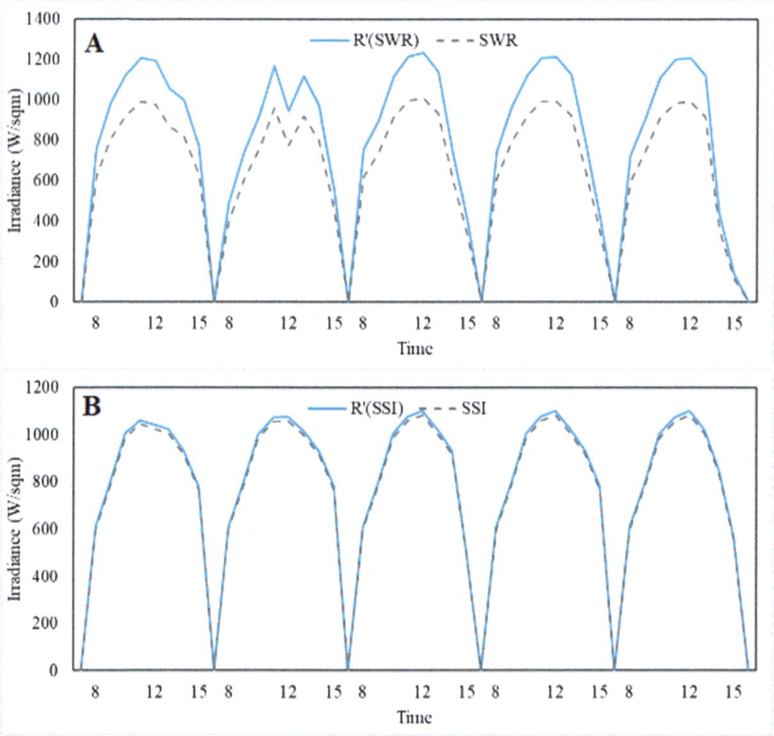

Fig. 14.2 Irradiance data plots: (**a**) R'_{SWR} and R_{SWR}, (**b**) R'_{SSI} and R_{SSI}

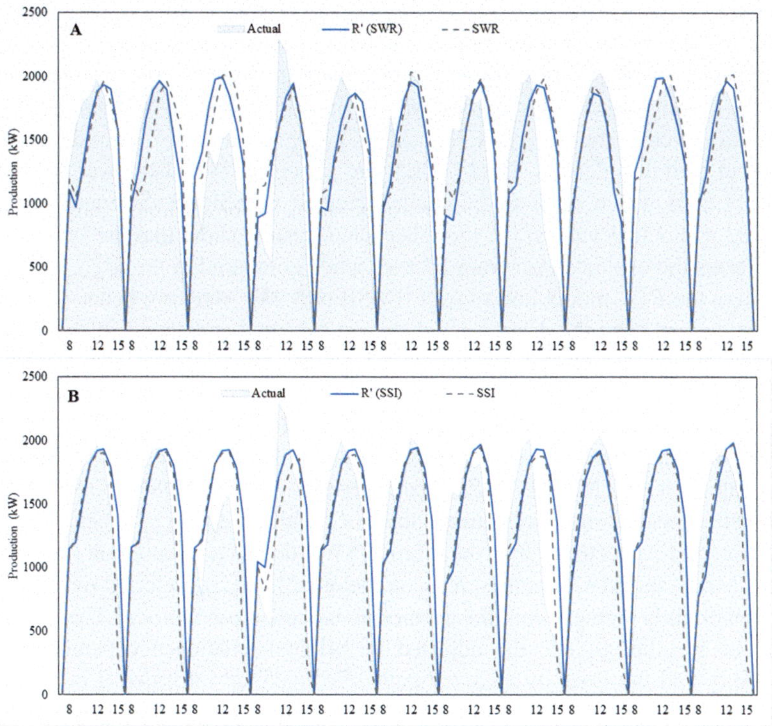

Fig. 14.3 Validation data plots: (**a**) production vs R'_{SWR} and R_{SWR}, (**b**) production vs R'_{SSI} and R_{SSI}

Table 14.3 Summary of model validation accuracies

Irradiance	RMSE	MAE	MAPE
R_{SWR}	9.078	4.362	11.824
R'_{SWR}	8.310	3.960	10.683
R_{SSI}	8.237	3.704	8.987
R'_{SSI}	8.302	3.917	10.937

Note: RMSE and MAE in kW, MAPE in %

adjustment had a greater impact on R_{SWR} than on R_{SSI} which could be attributed to the use of CLOT derived from AHI-8/9.

From Fig. 14.3a, results show that R'_{SWR} performs better than R_{SWR} when used in PPV forecasting. This suggests that the adjusted solar radiation data from AHI-8/9, R'_{SWR}, leads to better solar PV output forecasting accuracy. Adjustments to the original SWR data may have helped recognize more important patterns and characteristics, leading to more accurate predictions. Meanwhile, Fig. 14.3b shows that R_{SSI} outperforms R'_{SSI}, where the changes made to the original R_{SSI} data translates to negative impact on forecasting accuracy. Table 14.3 reports a detailed breakdown of the performance metrics of the PPV forecasting models. Results show that the adjusted R'_{SWR} consistently outperforms the R_{SWR} across a range of metrics.

With a lower RMSE of 8.310 as opposed to 9.078 of the original R_{SWR}, the adjusted R'_{SWR} shows improvements in forecasting accuracy. This suggests that the changes made to R'_{SWR} help lower the average gap between predicted and actual solar PV output values. The adjusted R'_{SWR}, having a lower MAE value of 3.960 compared to that of the R_{SWR} at 4.362, shows improved accuracy in predicting solar PV output with fewer errors. The MAPE for R'_{SWR} (10.683) is also lower than that of R_{SWR} (11.824) which indicates a smaller average percentage difference between predicted and actual values. On the other hand, results show that the original R_{SSI} outperforms the evaluated accuracy metrics when compared to the R'_{SSI}. Compared with R'_{SSI}, the R_{SSI}, model has a lower RMSE of 8.237, showing better forecasting accuracy in capturing the discrepancy between expected and actual solar PV output values. Similarly, R'_{SSI} and R_{SSI} produced a PPV forecasting model with MAE values of 3.917 and 3.704, respectively. The MAE values indicate that R_{SSI} has fewer errors and higher prediction accuracy for solar PV output than R'_{SSI}. Additionally, R_{SSI} shows a smaller average percentage difference between predicted and actual values, with a MAPE of 8.987 compared to R'_{SSI} at 10.937.

Therefore, results suggest the use of the CLOT-adjusted R'_{SWR} for better solar PV output forecasting accuracy than the original SWR data, R_{SWR}, based on the reported RMSE, MAE, and MAPE values. R_{SSI}, on the other hand, is preferred over the R'_{SSI} as it consistently outperforms the former across all error metrics. Therefore, the choice of whether to use the adjusted or original irradiance depends on data availability and the level of accuracy gains. To make an informed decision based on the particular needs and objectives of solar PV output forecasting, it is essential to carefully evaluate and compare the performance of various satellite-derived solar irradiance data.

14.4 Conclusions and Recommendations

This study analyzed the performance of solar PV output forecasting models using the original and adjusted satellite-derived solar irradiance data. The performance of the model that used the adjusted SWR data (R'_{SWR}) was better than its original version (R_{SWR}). Meanwhile, R_{SSI} performs better than its adjusted version (R'_{SSI}). The adjustment in solar radiation data using CLOT correction factor has improved forecasting model accuracy for solar PV output when using solar radiation data from AHI, but the same is not true for FY4A-derived data. It is therefore recommended that solar PV output forecasting models include adjusted solar irradiance data from AHI but must be very careful to apply the same adjustment for the FY4A product. Nonetheless, the adjustment applied to the solar radiation products produced more precise predictions by capturing key patterns and characteristics related to cloud effects. Grid managers and operators of solar PV systems benefit from using R'_{SWR} by achieving the optimal use of solar resources and preventing potential power supply-demand imbalances in energy generation and distribution. Furthermore, to improve the performance of R'_{SSI} in forecasting solar PV output, future study may explore the derivation of CLOT-based adjustment of satellite-derived solar radiation data from various data sources, such as FY-4A. Forecasting solar PV output using alternative correction techniques and including more meteorological factors may also help increase forecast accuracy. Additionally, to validate the applicability of the results and improve the adjustment technique for specific locations, similar studies could be carried out in various regions due to geographic and climatic variations.

The use of adjusted solar irradiance products derived from satellite data showed promising results for the accuracy improvement of solar PV output forecasting models. The integration of solar PV systems into the electrical grid can be improved through continuous enhancements in forecasting algorithms and adjustment techniques, leading to a more reliable and efficient renewable energy infrastructure.

Acknowledgments This study is implemented under Project SINAG (Solar PV Resource and Installation Assessment Using Geospatial Technologies). The authors wish to thank the Philippine Department of Science and Technology (DOST) as the funding agency of Project SINAG.

References

Abdel-Nasser, M., & Mahmoud, K. (2019). Accurate photovoltaic power forecasting models using deep LSTM-RNN. *Neural Computing & Applications, 31*(7), 2727–2740. https://doi.org/10.1007/s00521-017-3225-z

Akhter, M. N., et al. (2022). An hour-ahead PV power forecasting method based on an RNN-LSTM model for three different PV plants. *Energies, 15*(6), 6. https://doi.org/10.3390/en15062243

Alcañiz, A., Grzebyk, D., Ziar, H., & Isabella, O. (2023). Trends and gaps in photovoltaic power forecasting with machine learning. *Energy Reports, 9*, 447–471. https://doi.org/10.1016/j.egyr.2022.11.208

Antonanzas, J., Osorio, N., Escobar, R., Urraca, R., Martinez-de-Pison, F. J., & Antonanzas-Torres, F. (2016). Review of photovoltaic power forecasting. *Solar Energy, 136*, 78–111. https://doi.org/10.1016/j.solener.2016.06.069

Benitez, I., Gerna, L., Ibañez, J., Principe, J., & De Los Reyes, F. (2022a). Use of SARIMAX model for solar PV power output forecasting in Baguio City, Philippines. *International Conference and Utility Exhibition on Energy, Environment and Climate Change (ICUE), 2022*, 1–7. https://doi.org/10.1109/ICUE55325.2022.10113538

Benitez, I., Sotto, M., Bauzon, M. D. A., Santos, J. A., & Principe, J. (2022b). Preliminary analysis on the difference between solar PV power output forecasts derived from remotely sensed and in-situ solar radiation data. 宇宙からの地球環境・災害のモニタリングとリスク評価：生研フォーラム論文集 = *Monitoring of Global Environment and Disaster Risk Assessment from Space: the IIS Forum proceedings, 30*, 47–54.

Benitez, I. B., Ibañez, J. A., Lumabad, C. D., III, Cañete, J. M., De los Reyes, F. N., & Principe, J. A. (2023). A novel data gaps filling method for solar PV output forecasting. *Journal of Renewable and Sustainable Energy, 15*(4), 046102. https://doi.org/10.1063/5.0157570

Bonkaney, A., Madougou, S., & Adamou, R. (2017). Impacts of cloud cover and dust on the performance of photovoltaic module in Niamey. *Journal of Renewable Energy, 2017*, e9107502. https://doi.org/10.1155/2017/9107502

Das, U. K., et al. (2018). Forecasting of photovoltaic power generation and model optimization: A review. *Renewable and Sustainable Energy Reviews, 81*, 912–928. https://doi.org/10.1016/j.rser.2017.08.017

Haegele, A. (2020). Cloud cover and PV intermittence: monitoring, forecasting and its economical value. *Master's Theses*. [Online]. Available: https://opencommons.uconn.edu/gs_theses/1534

Harrou, F., Kadri, F., Sun, Y., Harrou, F., Kadri, F., & Sun, Y. (2020). *Forecasting of photovoltaic solar power production using LSTM approach*. IntechOpen. https://doi.org/10.5772/intechopen.91248

He, H., Hu, R., Zhang, Y., Zhang, Y., & Jiao, R. (2018). A power forecasting approach for PV plant based on irradiance index and LSTM. In *2018 37th Chinese Control Conference (CCC)* (pp. 9404–9409). https://doi.org/10.23919/ChiCC.2018.8483960

Hochreiter, S., & Schmidhuber, J. (1997). Long short-term memory. *Neural Computation, 9*(8), 1735–1780. https://doi.org/10.1162/neco.1997.9.8.1735

Kenny, D., & Fiedler, S. (2022). Which gridded irradiance data is best for modelling photovoltaic power production in Germany? *Solar Energy, 232*, 444–458. https://doi.org/10.1016/j.solener.2021.12.044

Kim, M., Song, H., & Kim, Y. (2020). Direct short-term forecast of photovoltaic power through a comparative study between COMS and Himawari-8 meteorological satellite images in a deep neural network. *Remote Sensing, 12*(15), 15. https://doi.org/10.3390/rs12152357

Kim, B., Suh, D., Otto, M.-O., & Huh, J.-S. (2021). A novel hybrid spatio-temporal forecasting of multisite solar photovoltaic generation. *Remote Sensing, 13*(13), 13. https://doi.org/10.3390/rs13132605

Liberti, L., Lavor, C., Maculan, N., & Mucherino, A. (2014). Euclidean distance geometry and applications. *SIAM Rev, 56*(1), 3–69. https://doi.org/10.1137/120875909

Lorenz, E., Kühnert, J., Heinemann, D., Nielsen, K. P., Remund, J., & Müller, S. C. (2016). Comparison of global horizontal irradiance forecasts based on numerical weather prediction models with different spatio-temporal resolutions. *Progress in Photovoltaics: Research and Applications, 24*(12), 1626–1640. https://doi.org/10.1002/pip.2799

Mahachi, T., & Rix, A. (2016). *Energy yield analysis and evaluation of solar irradiance models for a utility scale solar PV plant in South Africa*. https://doi.org/10.13140/RG.2.2.17960.72969

National Satellite Meteorological Center (National Center for Space Weather), "FY-4A," *FENGYUN Satellite Data Center*, Dec. 11, 2016. https://satellite.nsmc.org.cn/PortalSite/Data/Satellite.aspx. Accessed 14 Jun 2023.

pandas.DataFrame.resample—pandas 2.0.3 documentation. https://pandas.pydata.org/docs/reference/api/pandas.DataFrame.resample.html. Accessed 8 Aug 2023

Park, S., Kim, Y., Ferrier, N. J., Collis, S. M., Sankaran, R., & Beckman, P. H. (2021). Prediction of solar irradiance and photovoltaic solar energy product based on cloud coverage estimation using machine learning methods. *Atmosphere, 12*(3), 3. https://doi.org/10.3390/atmos12030395

Pawar, P., Mithulananthan, N., & Raza, M. Q. (2020). Solar PV power forecasting using modified SVR with Gauss-Newton method. In *2020 2nd Global Power, Energy and Communication Conference (GPECOM)* (pp. 226–231). https://doi.org/10.1109/GPECOM49333.2020.9247935

Sharma, J., et al. (2022). A novel long term solar photovoltaic power forecasting approach using LSTM with Nadam optimizer: A case study of India. *Energy Science & Engineering, 10*(8), 2909–2929. https://doi.org/10.1002/ese3.1178

Solar PV—Analysis, *IEA*, Sep. 2022. https://www.iea.org/reports/solar-pv. Accessed 8 Mar 2023.

SolarEdge, *SolarEdge monitoring*, 2023. https://monitoringpublic.solaredge.com/solaredge-web/p/home. Accessed 14 Jun 2023.

Sotto, M., Bauzon, M., Cañete, J., & Principe, J. (2023). *AHI-8 SWR adjustment using CLOT-derived correction factor for solar PV power potential assessment in the Philippines*. Proceedings of the 31st IIS Forum Earth Observation, Disaster Monitoring and Risk Assessment from Space.

Suri, M., et al. (2014). *Cloud cover impact on photovoltaic power production in South Africa*. Proceedings of the Southern African Solar Energy Conference.

Zhen, Z., Wang, Z., Wang, F., Mi, Z., & Li, K. (2017). Research on a cloud image forecasting approach for solar power forecasting. *Energy Procedia, 142*, 362–368. https://doi.org/10.1016/j.egypro.2017.12.057

Zhu, R., Guo, W., & Gong, X. (2019). Short-term photovoltaic power output prediction based on k-fold cross-validation and an ensemble model. *Energies, 12*(7), 7. https://doi.org/10.3390/en12071220

Chapter 15
Case Study Analysis of Improving Environmental Ethics Using a Collaboration Toolkit

Jason Challender

15.1 Introduction

> Environmental ethical behaviour is doing the right thing when no one else is watching, even when doing the wrong thing is legal. (Aldo Leopald)

> More and more companies are reaching out to their suppliers and contractors to work jointly on issues of sustainability, environmental responsibility, ethics and compliance. (Simon Mainwaring)

15.1.1 Introduction to Environmental Ethics

The above quotations highlight both the importance of environmental ethics and the current trend for organisations in the engineering and construction industries to embrace all aspects of environmental sustainability. To respond to this, the paper will set the scene for discussions around environmental ethics and encouraging sustainability in the design and construction processes and estates strategies. It will outline the recent trends in climate change and government pledges and targets to reduce greenhouse gases and carbon generation and the associated challenges around meeting such commitments. Furthermore, it will outline what we mean by 'environmental ethics' and define what sustainability is within the context of the construction and engineering sectors. The paper will then explain the importance of sustainability and the ethical case for participation, articulating the benefits for engaging with the environmental agenda. In addition, the potential disadvantages and risk for organisations from non-participation in environmental sustainability reforms and improvements will be covered including the irreprovable damage that

J. Challender (✉)
Salford University, Manchester, UK

A. Sayigh (ed.), *Transition Towards a Carbon Free Future*, Innovative Renewable Energy, https://doi.org/10.1007/978-3-031-61660-0_15

can be enacted for the planet through climate change. It will then outline the global and national agenda on sustainability and articulate world targets for issues around reducing carbon generation. This will then lead to the discussion around the UK Government's influence in environmental ethics and the many initiatives including grants and regulations that in recent years have been introduced to curb waste and emission levels specifically.

Environmental social responsibility will be covered in this paper and the mechanisms by which organisations can make a positive impact and example for others to follow. In this regard initiatives and commitments to waste, energy, carbon management and transport will be discussed and analysed. This will then lead into the many dilemmas around environmental ethics and how sometimes the financial aspects of introducing sustainability initiatives can sometimes present barriers to implementation. Awareness of the benefits of environmental commitments will be explained and how these can lead to advantageous outcomes for organisations in terms of aspects such as public relations, corporate responsibility and reputation. Finally, the paper will conclude by looking at two case studies in the education sector where commitments and delivery strategies related to environmental sustainability plans are being implemented, which are intended to present models of good practice for other organisations to follow.

15.1.2 Climate Change and Sustainability

The UK government has committed to cut greenhouse gas emissions by 80% by 2050 compared to 1990 levels and to halve them by 2025. In 2020 buildings in the UK accounted for nearly half of all the total carbon emissions. Accordingly, if the government pledge and commitments are to be achieved, buildings must become significantly more efficient, not just in the construction process but in terms of energy usage during their life cycle. This challenge has become more difficult owing to the fact that 80% of the UK building stock that will exist by 2025 has already been built, along with approximately 70% of the housing that will be occupied in 2025. Not only does the UK's building stock need to become more efficient, but it will also need to become more resilient if the climate changes as it is predicted. Despite this, government policies which include scrapping the Green Deal, withdrawing plans for zero-carbon homes and substantially reducing financial incentives for renewable energy would appear to undermine these commitments.

15.1.3 Context for the Discussion on Environmental Ethics

Construction professionals in their respective leadership roles, when procuring projects, should be aware of the growing importance of environmental ethics, as part of their organisations' social responsibilities. Increasing awareness of environmental ethics has in recent years occurred owing to the potential environmental problems including resource depletion and climate change. Environmental ethics reflects the impact of policies and actions on environmental matters and future individuals' well-being. There is an argument that the construction industry plays an important part in sustainability, conservation and environmental management, and environmental ethics should accordingly be particularly relevant to the ethos of projects (Tagawa, 2005, p. 12). Owing to the above factors, there is a strong argument that construction professionals should be increasing political influence in this area and therein drive the agenda forward. This could take the form of instructing their project teams to adopt environmental and sustainable design measures into their projects as one example. This could lead to changes on practice and have reputational benefits for their organisations. Notwithstanding the above premise, there has been much scepticism in the past about environmental ethics and whether simply enough has been done to curb the rise in carbon emissions into our atmosphere leading to global warming and potentially compromising the lives and livelihoods of the world's population. Construction and buildings in use account for a huge proportion of the problem, and some would argue that the industry has ignored some of the issues with carbon emissions for many years now. Strong arguments have emerged in the past that if construction professionals continually adopt policies which encourage use of non-sustainable construction strategies, then this could have negative connotations in their environmental credentials and lead to reputational damage. Such is the growing importance of environmental ethics, over recent years; this paper will focus on organisational sustainability strategies and with particular reference to two organisational case studies.

15.2 Literature Review

15.2.1 What Are Environmental Ethics and Sustainability?

In considering what constitutes environmental ethics, one definition is that:

> Environmental ethics concerns formulation and moral obligations regarding the environment. (Tagawa, 2005, p. 12)

According to the Chartered Management Institute (CMI) Codes of Ethics and Professional Conduct, the field of environmental ethics covers the ethical relationship that people have on their natural environment. The role of environmental ethics is to identify the moral obligations in terms of concerns of effects on the atmosphere, climate change, depletion of natural resources, loss of wilderness, dwindling animal and plant biodiversity. These have been in the past referred to as 'green issues' and are becoming more important for economies and organisations to recognise the effects on the environment from their actions. This has emanated in policies linked to social and environmental responsibilities. From a moral standing position, it is not always straightforward to identify what obligations we have for them. Some would argue that our obligations should be focused on ensuring that future generations are not disadvantaged and prevented from having their basic needs met.

Given the increasing concern for environmental issues and the impact that our actions will have on the environments that we live in, the field of environmental ethics will become more important in the future. Political efforts will seek to limit or in some cases reverse the environmental effects of certain delirious actions that have caused harm to the environment. There may be more interventions on a global stage, such as the Kyoto Protocol, that could attempt to deal with climate change. Such measures could be focused on large polluters to the environment with targeted reduction imposed on them in terms of carbon generation and greenhouse gases. This could present a challenge for developing economies who may find it difficult to meet such targets. Environmental ethics should be focused on not simply apportioning blame for what could be regarded as failures in this regard but suggest means for resolution of such problems. This raises the issue of whether countries should hold overall responsibilities in preserving the environment or, adopting a bottom-up approach seek to hold organisations responsible for putting forward solutions. This is becoming more topical in recent years, and the 'court of public opinion' on many global organisations in food and beverage for packaging linked to plastics has turned the tide in some respects in forcing change.

In understanding what sustainability is, there have been many different definitions over previous years. The United Nations Brundtland Commission, as cited in Appleton (2006) defined sustainable development as 'that which meets all the needs of the present without compromising the ability of future generations to meet their own means.' Principle 3 of the Rio Declaration on Environment and Development (Harvard Institute, 1992), however, embedded a revised definition of sustainable development as 'to equitably meet developmental and environmental needs of present and future generations.'

To assist in considering these definitions, sustainability could mean different things to many people when taking into account of the wide-ranging contexts and environments which construction management operates within. For instance, a project to utilise energy for renewable low carbon sources such as timber could only be regarded as truly sustainable if the environmental costs of transport to its intended destination do not outweigh the benefits. Whichever definition we accept, we need to establish how it applies to an organisation. For this we need to comprehend the scope of sustainability, what it involves and what is encompassed within it. Ding (2006, p. 463) explained that:

> It involves the efficient allocation of resources, minimum energy consumption, lower embodied energy intensity in building materials, reuse and recycling and other mechanisms to achieve effective and efficient short- and long-term use of natural resources.

15.2.2 The Importance of Sustainability and the Ethical Case for Participation

One could argue that it is important for all responsible organisations to uphold a sense of corporate social responsibility. This in turn will have far-reaching benefits for their businesses in terms of public perception, marketing, operational efficiency, adherence to ever-increasing regulations and maximising future government funding opportunities. In addition, organisations must not lose sight of the moral and ethical case for intervention on sustainability. The IPCC (Intergovernmental Panel on Climate Change) predicted that unless emissions of greenhouse gases decrease, there will be a temperature increase of between 1.4 C and 5.8 C by 2100 (IEA, 2006). More recently the Carbon Plan (HM Government, 2011) set out the government's plans for achieving the emission reductions and set down targets of such cuts of at least 80% of 1990 levels by 2050 and revealed that:

> Climate change is one of the greatest threats to both UK and global security and prosperity. There is an overwhelming scientific consensus that climate change is happening, and that it is very likely to be primarily the result of human activity.

The Carbon Plan articulated that all organisations should lead the way and be drivers of the transition to reducing emissions and they must appreciate and capitalise on the opportunities and challenges that this presents. In this regard, the Plan set out that innovation and cultural changes within organisations will be important in achieving government targets.

15.2.3 The Global and National Agenda on Sustainability

It could be argued moving global agendas including the international targets set at the Rio Summit (1992), the Kyoto Protocol (1997) and the Cancun Climate Summit (2010) will continue to have little direct influence on most organisations. Furthermore, the international agenda that has not been adopted at national or local level may have no impact whatsoever. The international agenda, in particular, is often criticised for not being proactive enough or concrete enough at an operational level. A converse argument is, however, that it is the indirect effect of these that is important as national agendas and regulations are heavily influenced by international politics. An example of this is the Climate Change Act 2008 that required the government to cut emissions by at least 34% by 2020 and 80% by 2050. The Carbon Plan (2011) is designed to ensure that action at a national level addresses barrier to energy efficiency and effectively supports and complements action at a European Union (EU) level. This demonstrates that there is government pressure to rebalance organisations' traditional key objectives for corporate growth, profits and increased share prices against environmental issues and the growth and welfare of our communities.

15.2.4 Government Influence in Environmental Ethics

Government has in the past contributed to encouraging environmental ethics into the design and construction of projects in the UK. The introduction by the Department of Trade and Industry (DTI) for Sustainable Progress Reports, which covers principles such as waste, energy efficiency, carbon generation, safety, skills, climate change and costs, has been used in the past for measuring how sustainable projects are. Sometimes government and international funding for construction projects stipulate minimum environmental performance criteria for buildings, which could be a *very good* or even *excellent* rating. Accordingly, any compromise on the environmental performance of buildings under such stipulations could jeopardise funding grants that may be available. In the UK it is quite common for grant funding on public buildings being conditional on gaining a minimum BREEAM rating. BREEAM is the world's leading sustainability assessment method for masterplanning projects, infrastructure and buildings. Through third-party certification, it assesses an asset's environmental, social and economic sustainability performance, using standards developed by the British research establishment (BRE). This means BREEAM-rated developments are more sustainable environments, although that enhance the well-being of the people who live and work in them, help protect natural resources and make for more attractive property investments. Notwithstanding this premise, some designers have argued in recent years that BREEAM has not always moved with the pace of environmental technological advancements. An example of this relates to measures to achieve complete electrification of buildings to avoid the use of gas, which in 2020 is not recognised by BREEAM in the rating process.

In addition to the design and construction of commercial buildings, the UK Government have exerted pressures on reporting on the environmental ratings of buildings for sale or rent with the introduction of the compulsory Eco Home ratings. These need to form part of the buyer packs for buildings, and once again this represents an intervention by the government to encourage environmental improvements.

15.2.5 The Local Agenda on Sustainability

In looking at local sustainability agenda in the UK and in particular local authority policies, there is strong emphasis on the participation of organisations to encourage reduction of waste, increased recycling and the adoption of sustainable transport policies. Councils have, however, previously had to rely on enforcement of such policies through the planning system (Town and Country Planning Act 1990) for major developments and had limited participation accordingly. One example of this is the Stockport Metropolitan Borough Council Sustainable Transport Supplementary Planning Document (SMBC, 2011) which forms part the Council's Local Development Framework and encourages 'more sustainable and active transport options where practicable'. Participating organisations in the area have responded with initiatives linked with reducing the reliance that employees have on vehicular means of commuting to work. Measures will need to be taken to approach this issue not least because of the finite number of car parking spaces currently available at their workplace. It is important and productive for organisations to work more closely with local authorities in partnership, through liaison and consultation, to promote a more robust sustainability strategy for their local area as a whole. This would improve the external relations and the public perception of businesses.

15.2.6 Environmental Social Responsibility

Environmental social responsibility is referred to as environmental conservation social responsibility and curtails to the long-term problem of global warming, increased waste production and contamination on an international scale. With the global focus on addressing the 'climate emergency' brought to light by high profile campaigners such as Bill Gates, Richard Attenborough and more recently Greta Thunberg, this has become front page news over many years. It followed on from Al Gore's campaign and lobbying to global powers in his climate emergency depiction 'An Inconvenient Truth' (2006). Such has been the prominence and exposure of environmental problems and global climate change that it has become increasingly important for organisations to align themselves in concerted efforts to help the situation. Accordingly, companies, especially in those larger economies have become more willing to commit to measures and initiatives linked to minimising environmental problems. These are designed to reduce their carbon footprint and therein decrease greenhouse gas emissions.

The questions that have been posed for business organisations across the world are whether they have a credible recycling programme in place and have they programmes and technologies in place for reducing energy consumption. This could take the form of creation of cleaner energy creation through renewable technology such as solar power through photovoltaic technologies and wind power through wind turbines. Other questions business leaders are tasked with relate to waste recycling and the degree to which their waste is managed in a sustainable manner. This could include the reduction in plastic waste which has become a global dilemma and could be affecting the world's oceans and contaminating the global food chain. There are many 'greening initiatives' that companies can adopt which could make a difference and some of which carry funding opportunities. Commitment to sustainable policies such as these is not simply the responsibility of large corporation but can easily be relevant to small- and medium-sized companies. In the UK construction industry, there is much research being conducted in reducing embodied carbon as part of the construction processes. This has called for innovative solutions and challenges to more tradition construction methods and building materials such as concrete and steel which have been commonly associated in being carbon intensive in their production.

Those organisations that have shown responsibility in 'doing their bit' for the sustainability and environmental agenda have benefitted from positive media coverage and gaining environmental accreditation. In the UK there has been a concerted effort for public sector organisations to 'lead from the front' and make a positive impact and example for other organisations to follow. Many universities have adopted strategies to raising their ranking in the global 'People and Planet' ratings, and gaining accreditation through the Eco Campus initiates as a commitment to waste, energy, carbon management, transport and greening initiatives. All these steps can make a small but effective contribution to the wider agenda in improving the environment. It is important also for organisations to look at their supply chains and have influence on them for responsible procurement of supplies and services. This could become a deciding factor in suppliers being selected as part of the supply frameworks especially for large organisations. Accordingly, those suppliers who show limited or no commitment to environmental measures could be severely disadvantaged in selection processes for new work orders. In the past the question for businesses is 'can they afford to make pledges and commitments to environmental initiatives?' For the reasons previously referred to, this has in recent years been turned around into the overarching question; 'can businesses afford not to have such commitments in place?'

15.2.7 Dilemmas Around Environmental Ethics

Ethical dilemmas could be generated by issues related to sustainability and environmental awareness. As part of the ongoing global climate emergency, it could be reasonable to assume that when designing new or refurbishing existing buildings, the design process should take full account of renewable energies in a concerted effort to reducing their 'carbon footprint'. It is important for the whole project team including clients and designers to consider both the construction and occupation phases in this regard as part of the life cycle of buildings. In doing so environmental ethics should take into account the embodied carbon of a building. Embodied carbon is the carbon footprint of a material. It considers how many greenhouse gases are released throughout the supply chain and is often measured from cradle to grave. It therefore needs to consider the construction phase and energy efficiency measures as part of the occupation 'in use' period of buildings with a particular emphasis on energy consumption and maintenance. Accordingly, it should take account and include all the emissions from embodied carbon in the construction materials, the building process, all the internal fixtures and fitting and the deconstruction and disposal of the demolition material at the end of the building's lifetime. Reducing embodied carbon is important for reducing resources and associated costs but also in reducing longer-term risks of resource availability. Furthermore, the embodied carbon agenda is growing more importantly given an awareness that almost one third of the total lifecycle carbon dioxide emissions from buildings stem from the embodied carbon in materials and the construction process. Addressing these sources of carbon is regarded as critical to meeting government and global targets for carbon reduction.

Notwithstanding the global push for 'carbon-friendly' buildings, the promotion and commitment to environmental and sustainability building technologies can sometimes create ethical dilemmas for designers. If their clients feel that they do not want or cannot afford such environmentally friendly technologies or system for their projects, then should the designers simply accept their position or challenge accordingly? This is not an easy question to answer and will depend on context and the relationship which exists between designers and their employers. At the very least, the design consultants should make clients aware of the benefits to sustainable design in terms of reduced environmental impact and potentially lower running costs. It could simply involve the designers questioning whether a particular client requires an air conditioning solution for their internal environment when a more sustainable and cost-effective solution might be predicated on natural ventilation through openable windows. This could save on carbon emissions and lower future utility costs, therein reducing the carbon footprint of buildings. Furthermore, architects, engineers and other construction design team consultants should be articulating the implications and disadvantages of ignoring the climate emergency and how this could specifically be regarded as irresponsible, leading possibly to reputational damage. At least then, they would be able to uphold their position from an environmental ethical perspective, given they have at least tried to influence the decision making in this regard. Clients may not be fully aware of the longer- term implications of not pursuing a policy of environmental sustainability and simply opting for solutions that generate the lowest capital costs in the construction phase. It is expected of professionals and ethical appropriate for them to educate their clients and make them aware of the future benefits in this regard. Notwithstanding this premise, it will always be clients or their senior advisors who will make the final decisions in this regard.

15.3 Case Study Findings: Designing and Implementing the Collaboration Toolkit as Action Research

15.3.1 Examples of Environmental Ethics and Sustainability in Practice: Case Study One—A UK Northwest College

15.3.1.1 Introduction to Case Study

The College was operating against an agreed environmental and sustainability agenda and existing legislation. Following the college's new build construction and engineering workshop project in 2011, it had been successful in gaining an excellent BREEAM accreditation. BREEAM represents the world's leading sustainability assessment method for masterplanning projects, infrastructure and building. It recognises and reflects the value in higher performing assets across the built environment lifecycle, from new construction to in-use and refurbishment. BREEAM does this through third-party certification of the assessment of an asset's environmental, social and economic sustainability performance, using standards developed by the British Research Establishment (BRE). This means BREEAM-rated developments are more sustainable environments that enhance the well-being of the people who live and work in them, help protect natural resources and make for more attractive property investments.

The College did seek, alongside our designers, to promote sustainable construction methods, mindful that the construction industry contributes to almost half of the world's energy consumption (Smith, 2004) and that there is the possibility of running out of fossil fuels in the next 40 years (Roaf, 2001).

The College's proposed masterplan redevelopment plans for further new energy-efficient buildings on the campus were cancelled following the Learning Skills Council withdrawal of funding in 2011. However, this did not preclude the College from looking at simple, cost-effective solutions to increasing building energy efficiency. These included improving heating controls, increasing thermal insulation, installing energy-efficient light fittings/bulbs, fitting door closers, managing classroom utilisation more effectively, fitting automatic controls and displaying energy meters so that students and staff can measure and monitor improvements. In addition, a waste management strategy involving different types of waste for recycling could be introduced quickly and at relatively little cost. This would then hopefully comply with EU Directive 2006/12/EC of the European Parliament which states:

> In order to achieve a high level of environmental protection, Member States should, in addition to taking responsible action to ensure disposal and recovery of waste, take measures to restrict the production of waste particularly by promoting clean technologies and products which can be recycled and reused.

15.3.1.2 Implications and Impacts for the College and the Need for Change

The College considered that unless it was to adopt a sustainable waste management policy, the changing agenda could present negative outcomes for the College on ethical, public relations, political and social grounds. Furthermore, on an economical basis, the financial operating costs for the College could increase as time passes through the introduction of government initiatives such as the Landfill Tax Escalator (Department of Trade and Industry, 2010). This taxation point is reinforced that 'further adoption of the "polluter pays principle" could see future differentiated taxes introduced by government seeking to remove or reduce emissions of carbon dioxide into the atmosphere'.

The dilemma for this organisation was not to assess the affordability and feasibility of sustainability as a choice of whether they chose to participate or not. The real question for the College was whether they could afford not to change the way they were currently operating and place sustainability high on their agenda.

15.3.1.3 The College's Corporate Social Responsibility

Given the importance of the sustainability agenda, the College looked for solutions that present environmental and social advantages and that will benefit the College in doing so. Some would argue that this corporate social responsibility does not exist in the real world. Banerjee (2007), supported this view, presents a cynical representation of this ideology for most organisations and explained:

> This ideology is a convenient, yet dangerous myth and this 'win-win' rhetoric is nothing but a mask, made up of glossy brochures, beguiling speeches and 'media-savvy' partnership initiatives.

One could, however, argue that the College is unlike most organisations as its primary function is not to make a profit but to serve the community that we work within through education and training, and this alone presents an important reason for our active participation in the sustainability agenda. On ethical grounds they considered that as a major public sector organisation, they should be paving the way for cultural change and setting a good example for other institutions in the process.

15.3.1.4 The Need to Embed Sustainability

The College had for many years now concentrated more on economic factors (especially in light of the real reductions in funding over the years) and educational excellence in meeting government targets on student achievement and retention rather than the bigger global issues of the environment. Michael Gove, Former UK Education Secretary, conceded this point and argued that 'we need to embed sustainability into the heart of our academic establishments' (Guardian, 2011). This view was underlined at the time by the Former Secretary of State for the Environment, Caroline Spelman:

> Genuine sustainability will never again be an add on or afterthought. It will be at the heart of everything we do, from making policy to running government buildings or purchasing goods and services. (DEFRA, 2011)

The College accepted that a long-term cultural change at the highest level was required within the organisation rather than consider the initiatives as individual short-term projects. This view is supported by the work of Hjorth and Bagheri (2006) who opined that 'sustainability should not be perceived as a 'project' that has an end point, but as an ongoing process that needs to be regarded as part and parcel of everyday work'.

15.3.1.5 Potential Measures to Address the Sustainability Agenda

Table 15.1 presents some potential recommendations that were designed to allow the College to address the sustainability areas. It lists a combination of *dematerialisation*

Table 15.1 Potential recommendations to address the sustainability agenda

Activity of college	Creative recommendations to address legislation and mitigate impacts of environmental risks
Environmental management	The College could: Promote sound environmental management by communicating procedures, guidelines and regulations via induction, staff training programmes and tutorials Encourage and facilitate feedback and suggestions on ensuring good practice via a network of operational groups and team meetings across College Develop a management system based on a sound understanding of current performance and the adoption of realistic targets to improve environmental performance
Waste reduction	The College could: Implement sustainable resource management practices based on 'reduce, reuse, recycle' principles Re-use and recycle materials wherever practicable Reduce the amount of waste being disposed of in landfill sites, by segregating different types of waste Develop a programme of target setting to reduce waste Prevent pollution Comply with the legislation and meet the requirements of the waste regulators Reduce litter
Water consumption	The College could: Make efficient and responsible use of water, including making opportunities for water re-use and rainwater harvesting
Carbon management	The College could: Implement a carbon management strategy to ensure the efficient use of energy. Ensure the uptake of low-carbon technology in buildings and equipment. Improve the Energy Performance Certificate assessment grade of buildings from the base year, October 2008
Transport	The College could: Implement sustainable transport practices across all activities, as indicated in College Travel Plan Encourage the use of public transport, walking, cycling and vehicle sharing, where appropriate Possible discounts to incentivise train and bus travel Charging students and staff for car parking Reduce the need to travel especially by car, e.g. working from home. Extension of bus routes and increasing bus frequency Encouraging multipurpose/linked trips Engage with public transport providers to maximise use of existing transport infrastructure, including Designing our campus to encourage cycling to work, e.g., more cycle storage, dedicated changing areas and showers
Ethical and sustainable procurement	The College could: Promote the use of sustainable resources whenever practicable Promote life cycle thinking in the procurement of goods and

(continued)

Table 15.1 (continued)

Activity of college	Creative recommendations to address legislation and mitigate impacts of environmental risks
	services Promote the development of central purchasing strategies to ensure efficient and sustainable purchasing practices. Embed sustainability into selection criteria, processes, procedures, business cases, budget allowance and early market engagement when purchasing goods and services Manage relationships with suppliers to raise awareness, and allow suppliers to respond to additional or different measures in some cases
The College estate	The College could: Ensure that, wherever practicable, all new building works meet the sector standards for sustainability Promote a responsibility culture to improve the internal environment of all College rooms Promote a responsibility culture to improve the external environment of the College Promote the efficient use of all buildings and facilities to meet sector standards for space utilisation Manage the College estate to enhance biodiversity wherever possible
Education for sustainable development	The College could: Develop links with local and national groups which encourage education for sustainable development Endeavour to embed education for sustainable development into all curriculum areas Encourage students to participate actively in environmental projects, locally and globally

and *substitution* measures. Dematerialisation would include reduction in energy usage, waste, paper, etc., and substitution could be achieved through alternative energy sources. In some cases, innovative solutions could fulfil both objectives at the same time. For instance, the College could have reduced the extent of wood waste through recycling (dematerialisation) and utilised the waste to fuel biomass boilers for heating as an alternative zero carbon energy solution (substitution).

15.3.1.6 Economic, Political and Social Factors Influencing the Evaluation of the Measures

Detailed evaluation of the financial viability and environmental benefits of each measure was required to be undertaken by the College. This approach to business decisions around sustainability is advocated by Ding (2006) who explained that 'Environmental issues and financial considerations should go hand in hand as parts of the evaluation framework of a project'.

Some of the recommendations in Table 15.1 were intended to save money especially when considering life cycle costs. Ayres (2008, p. 289) supported this point and states that '...it may be possible, in principle to cut costs and pollution at the same time in some sectors...although it is rarely accomplished in practice'. Sant and Carhart (1981) also endorsed this view and suggest evidence through 'least cost' studies that very significant opportunities for savings through sustainability measures have been neglected at the national level. The converse argument is made by Appleton (2006, p. 3) who claimed:

> An increasing number of corporations and governments have recognized that practices like green buildings, alternative ways of storm water management and certain change to land use practices offer them chances to make additional profits at little downside cost or risk.

The College was required to identify projects which should take priority over others and develop overall timescales for instigation of our strategies. Factors included complexity, cost resources, funding incentives and the political appetite for change in certain areas. One argument could have been made for priority to be given to those that stand a relatively good chance of yielding a good return on investment. Graham, (2003, p. 54) presented a converse argument in that 'sustainability should not be influenced to such an extent by cost alone but by all the environmental and social benefits that generate from it'.

15.3.1.7 A Sustainable Procurement Process

Using the sustainable procurement processes for the purchase of goods and services, the College attempted to implement measures and realise benefits from those initiatives very quickly. The Office for Government and Commerce (OGC) announced in 2015 new commitments for 'greening' public sector procurement to understand and reduce the impacts of its supply chain. It is hardly surprising with an annual public sector spend in excess of £236 billion that government regards procurement as a powerful lever that can reduce emissions and drive innovation. Government also supports the National Procurement Programme (OGC, 2011) which is delivering training on sustainable procurement to public procurers including the FE and HE sectors as well as interest-free loans for setup costs. In addition, there were already procedure, processes and documentation from previous models of good practice as templates that could be utilised for this purpose, which would have precluded the need for a bespoke solution. Financial and resourcing demands would therefore have presented few barriers to changing the procurement strategy of the College in the short term, given the above incentives for change.

15.3.1.8 A Sustainable Transport Policy

There are other examples of sustainability initiatives such as changes to transport and the introduction of a green travel plan which could take much longer to initiate. Any changes would have had to involve a considerable consultation period and extensive negotiations with partners such as the local authority (as part of their Agenda 21 Sustainable Transport 1 initiative) and public transport providers in the borough. In addition, the need to undertake physical alterations and build new additional facilities on the campus was a consideration. Such measures could have included creating additional changing rooms, bike storage and showers for cyclists which could take some time in planning and construction. Whilst this was a valid argument, the transport agenda was a major focus for the College. After all, according to the DTI (2010), domestic transport accounted for 22% of UK greenhouse gas emissions. In addition, the introduction of the Government's Local Transport White Paper in 2010 (HM Government, 2010) detailing £560 million enabled local bodies to deliver sustainable transport solutions. This should have provided an incentive to the College to start the process and planning earlier in our programme, to ensure implementation within a reasonable timescale.

15.3.1.9 The Need for Effective Consultation and Communication

In planning long-term changes to the College's sustainability strategy, their success was dependent on undertaking extensive consultation and dialogue with their students and staff as well as third-party stakeholders such as their local authority. Innovative ideas were generated to assist the College in developing environmental strategies and sought to discard the former 'identify-and-repair' approach and instead adopt an 'anticipate-and-prevent' approach (Al-Sharrah et al., 2010).

At each stage the College shared and communicated their sustainability strategy as it developed and unfolded. This was aimed at generating maximum support and to be as transparent as possible on the issues and their sustainability objectives. This improved public relations for the College and was intended to enable transparency which could open up avenues of possibilities that would be difficult to achieve without it.

15.3.1.10 The Need for Commitment and Cultural and Organisational Change

The College decided that success of their sustainability strategy was predicated on securing full commitment from the College's Senior Management Team and Board of Governors. For this reason, all college managers were given responsibility and duties to support the implementation and monitoring of the policy. It was recommended that a new role should be created as Head of Environmental Services, who would be required to report to the principal, senior management and the Governing Board and provide regular progress reports against sustainability programmes and implement an annual review of all environmental policies.

Finally, in terms of cultural change, educational institutions, such as the College, did not simply regard themselves as academic institutions that enabled students to obtain qualifications and training. Instead, they proclaimed that they offered and created an excellent opportunity to encourage people to adopt more sustainable lifestyles. In addition, the College believed that the increasing regulation and sustainability agenda affecting organisations like theirs should not be regarded as a negative influence to adversely affect their business operations. Moreover, they professed that they should be encouraged to push the boundaries in the creation of more environment-friendly communities (Architecture Week, 2011).

15.3.1.11 Summary of Case Study

This case study hopefully provides an insight into an organisational sustainability agenda, which in this instance was a college. The example articulated the analysis and evaluation of the implications and impacts from ever-increasing regulation alongside the benefits that could be realised from participation which suggested the case for adopting sustainability principles and embedding these within their organisation and everyday activities. The College's corporate social responsibility on ethical grounds alone in promoting the agenda provided many benefits. Furthermore it was intended that the proposed innovative measures would provide the tools and motivation for implementing the necessary sustainability strategies and policies towards promoting a cleaner sustainable environment.

15.3.2 Examples of Environmental Ethics and Sustainability in Practice: Case Study Two—A UK Northwest University

15.3.2.1 Introduction to Case Study

This case study relates to achieving sustainability and environmental enhancements through a collaboration and partnering 'toolkit.' The measures undertaken and successful outcomes at the University will be explained and how the development of a carbon management plan assisted the organisation in achieving its environmental target.

The author whilst employed at a university in the Northwest of England worked alongside a sustainability team around improving environmental performance of the organisation. A 'collaboration toolkit' was formulated through the author's research to assist client bodies to adopt a more collaborative model of procurement that integrates the supply chain into the environmental agenda at early stages of projects. Plans were in train for the toolkit to be piloted on a major project as part of the University's masterplan and capital programme.

The focus for the toolkit was finding environmental solutions through collaboration with industry and delivering a new sustainability strategy for the University. In this regard the following measures and initiatives were developed and planned to be deployed as part of the case study from many different environmental perspectives including:

- Pursuing a holistic approach across the supply chain to meet the demands of the carbon-neutral agenda. All bodies (client end users, design consultants, contractors, subcontractors and other external stakeholders) are being encouraged to collaborate early in design stages on alternative construction methods, systems or solutions to generate more sustainable solutions.
- Reducing embodied carbon in buildings, with academics and industry working collaboratively to formulate pragmatic solutions through workshops.
- Introducing an energy partner to work with the University to assess the ways and means of them reducing its gas, electricity and water consumption.
- Reviewing which renewable technologies are going to give a significant and practical reduction in carbon and are cost-effective, respective to site logistics decanting and appraisal.
- Creating a new sustainable transport policy linked to reducing car travel.

An energy, water, and carbon management plan was developed by the University and this is in Appendix A. It gives context and background to the plan and contains an energy and water review. It also incorporates baseline data and key targets for future energy and water usage alongside an action plan and risk assessment. Furthermore, it provides month upon month results from energy audits conducted with data on water and energy usage. In addition to this document, an environmental sustainability strategic plan was prepared by the University which sets out further targets linked to wider sustainability aims and objectives, including procurement, transport and carbon generation (included in Appendix B). The targets linked with these measures as part of the case study in 2020, 3 years after implementing the plans, were as follows:

- Carbon emissions on target for reduction of 81% by 2030 (from a baseline of 2012).
- Reductions of energy and water consumption by an average of 5% over the last 3 years.
- Environmental accreditations achieved in the last 2 years.

    1. Green Flag.
    2. Eco Campus Platinum Status.

- Reduction in waste of 20% over the last 3 years through recycling.
- Increase of EV charge points by 30%.
- Flexible (Sustainable Procurement) Framework Level 4 achieved.
- Robust energy, water and carbon management plan developed consisting of:

    1. Energy audits.
    2. Training.
    3. Automated systems.
    4. Measurement.

5. Switch it off campaign.
6. Water leak surveys.
7. Low flow sanitary fixtures.
8. Rainwater and grey water harvesting.
9. Electric vehicles.
10. LED lighting.
11. Green IT.
12. Cycle to work initiatives.
13. BREEAM Excellent for new builds and refurbishments.

- Communication and engagement on sustainability measures and outcomes.
  1. Sustainability linked into the curriculum.
  2. Active member of the Greater Manchester Green Impact Committee.
  3. Active member of Go Green Salford.
  4. Green Campus Group at the University.

The above successes highlight the measures and consequential beneficial outcomes for the University and how the collaboration toolkit played an integral part in delivery. This will be no doubt of interest to those organisations that are planning for long-term transformation change and raising their sustainability credentials and accreditations such as the Eco Campus Scheme.

15.3.2.2 Practical Use of the Collaboration Toolkit to Target and Deliver Environmental Sustainability Objectives

The collaboration toolkit provides the practical guidance and the platform for systematic partnering at all design and construction stages with the supply chain. This is intended to provide the conduit for improving outcomes on projects as part of strategies designed around environmental sustainability. Notwithstanding this premise, the following considerations need to be integrated into environmental sustainability strategies and policies.

(i) The Need for an Effective Environmental Management System

As ‘part and parcel’ of the sustainable procurement of construction and engineering projects, an Environmental Management System should be put in place as a way for organisations to demonstrate that they have addressed and minimised the environmental impacts of development. In this way it can oversee and manage legal compliance of development whilst striving for continuously improvement on environmental sustainability performance. Furthermore Fig. 15.1 shows the environmental measures which should be incorporated into the Environmental Management System as part of the project management process.

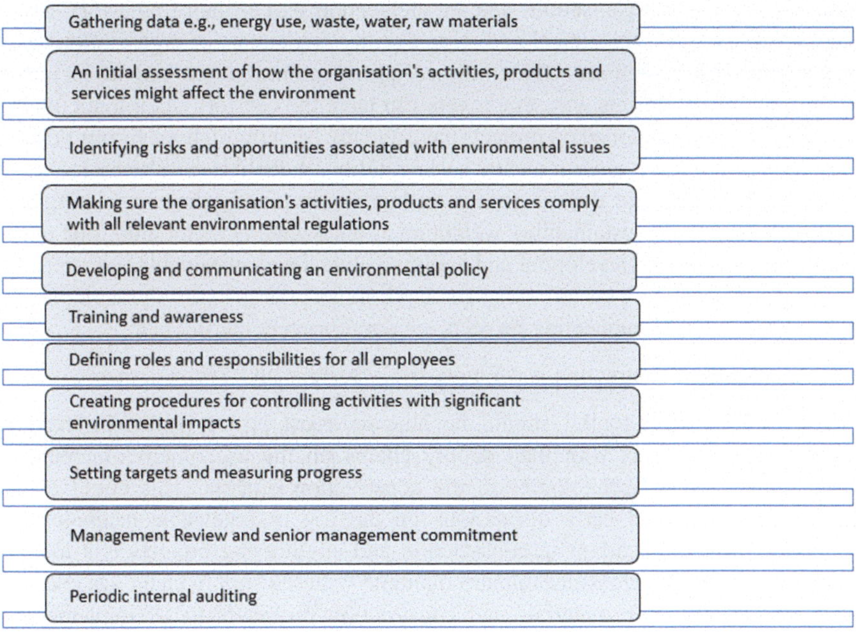

Fig. 15.1 Environmental measures which should be incorporated into the Environmental Management System

In addition to the above measures, having energy, water and carbon management plans in place can greatly improve the success of environmental sustainability projects. The collaboration toolkit can be tailored to ensure that these plans are developed by the projects team, assisted by the supply chain and then become embedded into the construction and engineering programmes.

(ii) Waste Reduction

On all projects through the collaboration toolkit, there should be integrated measures to ensure that waste from construction and engineering activities is minimised. Every effort should be made to recycle waste materials and thus reduce the overall amount of arisings disposed off-site. This will serve to reduce the embedded carbon of the build process, thus lowering the carbon footprint. Furthermore, measures to improve the green credentials and status of projects will have many advantages for those organisations involved in them, and these revolve around corporate social responsibility, community relations and reputation.

(iii) Sustainable and Ethical Procurement

In a construction and engineering context, sustainable procurement philosophies should be deployed as part of the collaboration toolkit when appointing consultants, main contractors, sub-contractors, suppliers and other specialists. In the UK there are various purchasing consortiums, who ensure that 'checks and balances' are conducted on those organisations that are included as part of their framework. Such checks will ascertain the ethical and sustainability credentials of organisations and ensure that only those firms which meet a required minimum standard are included on the framework. In this way, employers can have the comfort factor that companies who are commissioned on projects have already been through a rigorous vetting process to assess their environmental sustainability credentials. Furthermore, organisations should ensure that all tender evaluations required when appointing the supply chain include sustainability weighting as a key quality factor alongside other criteria such as price. Developing and applying ethical and sustainable procurement policies in this way will ensure social and environmental impacts of those employed on construction and engineering projects are appropriately considered.

(iv) Sustainable Construction and Engineering Designs and Technologies

The collaboration toolkit should be able to assist project teams in making collaborative decisions with their supply chains on the use of environmentally sustainable building technologies in the construction process. This could entail careful and protracted early discussions on the use of renewable technologies. Reducing embedded carbon in construction and engineering projects is a topical issue in the UK, and cross-disciplinary dialogue with the supply chain has encouraged more sustainable materials and construction processes to be implemented within projects. Accordingly, sustainability is the core principle of the any project and should embed a high standard of adherence to best practice philosophies.

(v) Community Engagement

Communication and engagement, encouraged and facilitated through the collaboration toolkit at all construction and engineering stages, are regarded as good practice on most construction and engineering projects. Processes around consultation with the local community into environmental issues and corporate social responsibility to maintain or enhance their well-being, environment and facilities can prove extremely fruitful in most cases. In addition, becoming accredited with schemes such as the Considerate Contractors Scheme in the UK can pay dividends and enhance the reputation of the project team.

Once again, the collaboration toolkit can play an important part on bringing parties together, formally or informally to present innovative ideas and solutions in this regard.

(vi) Education Around Environmental Sustainability

Engineering and construction projects should ‘promote sustainability and social responsibility’ through teaching, learning and research. The collaboration toolkit encourages industry partnerships and therein can encourage teaching and learning. As part of construction and engineering projects using the toolkit, strategies should be focused on the environmental sustainability of all operations and teaching, learning and research around this agenda through education programmes and continuous professional development should be encouraged. In this way construction and engineering projects could be regarded as a 'Living Laboratories' especially when partnering with specialist energy partners.

15.3.2.3 Successful Outcomes That the Collaboration Toolkit Has Had on Delivering Environmental Sustainability

Although projects where the collaboration toolkit is being piloted are at a relatively early stage in terms of delivery, they are proving to have many beneficial outcomes from an environmental sustainability perspective. For the client body, these include those benefits shown in Fig. 15.2.

15.3.2.4 Conclusions and Recommendation from the Case Study

The collaboration toolkit has been presented and as previously outlined forms part of action research as part of an ongoing case study where it is being piloted on UK construction projects. It has so far, since it was introduced in 2018, achieved very

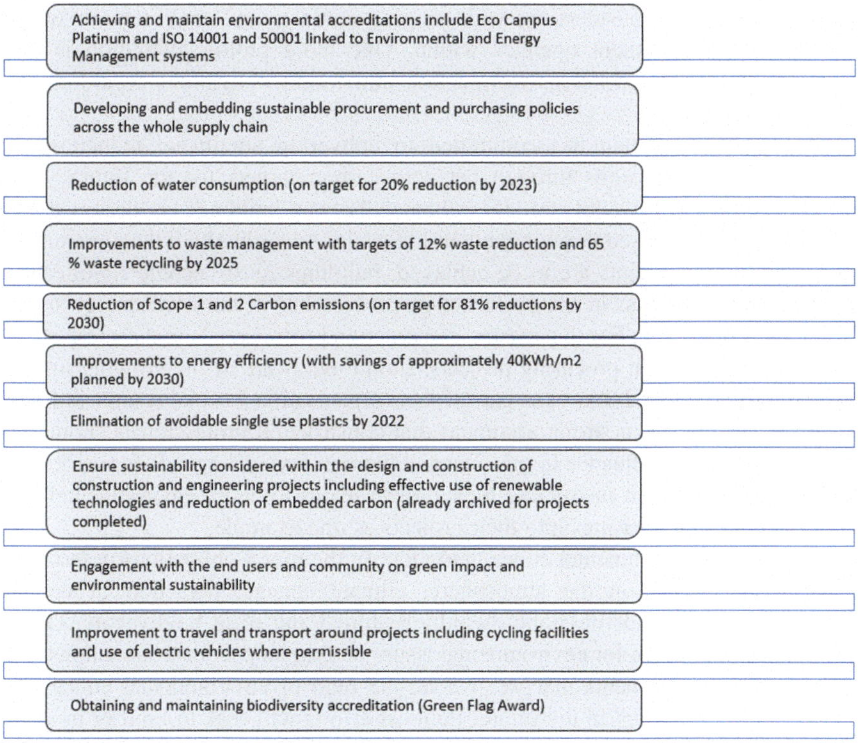

Fig. 15.2 Beneficial outcomes from an environmental sustainability perspective

successful outcomes for two client organisations in the UK and facilitated and encouraged many different environmental sustainability initiatives and policies to be brought forward. In this way it has assisted the organisations to keep on target with their environmental sustainability objectives. It has created the collaboration and partnering ethos that is so important for innovation and delivery strategies linked to environmentally sustainable construction and engineering projects and processes that could otherwise have been problematic.

The benefits of utilising the toolkit have been clearly articulated in this paper. Accordingly, it is hoped that the toolkit can provide the right platform for construction and engineering projects and therein achieve similar environmentally beneficial outcomes for many other countries other than the UK. For this reason, it is recommended that the principles and mechanics of the toolkit be deployed when embarking on environmental projects as part of global aspirations for a more sustainable future.

15.3.2.5 Summary

In understanding what sustainability is, there have been many different definitions over previous years. For this reason, sustainability could mean different things to many people. This considers the wide-ranging contexts and environments which construction management operates within. One more prolific definition is that environmental ethics concerns formulation and moral obligations regarding the environment.

The UK Government has committed to delivering significant reductions in greenhouse gas emissions through net zero carbon targets for the future years ahead. In doing so this will strongly influence the way buildings are designed and constructed but also used throughout their life cycle. Accordingly, if the government pledge and commitments are to be achieved, buildings must become significantly more efficient, not just in the construction process but in terms of energy usage during their operation. For this reason, construction professionals in their respective leadership roles, when procuring projects, should be aware of the growing importance of environmental ethics, as part of their organisations' social responsibilities. Accordingly, there is a strong argument that construction professionals should be increasing political influence in this area and therein drive the agenda forward. This could take the form of instructing their project teams to adopt environmental and sustainable design measures into their projects as one example.

The role of environmental ethics is to identify the moral obligations in terms of concerns of effects on the atmosphere, climate change, depletion of natural resources, loss of wilderness and dwindling animal and plant biodiversity. Given the increasing concern for environmental issues and the impact that our actions will have on the environments that we live in, the field of environmental ethics will become more important in the future. Political efforts will seek to limit or in some cases reverse the environmental effects of certain delirious actions that have caused harm to the environment. In this regard, it is important for all responsible organisations to uphold a sense of corporate social responsibility. This in turn will have far-reaching benefits for their businesses in terms of public perception, marketing, operational efficiency, adherence to ever-increasing regulations and maximising future government funding opportunities. However, the global and national agenda on sustainability has in the past had little direct influence on most organisations. Notwithstanding this premise, the government in recent years has contributed to encouraging environmental ethics into the design and construction of projects in the UK. One example of a UK government initiative in this regard is the introduction by the Department of Trade and Industry (DTI) for Sustainable Progress Reports. This covers principles such as waste, energy efficiency, carbon generation, safety, skills, climate change and costs which have been used in the past for measuring how sustainable projects are. Furthermore, in looking at local sustainability agenda in the UK, and in particular local authority policies, there is strong emphasis on the participation of organisations to encourage reduction of waste, increased recycling and the adoption of sustainable transport policies. In some cases, county councils have previously had to rely on enforcement of such policies through the planning system (Town and Country Planning Act 1990) for major developments. In addition, it is important and productive for organisations to work more closely with local authorities in partnership, through liaison and consultation, to promote a more robust sustainability strategy for their local area. This would improve the external relations and the public perception of businesses.

Environmental social responsibility is referred to as environmental conservation social responsibility and curtails to the long-term problem of global warming, increased waste production and contamination on an international scale. Such have been the prominence and exposure of environmental problems and global climate change that it has become increasingly important for organisations to align themselves in concerted efforts to help the situation. The questions that have been posed for business organisations across the world are whether they have a credible recycling programme in place and have they programmes and technologies in place for reducing energy consumption. This could take the form of creation of cleaner energy through renewable technology such as solar power through photovoltaic technologies and wind power through wind turbines. There are many 'greening initiatives' that companies can adopt which could make a difference and some of which carry funding opportunities. In the UK construction industry, there is much research being conducted in reducing embodied carbon as part of the construction processes. This has called for innovative solutions and challenges to more tradition construction methods and building materials such as concrete and steel which have been commonly associated in being carbon intensive in their production. Notwithstanding the global push for 'carbon-friendly' buildings, the promotion and commitment to environmental and sustainability building technologies can sometimes create ethical dilemmas for designers especially. Such dilemmas have revolved around short-term client focus on reducing capital costs at the expense of longer-term outcomes. Accordingly, consultants should be articulating the implications and disadvantages of ignoring the climate emergency and how this could specifically be regarded as irresponsible, leading possibly to reputational damage for organisations.

In this way clients could then be persuaded to adopt environmentally sustainable policies and not simply discount them on the grounds of affordability in the short term. This drive to increase client awareness is important as they may not be fully aware of the longer-term implications of not pursuing a policy of environmental sustainability and simply opting for solutions that generate the lowest capital costs in the construction phase.

Two case studies have been articulated and discussed in this paper relating to environmental sustainability plans and strategies which have been adopted by both a UK college and university. Measures undertaken by these further and higher education establishments have included the introduction of effective environmental management systems, waste reduction initiatives, sustainable and ethical procurement processes, sustainable construction and engineering designs and technologies, community engagement and education around environmental issues. Successful outcomes on delivering environmental sustainability in these case studies have included reduction in energy and water consumption, the award of many environmental accreditations, improvements to travel and transport policies, reduction in waste in carbon generation and waste, decrease in the use of single-use plastics and the introduction of more robust environmentally sustainable procurement policies.

15.4 Conclusions and Recommendations

This study, conducted in the UK, provides an important piece of action research for projects linked to environmental ethics and using a 'collaboration toolkit'. The two case studies articulate the benefits of using the toolkit in developing successful sustainability outcomes with environmental ethics at the heart of decision-making. In this regard there have been many policies and initiatives developed which have assisted the two case study organisations in meeting their environmental sustainability aims and objectives and to be able to constantly monitor and track their progress against predetermined goals and targets. Accordingly, it has created the partnering and collaboration ethos which is considered vitally important for delivery strategies and innovation, linked to engineering processes and projects around environmentally sustainable construction. Without this it could otherwise have been problematic and challenging in attaining the successful environmental sustainability outcomes. It is hoped that the toolkit can provide the right platform for construction and engineering projects globally and therein achieve similar environmentally beneficial outcomes for many other organisations.

References

Al-Sharrah, G., Elkamel, A., & Almanssoor, A. (2010). Sustainability indicators for decision-making and optimisation in the process industry: The case of the petrochemical industry. *Chemical Engineering Science, **65***(2), 1452–1461.

Appleton, A. F. (2006). Sustainability: A practitioner's reflection. *Technology in Society, **28***(1–2), 3–18.

Architecture Week (2011), *Architectural Global Warming*. Online Available: http://www.architectureweek.com/2011/0218/environment_1-1.html. Accessed 5th Apr 2011

Ayres, R. U. (2008). Sustainability economics: Where do we stand? *Ecological Economics, **67***(2), 281–310.

Banerjee, S. B. (2007). Corporate social responsibility, the good, the bad and the ugly. *Journal of International Management.*, Edward Elgar Publishing, ***15***(1), 118–119.

DEFRA. (2011). *Embedding sustainability in Government*. (internet). Available: http://www.sd.defra.gov.uk/2011/02/embedding-sustainability-in-government. Accessed 28 Apr 2011

Ding, K. C. (2006). Sustainable construction—The role of environmental assessment tools. *Journal of Environmental Management, 86*(6), 451–464.

Graham, P. (2003). *Building ecology. First principles for a sustainable built environment*. Blackwell Science Ltd.

Guardian. (2011). *The Guardian home page* (Internet). Available from: http://www.guardian.co.uk/politics/2011. Accessed 3 May 2011

H M Government. (2011). *Carbon plan* (p. 2011). DECC Publications.

Harvard Institute for International Development (1992). *Report of the United Nations Conference on Environment and Development*. UNCED Report A/CONF.151/5/Rev.1 13 June 1992.

Hjorth, P., & Bagheri, A. (2006). Navigating towards sustainable development: A system dynamics approach. *Futures, 38*(1), 74–92.

IEA. (2011). *A workshop on energy efficiency in buildings*, Paris. Online Available: http://www.iea.org/Textbase/work/2011/cert_slt/announcement.pdf. Accessed 15 Apr 2011

Roaf, S. (2001). *A design guide* (p. 2001). Butterworth-Heinemann.

Sant, R. W., & Carhart, S. C. (1981). *8 great energy myths: The least cost energy strategy*. Carnegie-Mellon University Press.

Stockport Metropolitan Borough Council. (2011). *Local Agenda 21 Sustainable Transport Supplementary Planning Document* 2007. (Internet). Available: http://www.stockport.gov.uk. Accessed 16 Apr 2011

Tagawa, S. (2005). Environmental ethics and project management of architecture. *The 2005 World Sustainable Building Conference* (27–29 Sept 2005). (pp. 4529–4531). Tokyo.

Chapter 16
Wind Energy in Morocco: Tangier and Lagouira in Comparative Perspective

H. Nfaoui and A. Sayigh

16.1 Introduction

From 1963 to 1996, of thermal and hydraulic origin, almost all electricity production is provided by the National Electricity Office (NEO). The gradual liberalization of the electricity sector began in 1997, through the concessional electricity production.

On the other hand, the development of wind electricity has undergone a remarkable evolution, in particular after the adoption of Law No. 57-09 establishing the company "Moroccan Agency for Sustainable Energy" (MASEN) in 2010. Among the objects of this company, we quote: the wind potential evaluation, the wind park development in the national territory zones qualified to shelter such parks, etc. Thus, Law No. 2015-48, adopted in 2016, relates to the regulation of the electricity sector. In addition to fixing the tariff for the use of medium voltage electricity distribution networks, this law mainly targets the competition between various private actors (producers or consumers) and encourages them to invest more in the electricity and energy sectors.

Morocco not only has the largest wind farm in Africa, located in Tarfaya (300 megawatts (MW)) in the south, but also a huge potential of offshore and onshore wind electricity productivity in the Tangier region in the north and in the southern part from the Atlantic coast from Dakhla to Lagouira, areas where annual average wind speeds can exceed 8.83 m/s (2005–2007) and 8.02 m/s (1993–2003), respectively.

H. Nfaoui (✉)
Solar Energy & Environment Laboratory, Mohammed V University of Rabat, Rabat, Morocco
e-mail: h.nfaoui@um5r.ac.ma

A. Sayigh
Chairman of WREC & Director General WREN, Editor-in-chief Renewable Energy, Brighton, UK

A. Sayigh (ed.), *Transition Towards a Carbon Free Future*, Innovative Renewable Energy, https://doi.org/10.1007/978-3-031-61660-0_16

Thanks to its energy strategy, adopted in 2009, bringing the installed electrical power of renewable energy to 52% in 2030, Morocco has become a model for many African countries in the field of renewable energies. Morocco ranks second in the field of wind energy on the African continent after South Africa.

The development of wind electricity has undergone a remarkable evolution, particularly after the adoption of Law No. 13-09 on the development of renewable energies in 2010. The electrical power installed in 2021 reached 1466 MW, or 13% of total power, compared to only 50 MW, the capacity of the first wind farm installed in 2000. The contribution of wind electricity has increased from 100 GWh in 2000 to 5024 GWh in 2021, contributing with 12% in the total electricity consumption. It should be noted that for the electricity production, coal-fired power plants remain predominant with 68% of the total production in 2021. With a share of electricity production exceeding that of natural gas (8%), wind power is now the second most important source of electricity production in Morocco.

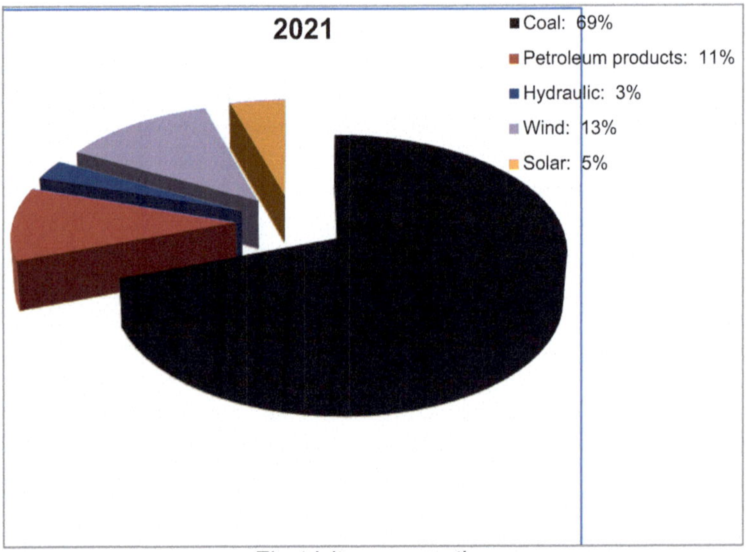

Electricity consumption

As part of the ambitious integrated wind energy project, Morocco aims to install 850 MW, spread over five wind farms, including that of Boujdour in the south (320 MW) and that of Tangier (70 MW). These projects attract billions of dollars in investment. In addition to reducing its fossil fuel bill, these investments contribute to job creation and other economic benefits to local economies through their construction and operation, particularly in the context of supporting the global clean energy transition, energy security, and energy price stability.

In addition to the economic benefits and energy savings, accelerating the deployment of wind energy could be a major tool for Morocco wishing to achieve climate objectives (reduction of greenhouse gas emissions) within the framework of international commitment to them and to achieve a balance between its economic growth and its energy security.

The objective of this work is to study the characteristics of the wind potential of Tangier and Lagouira in comparative perspective, as well as to respond mainly to requests from users of wind energy (engineers, technicians, and research groups) for the regions of Tangier and Lagouira. Indeed, the design and sizing of a wind farm as well as the establishment of its energy balance require prior knowledge of the wind energy it receives. This energy will be characterized by several factors, such as seasonal variations and frequency distributions of wind speed for a long period of measurements.

16.2 Sources of Wind Speed Data Used

The Directorate General of Meteorology (DGM) is responsible for most of the synoptic stations, including those located at airports. Among the meteorological parameters that the DGM has available, there are observations and hourly average wind speeds (HAWS). For some stations, the wind speed data goes back to 1948 (Knidiri & Laaouina, 1986; https://www.marocmeteo.ma/).

We are satisfied in this work with the three sites, namely, Tangier, Rabat, and Lagouira, since we have available 12 years (1978–1989) and 5 years (1978–1982) of measured HAWS for Tangier and Rabat, respectively (https://www.marocmeteo.ma/ ; Nfaoui, 2004), as well as estimated HAWS for Lagouira from 1985 (https://www.meteoblue.com/fr/meteo/historyclimate/climatemodelled/lagouira_maroc_6546558). On the other hand, Tangier is located in the extreme north of Morocco, Rabat in the center, and Lagouira in the extreme south.

For Tangier and Rabat, measurements are taken by DGM. For these two sites, the anemometers are installed at the airports at a height of 10 m (Knidiri & Laaouina, 1986; https://www.marocmeteo.ma/). Generally, airports are located in open terrain where the wind force is low for the safety of aircraft landing.

For Lagouira, the "meteoblue" service in Switzerland permits us to access its website: www.meteoblue.com. In the section history+, we have available historical HAWS for every point on earth. The estimation of the data is obtained by the application of the NEMS and ERA5 numerical weather models, including the integration of measurements and observations from several different meteorological worldwide sources. For wind potential, the "meteoblue" service recommends the wind speed data established from the ERA5 model (https://www.meteoblue.com/fr/meteo/historyclimate/climatemodelled/lagouira_maroc_6546558).

16.3 Comparison of the HAWS Measured by the DGM and Estimated by "Meteoblue" Service for Dakhla

For Tangier, which is a windy site, we have shown how to properly assess the wind potential. It is necessary to have at least nine available years of measurements with four measurements per day (Nfaoui, 2004; Nfaoui & Sayigh, 2017). However, for Lagouira, we do not have access to the HAWS measured for a long period, which is the opposite for Dakhla, a site located not far from this one where we have 10 years (1980–1989) of the HAWS.

To obtain more information on the characteristics of the wind speed and to estimate the wind potential for Lagouira, we compared the interannual and seasonal variations of the monthly means, the monthly maxima, and the annual distributions of the frequencies of the HAWS, estimated by the ERA5 model, with those of Dakhla obtained from the measurements and those estimated from the considered model for the period 1985–1989 (5 years) (Table 16.1, Figs. 16.1, 16.2, 16.3 and 16.4).

Table 16.1 Measurement periods, annual averages of wind speed, and wind potential

Stations	Period	Wind speed: measured/ estimated	Wind speed (m/s)	Wind power (W/m^2)
Tanger	1978–1989 (12 years)	Measured[a]	5.90	366
Dakhla	1980–1989 (10 years)	Measured[a]	7.70	472
Dakhla	1985–1989 (5 years)	Measured[a]	7.03	307
Dakhla	1985–1989 (5 years)	Estimated[b]	5.88	162
Lagouira	1985–1989 (5 years)	Estimated[b]	6.27	193
Lagouira	1913–1922 (10 years)	Estimated[b]	6.43	204

[a]Measured by the Directorate General of Meteorology (Morocco)
[b]Estimated by "meteoblue" service with the ERA5 model (Switzerland)

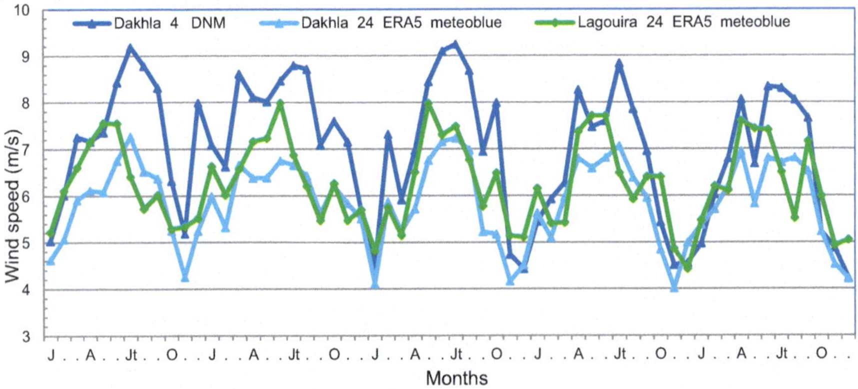

Fig. 16.1 Interannual variations of monthly averages

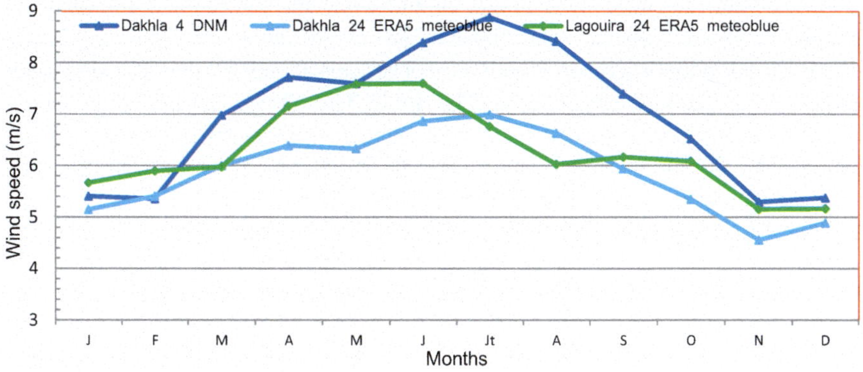

Fig. 16.2 Seasonal variations of monthly averages

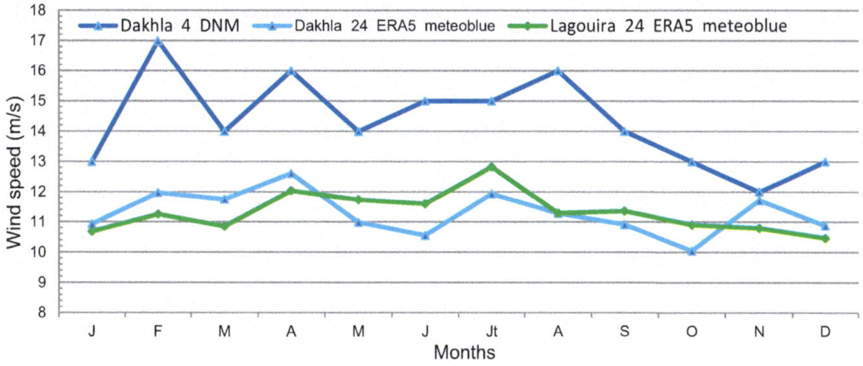

Fig. 16.3 Seasonal variations of monthly maxima

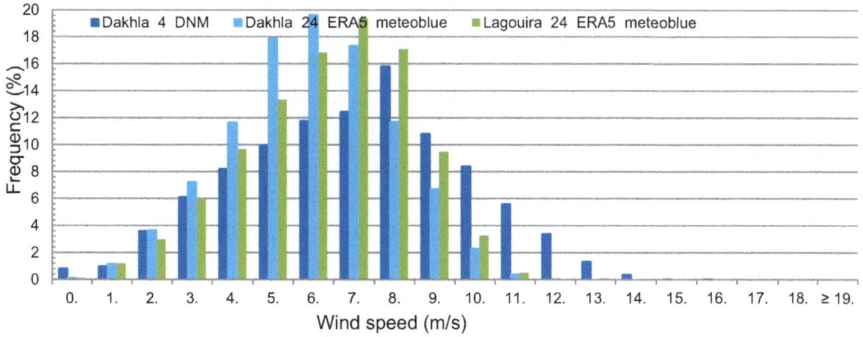

Fig. 16.4 Annual frequency distributions

For Dakhla, the monthly averages of the wind speed, which is the absolute maxima obtained from the measurements, are much higher than those estimated by the ERA5 model (Figs. 16.1 and 16.2). The frequencies for the estimated wind speeds greater than 10 m/s are zero (Fig. 16.3). This shows that the ERA5 model does not predict strong winds. The annual average wind speed and wind potential estimated by the ERA5 model compared to those obtained from the measurements are underestimated by 16% and 47%, respectively (Table 16.1).

Dakhla and Lagouira, two Saharan sites located in a windy region in southern Morocco, are subject to the same system of prevailing winds, the trade winds from the northeast. They overlook the Atlantic coast and are located on peninsulas.

We would expect the characteristics of the wind speed for Lagouira, which is the wind potential, to be comparable to those of Dakhla, obtained from the measurements. But this is not the case. For Lagouira, it can be argued that the annual average wind potential is underestimated by at least 47%. Hence, there is a need to have wind speed measurements for a long period with a smaller measurement step to assess the wind potential with good precision for a given site.

16.4 Statistical Analysis of HAWS Data

The earth is subject to relatively regular strong prevailing winds (general circulation) due to changing pressure systems in the atmosphere. These winds are, however, interrupted and modified by atmospheric disturbances and other smaller-scale pressure systems, such as the midlatitude winds in the north of Morocco and the trade winds from the northeast to the south (Lysen, 1983; Johnson, 1985). Local thermal winds also make an important contribution, for example, the sea/land breeze, especially coastal sites located in the Saharan regions such as Lagouira and Dakhla.

16.4.1 Average Annual Speeds and Seasonal Variations

The annual wind speed averages for Tangier and Lagouira are comparable, 5.90 m/s and 6.07 m/s, respectively, double that of Rabat (2.90 m/s). This shows that the north and south of Morocco are windier areas than the center (Table 16.2).

The monthly averages for Tangier fluctuate around the annual average. On the other hand, for Dakhla, for summer, they are above the annual average and those for winter are below. Moreover, the influence of the sea/land breeze may be due to a greater warming of the atmospheric air for Dakhla, compared to that of Tangier. However, Tangier is a special case. In effect, the Strait of Gibraltar connecting the Atlantic Ocean and the Mediterranean Sea forces the winds to be mainly from the dominant directions: east/west or west/east. There is also a certain acceleration and consequently an increase in wind speed (Fig. 16.5).

Table 16.2 Meteorological characteristics, measurement periods, annual wind speed averages, and standard deviations

Stations	Altitude above sea level (m)	Measurement period	Annual wind speed averages (m/s)	Standard deviations σ	Interannual variation factor, I
Tanger	15	1978–1989 (12 years)	**5.90**	1.16	**0.20**[a]
Rabat	75	1978–1982 (5 years)	**2.90**	0.42[a]	**0.10**[a]
Lagouira	11	2013–2022 (10 years)	**6.43**	2.03	–

Stations	Air density (kg/m^3)	Rugosity (m)	Anemometer height (m)	Wind speed measurements per day	Zero wind speed frequency (%)
Tanger	1.18	0.017	**10**	**24**	**11.57**
Rabat	1.17	0.017	**10**	**4**	**4.40**
Lagouira	1.20	0.001	**10**	**24**	**0.09**

[a]Knidiri and Laaouina (1986)

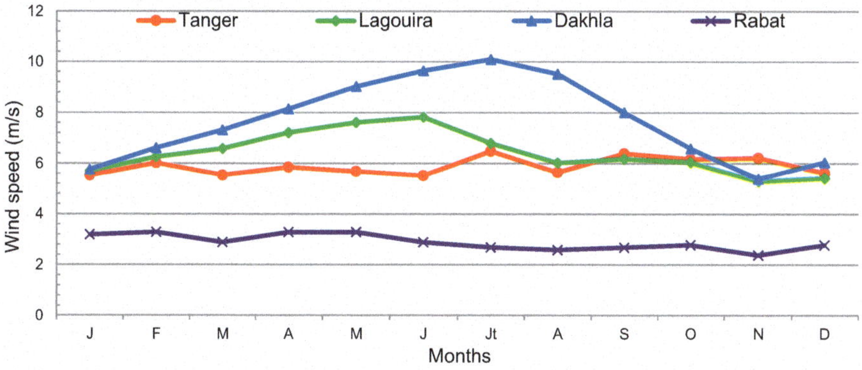

Fig. 16.5 Seasonal variations

The amplitude of the seasonal variation of the wind speed of Lagouira is smaller than that of Dakhla. The absolute maximum wind speed estimated for Lagouira varies between 10.5 m/s and 14 m/s. On the other hand, those for Dakhla and Tangier, obtained from the measurements, vary from month to month and can reach 25 m/s and 30 m/s, respectively (Fig. 16.6).

16.4.2 *Daily (Diurnal) Variations*

Just as the wind varies from season to season and from year to year, it varies between day and night. The diurnal variations of wind speed for Tangier, Dakhla, and Lagouira are greater than that for Rabat (Fig. 16.7). Apart from Lagouira, in general, the wind is strong during the day and reaches its maximum in the afternoon and its

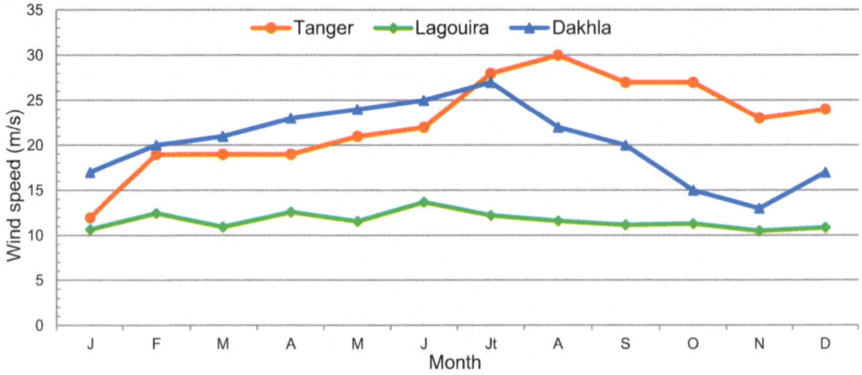

Fig. 16.6 Seasonal variations of absolute maxima

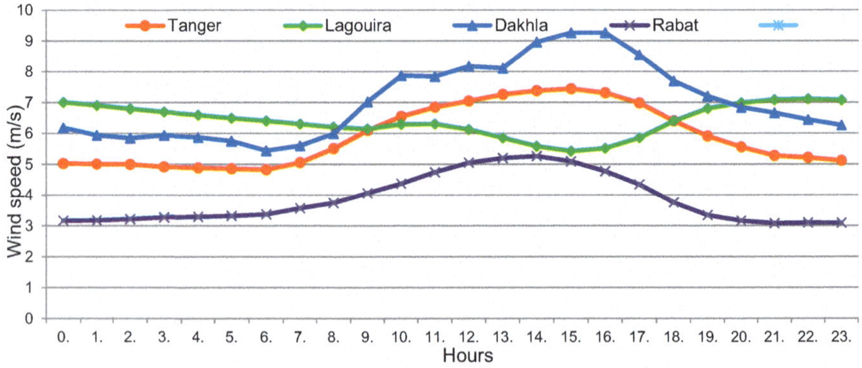

Fig. 16.7 Diurnal variations

minimum around midnight. These are due to the effect of the sea/land breeze which is greater during the day.

16.4.3 Interannual Variations

Interannual variations are a type of temporal variations that correspond to the differences observed from 1 year to another. The interannual variation factor for the *i*th month, I_i, is quantitatively defined for each month of the calendar over a period of several years by (Knidiri & Laaouina, 1986):

$$I_i = \frac{\sigma_i}{V_i} \tag{16.1}$$

where:

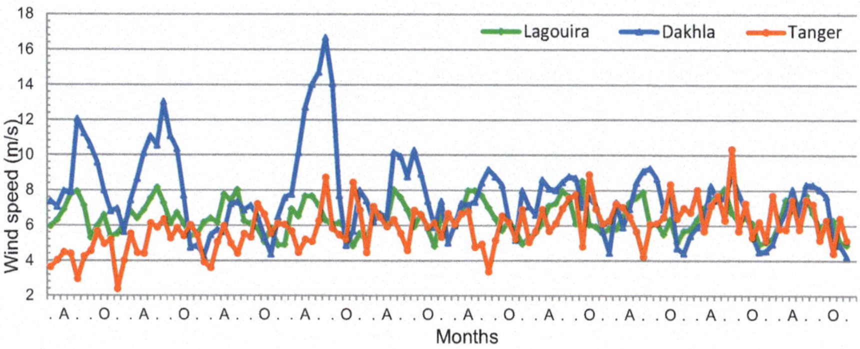

Fig. 16.8 Interannual variations

σ_i: standard deviation of monthly mean wind speeds for the *i*th month
V_i: global average speed for the *i*th month

Unlike Tangier, the periodicity of the seasonal variation is respected for Dakhla and Lagouira (Fig. 16.8). The amplitude of variation of the monthly averages of the wind speed for Dakhla is greater than that for Lagouira. Thus, the monthly averages for Dakhla vary between 4 m/s and 17 m/s, but for Lagouira, they vary between 4 m/s and 8 m/s. This shows that those obtained from the measurements are more accurate than the estimated ones.

The interannual variation I_i for Tangier is around 0.20, twice that of Rabat ($I_i = 0.10$). This shows that Tangier is located in an area of irregular winds compared to that of Rabat, which is in an area with stable winds (Table 16.1). This result is in agreement with the geographical location of Rabat in a transition zone of calm winds.

16.4.4 Frequency Distributions

To assess wind energy, the annual average wind speed alone is hardly sufficient. Indeed, the evaluation of the wind potential in a given site requires a detailed and precise knowledge of the wind speed data and their variations over time (Knidiri & Laaouina, 1986; https://www.marocmeteo.ma/; Nfaoui, 2004; https://www.meteoblue.com/fr/meteo/historyclimate/climatemodelled/lagouira_maroc_6546558; Nfaoui & Sayigh, 2017; Lysen, 1983; Johnson, 1985; Manwell et al., 2009; Nfaoui et al., 1995; Nfaoui, 2020; Nfaoui et al., 2022). The most important aspect of wind data analysis is the compilation and characterization of wind frequency distributions.

For the 0 m/s class, we see that the frequencies are around 12%, 2%, and 0% for Tangier, Dakhla, and Lagouira, respectively. It can also be noted that the frequencies corresponding to wind speeds greater than 10 m/s are zero for Lagouira, which is the opposite for Tangier and Dakhla. This confirms that strong winds are not predictable

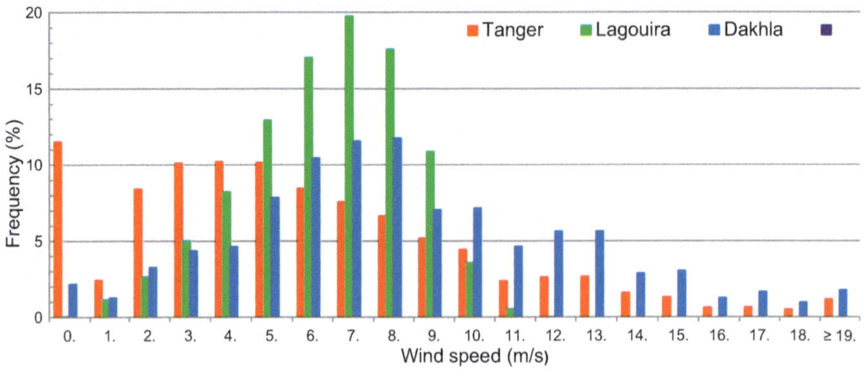

Fig. 16.9 Frequency variations

by the ERA5 model. We also notice that the frequencies of the wind speed varying from 4 m/s to 9 m/s are greater for Lagouira compared to those of Tangier and Dakhla (Fig. 16.9).

16.5 Mathematical Modeling of the Wind Speed Frequency Distribution

16.5.1 The Weibull Law

The frequency distribution of HAWS can be represented by mathematical equations such as the Weibull probability density. It is suitable for sites where calm wind frequencies are relatively small. It is given by the following relation (Knidiri & Laaouina, 1986; https://www.marocmeteo.ma/; Nfaoui, 2004; https://www.meteoblue.com/fr/meteo/historyclimate/climatemodelled/lagouira_maroc_6546558; Nfaoui & Sayigh, 2017; Lysen, 1983; Johnson, 1985; Manwell et al., 2009; Nfaoui et al., 1995; Nfaoui, 2020; Nfaoui et al., 2022):

$$p(v) = \left(\frac{k}{c}\right)\left(\frac{v}{c}\right)^{k-1}\exp\left[-\left(\frac{v}{c}\right)^{k}\right](k > 0,\ \ v > 0,\ \ c > 1) \tag{16.2}$$

where:

$p(v)$: probability density at wind speed, v (m/s)
K: shape factor (dimensionless)
C: scale factor (m/s)

16.5.2 Hybrid Weibull Law

In regions where calm wind frequencies are relatively high, the Weibull distribution is not always suitable. For this, we define a new probability density (Knidiri & Laaouina, 1986; Nfaoui, 2004; Nfaoui & Sayigh, 2017):

$$\begin{aligned} \mathrm{ph}(v) &= F_0 && \text{pour} \quad v = 0 \\ \mathrm{ph}(v) &= (1 - F_0)\ p(v) && \text{pour} \quad v > 0 \end{aligned} \tag{16.3}$$

where:

F_0 = calm wind frequencies

The value F_0 comes directly from the wind data. A relatively long, but extremely precise, method, called the maximum likelihood method, has been used, with good results for the calculation of the parameters K and C. It requires the resolution of the following two equations by the iterative method (Knidiri & Laaouina, 1986; Nfaoui, 2004; Nfaoui & Sayigh, 2017):

$$\begin{aligned} k &= \left[\left[\frac{\sum_{i=1}^{N} V_i^k \ \ln(V_i)}{\sum_{i=1}^{N} V_i^k}\right] - \left[\frac{\sum_{i=1}^{N} n_i \ \ln(V_i)}{N}\right]\right]^{-1} \\ C &= \left[\frac{\sum_{i=1}^{N} n_i \ V_i^k}{N}\right]^{\frac{1}{k}} \end{aligned} \tag{16.4}$$

where:

N = total number of nonzero wind measurements
n_i = number of measurements in the ith interval
V_i = center of the ith interval

For Tangier, the Weibull law adapts better to the observed frequencies. On the other hand, for Rabat and Lagouira, it is the Weibull hybrid distribution. The scale factors C for Tangier and Lagouira are large ($C = 7.14$ m/s), which shows that these two sites are windy. However, the form factor K for Lagouira is around 3.64, which is more than double that of Tangier. This means that the winds are widely dispersed for Tangier. A result is confirmed by the large value of the standard deviation ($\sigma = 4.3$) compared to that of Lagouira (Table 16.3)

Table 16.3 Statistical characteristics of the wind speed

Sites	Wind speed (m/s)	Standard deviation σ	Frequency (%) ($V = 0$ m/s)	Shape factor	Scale factor (m/s)
Tangier	**5.90**	4.38	11.57	1.62[a]	7.14[a]
Rabat	**2.89**	0.42	4.40	2.09[b]	3.43[b]
Lagouira	**6.27**	2.03	0.12	3.64[b]	7.14[b]

[a]Weibull
[b]Weibull hybrid

16.6 Evaluation of the Wind Potential Available in the Wind

The wind kinetic energy available in an elementary volume of air of mass m moving at speed v (m/s) in the x direction is:

$$\begin{aligned} P &= ½.m.v^2 \\ &= ½.\rho.\ Av^2.x \end{aligned} \tag{16.5}$$

However, the wind power is the derivation with respect to time of the kinetic energy:

$$P = \frac{1}{2}\ \rho\ Av^3 \tag{16.6}$$

where:

A: surface crossed by the air (m^2).
ρ: air density (kg/m^3).
v: wind speed (m/s),
x: thickness of the elementary volume (m).

For the three considered sites, Eq. (16.6) was used to assess the annual wind potential with $\rho = 1.18$ kg/m^3. The annual average wind power for Tangier is around 366 W/m^2, which is almost 12 times greater than that of Rabat (31 W/m^2). For Dakhla and Lagouira, the annual averages of the wind speed and the wind potential for a period of 10 years are higher than those corresponding to 5 years, and consequently, they are more representative (Fig. 16.10).

For Lagouira, the annual averages of the wind speed and the wind potential obtained from the ERA5 model for the period 1985–1989 (5 years) are higher than those for Dakhla, which shows that Lagouira is windier than Dakhla. The average annual wind potential obtained from wind speed measurements (307 W/m^2) is almost double that estimated by the ERA5 model (162 W/m^2) for Dakhla.

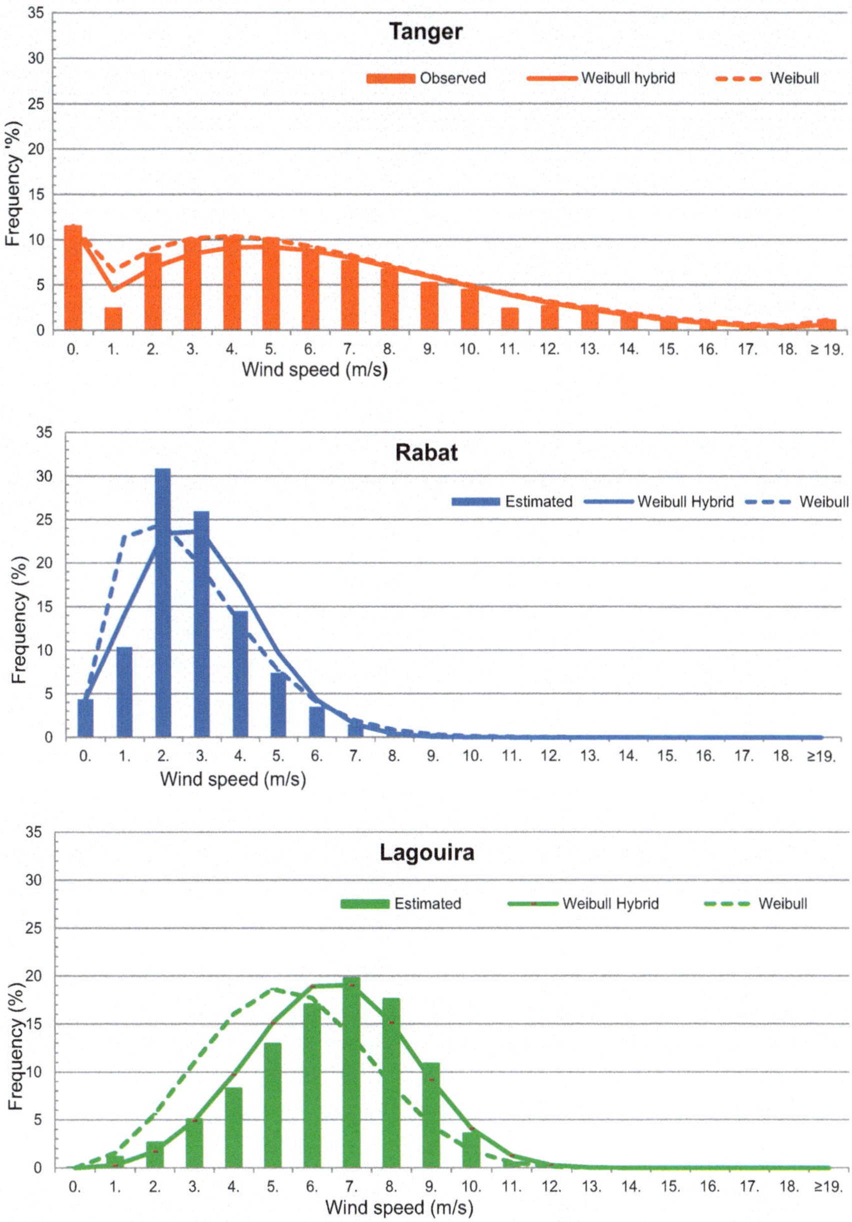

Fig. 16.10 Fit of annual wind speed distributions by Weibull hybrid and Weibull functions

Therefore, the estimated wind potential is undervalued by 47%. Since Lagouira is windier than Dakhla, we could consider that the wind potential for Lagouira is underestimated by at least a percentage greater than this.

16.7 Conclusion

The study of the wind potential assessment for Lagouira is based on the hourly average wind speed (HAWS), estimated by the ERA5 model, and those of Dakhla obtained from the measurements from 1980 to 1989 (10 years). For Dakhla, the measured absolute maxima are much higher than those estimated by the ERA5 model. The estimated wind speed frequencies greater than 10 m/s are zero for Lagouira. This shows that the ERA5 model does not predict strong winds. The annual averages of wind speed and wind potential estimated by the ERA5 model compared to those obtained from the measurements are underestimated by 16% and 47%, respectively.

The annual wind speed averages for Tangier and Lagouira are comparable, 5.90 m/s and 6.07 m/s respectively, double that of Rabat (2.90 m/s). This shows that the north and south of Morocco are windier than the center, a transition zone of calm winds.

The monthly averages of wind speed for Tangier fluctuate around the annual average. But, for Dakhla, they are above the annual average for summer. In addition, the influence of the sea/land breeze may be due to a greater warming of the atmospheric air. However, Tangier is a special case. In effect, the Strait of Gibraltar connecting the Atlantic Ocean and the Mediterranean Sea forces the winds to be mainly from the dominant directions: east/west. There is also a certain acceleration and consequently an increase in wind speed.

The estimated absolute maximum wind speed for Lagouira varies between 10.5 m/s and 14 m/s. But those for Dakhla and Tangier, obtained from the measurements, vary from month to month and can reach 25 m/s and 30 m/s, respectively.

Apart from Lagouira, the wind (diurnal variation) is strong during the day and reaches its maximum in the afternoon and its minimum around midnight. These are due to the effect of the sea/land breeze.

Unlike Tangier, the periodicity of the seasonal variation is respected for Dakhla and Lagouira. The monthly averages for Dakhla vary between 4 m/s and 17 m/s, but for Lagouira, they vary between 4 m/s and 8 m/s. This shows that those obtained from the measurements are more accurate than the estimated ones. The interannual variation Ii for Tangier is around 0.20, twice that of Rabat (Ii = 0.10). This shows that winds for Tangier are irregular.

For the 0 m/s class, we see that the frequencies are around 12%, 2%, and 0% for Tangier, Dakhla, and Lagouira, respectively. For Tangier, the Weibull law adapts better to the observed frequencies, but for Lagouira, it is the Weibull hybrid law. The scale factors C for Tangier and Lagouira are higher ($C = 7.14$ m/s), which shows that these two sites are the windiest. The winds are widely dispersed for Tangier because the form factor (K = 3.64) for Lagouira is almost double. A result is confirmed by the large value of the standard deviation ($\sigma = 4.3$).

The annual average of wind power for Tangier is 366 W/m^2, which is almost 12 times greater than that of Rabat (31 W/m^2). For Dakhla and Lagouira, the annual averages of wind speed and wind potential, which correspond to 10 years of measurements, are more representative. Those of Lagouira obtained from the ERA5 model for 1985–1989 are higher than those for Dakhla, which shows that Lagouira is the windiest. But that of Dakhla obtained from the measured wind speed (307 W/m^2) is almost double that estimated by the ERA5 model (162 W/m^2). Therefore, the estimated wind potential is undervalued by 47%. Since Lagouira is windier than Dakhla, we can argue that the wind potential for Lagouira is underestimated by at least a percentage greater than this.

References

Johnson, G. L. (1985). *Wind energy systems*. Prentice-Hall, Inc.

Knidiri, F., & Laaouina, A. (1986). *L'énergie éolienne au Maroc*. le Centre Développement des Energies Renouvelables (CDER), Marrakech.

Lysen, E. H. (1983). *Introduction to wind energy, CWD 82-1* (2nd ed.). Amerfoorst.

Manwell, J. F., Mcgowan, J. G., & Rogers, A. L. (2009). *Wind energy explained, Theory, design and application*. Wiley.

Nfaoui, H. (2004). *Caractéristiques du gisement éolien marocain et optimisation d'un système aérogénérateur/groupe électrogène pour l'électrification des villages isolés* (Thèse de Doctorat d'Etat). Faculté des Sciences de Rabat.

Nfaoui, H. (2020). Wind Energy in Morocco: Resources, potential, and progress. In *The age of wind energy* (pp. 219–282) https://www.springer.com/gp/book/9783030264451

Nfaoui, H., & Sayigh, A. (2017). Influence of the period of measurements on wind potential assessment for a given site. In A. Sayigh (Ed.), *Renewable energy and a cleaner environment, innovative renewable energy* (pp. 311–325). Springer.

Nfaoui, H., Buret, J., & Sayigh, A. (1995). Wind characteristics and wind energy potential in Morocco. *Solar Energy, 63*(1), 51–60.

Nfaoui, H., Abbou, A., Sayigh, A., & Elmalki, A. (2022). *Prospects for wind energy in Southern Morocco and Northern Mauritania*. Mediterranean architecture and the green-digital transition, Springer.

Chapter 17
Renewable Energy for Climate Protection and Energy Security: Lessons Learned from the European Green Deal and REPowerEU

Rainer Hinrichs-Rahlwes

17.1 Introduction

The *global climate crisis* is aggravating. Extreme weather events—extreme temperatures, more and stronger tropical storms, deadly floods, deadly droughts, more and larger wildfires, failed harvests and starving people—are becoming much more frequent with growing impact on the lives of thousands and millions of people. Among serious scientists, there is no doubt that global warming is accelerating more than most of them expected, and faster than in the last decades, resulting in more devastating weather events threatening more people in more places across the globe. Accelerated warming results in earlier than expecting kicking in of irreversible tipping points which is leading to faster warming and resulting in faster melting of glaciers—in the world's mountainous areas across the continents, and particularly in the Artic and Antarctic regions, leaving more and bigger parts ice-free during longer periods of the year. Figure 17.1 shows clearly how global temperature anomalies increased rapidly in the last few decades, reaching new records in the last 9 years, all of which were hotter than any other year so far. The temperature in 2022 was 0.89 °C above the average of 1951–1980. And 2023 is poised to add to this record temperature increase. July 2023 has already been confirmed as globally the hottest month ever recorded, and June, July and August were the hottest 3-month period on record.

The need for decisive, accelerated and unambiguous climate action is more evident than ever, and the targets of the Paris Climate Agreement (United Nations, 2015) seem to be increasingly difficult to achieve. In its 2023 Synthesis Report (Intergovernmental Panel on Climate Change, 2023), the International Panel on Climate Change (IPCC) sent the clear message that more needs to be done to limit global warming to just tolerable temperature levels. It finds that "mitigating climate

R. Hinrichs-Rahlwes (✉)
European Renewable Energies Federation (EREF), Brussels, Belgium

German Renewable Energy Federation (BEE), Berlin, Germany

A. Sayigh (ed.), *Transition Towards a Carbon Free Future*, Innovative Renewable Energy, https://doi.org/10.1007/978-3-031-61660-0_17

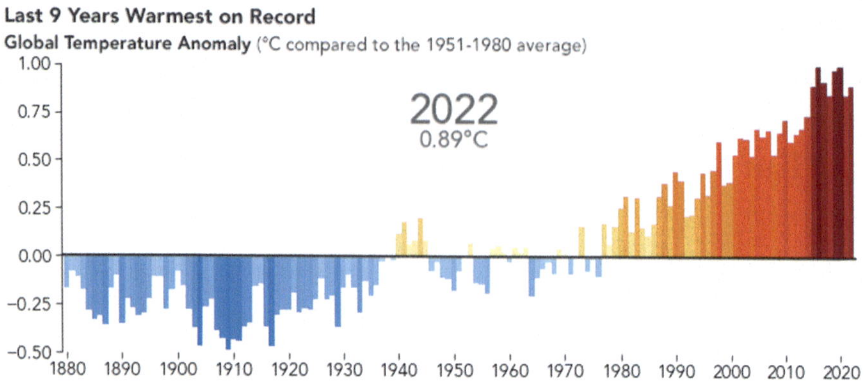

Fig. 17.1 Warmest years on record (NASA, 2022)

change to no more than 1.5 °C above pre-industrial levels by the end of this century will require complete elimination of all anthropogenic greenhouse gas emissions by 2050, or even earlier". And "rapid and deep, and in most cases, immediate greenhouse gas emissions reductions in all sectors in this decade" are needed to keep the chance of limiting the accelerating climate crisis. The United Nations Secretary General Antonio Guterres used more urgent language: "Climate change is here. It is terrifying. And it is just the beginning. It is still possible to limit global temperature rise to 1.5 °C [above pre-industrial levels] and avoid the very worst of climate change. But only with dramatic, immediate climate action" (Guterres, 2023).

The urgency of rapid and decisive climate action requires accelerated deployment of renewable energy across the globe and for all end-uses. Given the urgency of avoiding greenhouse gas (GHG) emissions, as well as recent cost developments of renewable energy, particularly in the power sector, make them the number one choice for all new capacity. According to a recent IRENA-Report (IRENA, 2022), the global weighted levelised costs of energy (LCOE) of utility-scale solar PV decreased by 3% year on year from 2021 to 2022, for onshore wind 5%, for bioenergy 13% and for geothermal energy 22%. Carbon removal technologies and nuclear are much more expensive and—in addition to being risky and immature—they need much longer permitting and construction times.

Accelerating GHG reductions and achieving GHG neutrality by 2050 requires a comprehensive set of legislative, regulatory and strategic action. Proposing, agreeing on and eventually enacting and implementing such decisive action is the main objective of the *European Green Deal* (European Commission, 2019), which was introduced by the European Commission (COM) in the beginning of the present COM's term in 2019. The various strategies and pieces of legislation needed to be discussed and agreed between the European Parliament (EP) and the Member States (represented by the European Council (EC)). Most of this was already finalised when this paper was written. The procedures on the remaining dossiers need to be concluded before the next elections for the European Parliament in June 2024. Chap. 2 will elaborate on this in more detail.

In early 2022, Russia's invasion of Ukraine added another major crisis to be addressed and tackled immediately. Given that Russia was a major supplier of energy, in particular oil and gas, the European Union (EU) and its Member States (MS) had to rapidly find ways to secure the bloc's energy supply at affordable costs while at the same time developing and agreeing on effective sanctions against Russia's war.

According to official statistics as shown in the *REPowerEU* Communication (European Commission, 2022a), more than 45% of the EU's fossil gas supply was imported from Russia, with some Member States importing more than 70% of their demand from Russia and others nearly nothing. Eighty-five percent of the EU's oil also came from Russia (Eurostat, 2023). This significant dependency from Russian oil and gas required to rapidly diversify supply and to accelerate the shift towards renewable energy. To tackle this challenge, COM presented the REPowerEU Communication (European Commission, 2022a) in March and the more detailed REPowerEU plan (European Commission, 2022b) in May. Chap. 3 addresses important elements in more detail.

17.2 The European Green Deal

The European Green Deal was presented as a lighthouse project of the newly appointed European Commission. The main elements were published (European Commission, 2019) in December 2019, only 10 days after they had officially taken office. The ambition was high: "The European Green Deal provides a roadmap with policies and measures to deliver the transformative change we need across all sectors". Figure 17.2 provides an overview of the main objectives of "the new growth strategy", which was also intended to be the basis for the EU's contribution to the United Nations' Sustainable Development Goals and to achieving the objectives of the Paris Agreement.

The Green Deal is designed to deliver on a broad range of targets while at the same time leaving no one behind (*Just Transition*) and striving for global leadership in the energy transformation. The most important elements in terms of fighting climate change and achieving a clean energy supply and demand are aiming at defining and agreeing on a higher than before GHG reduction target for 2030 and legally enshrining the objective of GHG neutrality by 2050. All these objectives should be reached by mobilising industry and citizens, by supplying clean, affordable and secure energy and increasing energy efficiency. The strategy also includes a zero-pollution ambition, a biodiversity and a farm-to-fork strategy.

The first major piece of legislation of the Green Deal was a *European Climate Law,* which legally enshrined the EU's objective of becoming climate-neutral by 2050. It set the legally binding target to reduce GHG emissions by 55% by 2030. A draft proposal was presented by COM in March 2020 (European Commission, 2020) and finally adopted in early 2021 (Official Journal of the European Union, Regulation (EU), 2021).

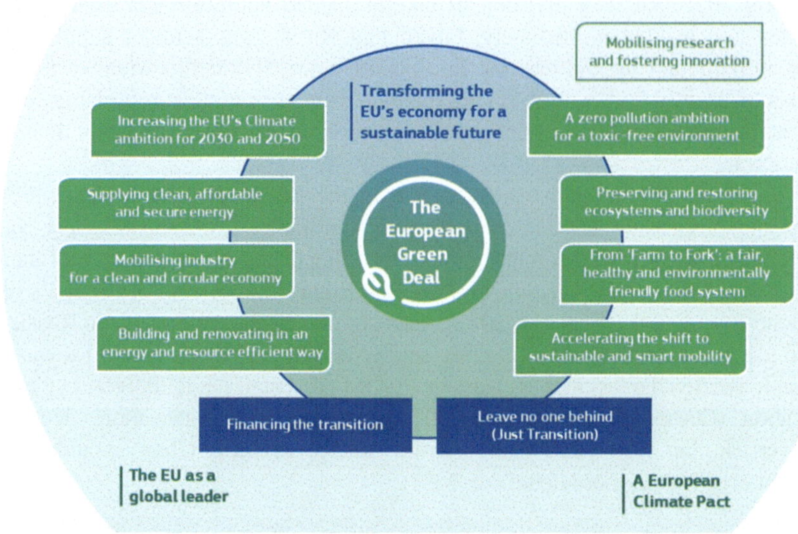

Fig. 17.2 The European Green Deal (European Commission, 2019)

The detailed objectives of the Green Deal shall be achieved by a number of strategies and legislative element. There is a *New Industrial Strategy for Europe,* a *Renovation Wave* and a strategy for *Smart Sector Integration* and for *Sustainable and Smart Mobility.* There is a strategy to protect *Biodiversity*, and there are strategies for *Offshore Renewable Energy,* for *Hydrogen* and for *Sustainable Finance.* And there is a *European Climate Pact*—to mention just some of the key building blocks.

The core of the Green Deal is a huge legislative package, branded as *Fit-for-55* to highlight the bloc's new GHG reduction target for 2030. This package was presented by COM in July 2021. Figure 17.3 shows the main elements of this package, which was presented by the European Commission in July 2021 (European Commission, 2021). The individual proposals for directives, regulations and other elements had then to discuss and eventually agree between EP and MS.

Fit-for-55 consists of proposals for a revised *Renewable Energy Directive,* a revised *Energy Efficiency Directive,* a revised *Emissions Trading Directive,* a new *Carbon Border Adjustment Mechanism* and some more. This paper concentrates on the main dossier for renewable energy—the revised Renewable Energies Directive (RED) (European Commission 2021 (2), 2021).

Like all the elements of the Fit-for-55 package, the revised RED was designed to deliver on a higher than before renewable energy target for 2030 in order to achieve the 55% GHG reduction target. COM therefore suggested to increase the renewable target for 2030 to at least 40%. An indicative target of 49% renewables in buildings was introduced as well as an increased target of 2.1 ppt of annual increase of renewables in district heating and cooling. Provisions were suggested to mainstream renewable energy in industry by increasing the share by 1.1 ppt/a. To accelerate

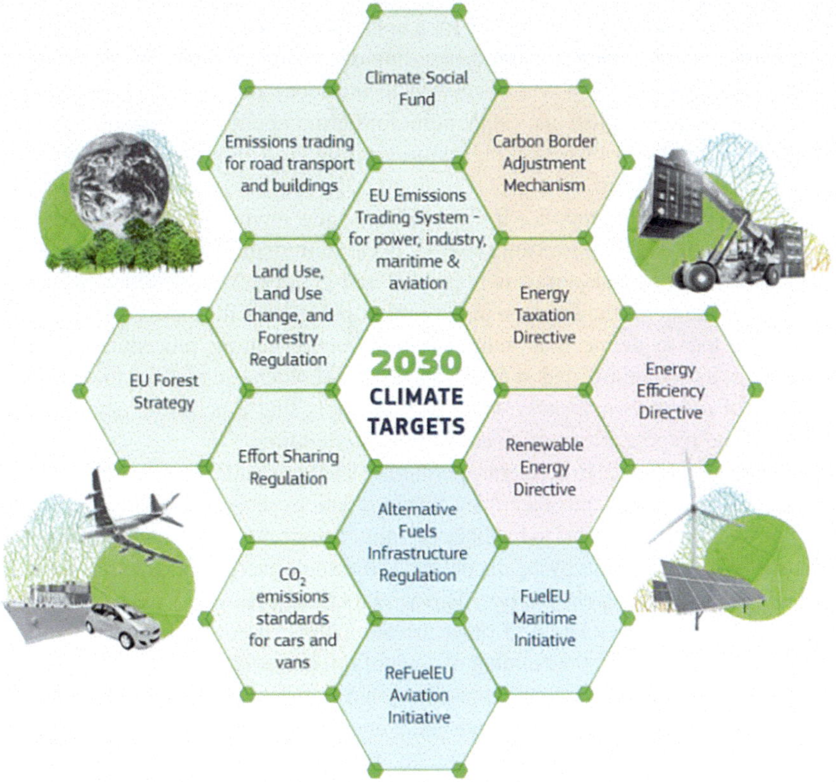

Fig. 17.3 Fit-for-55 (European Commission, 2021)

renewable energy-driven electrification, it was suggested to strengthen support for power purchase agreements (PPA). MS should be obliged to develop cross-border projects to enhance cooperation in the single market. Joint offshore energy planning should be facilitated in both Baltic and North Sea. In addition, the proposal aims at accelerating the switch from fossil fuel to renewable sources for heating systems, inter alia by adjusting qualification needs for installers. Furthermore, transmission and distribution system operators (TSOs and DSOs) should be obliged to make available information about renewable shares and GHG content of the power supply, battery capacity and smart charging capabilities and particularly not to discriminate against small and mobile storage systems.

The COM proposal was widely welcomed, but the discussions among stakeholders, in the committees of the European Parliament and among the Member States also revealed controversies that had to be solved to reach an agreement between the institutions. Renewable energy industry associations and environmental NGOs asked for a significantly more ambitious 2030 renewable target of up to 56%. A broadly supported campaign demanded an EU-wide target of at least 45% renewable energy in gross final energy consumption (GFEC) by 2030. In addition, various stakeholders asked for national binding targets for each Member State to facilitate and secure actual target achievement.

Another key topic in the discussion was the need for acceleration of permitting procedures, which takes up to 7 years for a wind farm in some countries and which is often extensively bureaucratic and demanding due to a variety of contact points for one permit. This is why it was suggested to introduce *one-stop shops*—requiring Member States to establish one single point for permit applications and also to allow for digital submission of permitting requests, instead of thousands of pages to be submitted in paper form.

It was suggested to renounce from mandatory new environmental impact assessments (EIA) for most repowering projects, because in-depth EIA are already undertaken when the original permit is requested and granted. This would also include setting time limits for the duration of permitting procedures for these projects. It was also suggested to define maximum duration for permitting procedures for new projects to avoid unjustified delays. Some stakeholders suggested to apply the principle of *affirmative silence*, which would imply that a permit is automatically granted if the responsible body does not react in due time.

Another key proposal to accelerate permitting and to provide the necessary areas for new wind or solar farms and other renewable installations such as anaerobic digesters or hydropower plants suggested to facilitate permitting and prioritising environmental and sustainability requirements by legally defining renewable energy projects as being in the *overring public and security interest.* This would create a relevant legal tool for many court cases when the violation of other private, public or environmental interests is brought forwards against the building of a windfarm or a solar park.

Another increasingly urgent discussion point is the need for strengthening and growing qualification schemes for installers and maintenance capacities for renewable energy installations, for example, by retraining workers of the incumbent energy sector. The need for more and more qualified workforce is becoming relevant in most EU MS and needs to be addressed in order not to delay the transformation.

The discussions and the informal political debates were ongoing, and the EP and EC committees were discussing compromise solutions, when Russia invaded Ukraine in early 2022—which dramatically highlighted the risks of energy import dependencies, in this case the strong dependence from Russia's oil and gas resources (and also nuclear fuels, but this was not in the focus of early reactions). Politicians and governments across Europe (and elsewhere) became aware of the risks of no longer available imports and/or significant price increases. Some agreements and votes including the final agreement on the revised RED were postponed to integrate higher ambition and stronger measures, which were proposed as part of the answer to Russia's aggression.

17.3 REPowerEU

Decisive and immediate action was required—on the side of the EU MS, but also at the EU level, particularly because union-wide energy policies and the single market require joint efforts and a coordinated response to mitigate potential energy shortages and looming price crises. It was obvious that the proposed targets and measures of the Fit-for-55 package were not sufficient to meet the new challenges.

On 8 March 2022, the COM presented a draft strategy to deal with the repercussions of the war on the EU's energy policies and security of supply—*REPowerEU: Joint European Action for more affordable, secure and sustainable energy* (European Commission, 2022a). It outlined a strategy to accelerate the diversification of energy supply—to a large extent by accelerating renewable energy deployment. Given that nearly half of the EU's fossil gas consumption and also high shares of the coal, oil, and uranium resources were imported from Russia, COM proposed measures to mitigate these risks as soon as possible. The document suggested—inter alia—frontloading some measures of the *Fit-for-55 package* and at the same time increasing some of the ambition levels and targets. It also suggested removing administrative barriers, developing a solar industry strategy and diversifying gas supplies including towards biogas and biomethane. Accelerated permitting procedures for renewable energy projects were temporarily established in an *emergency regulation (*European Council, 2022*)*, which was formally adopted on 19 December 2022, before most of these emergency measures will be evaluated and eventually become permanent rules in the revised (RED), which still had to be formally adopted when this paper was written.

In the REPowerEU strategy, the "Commission calls on Member States to ensure that the planning, construction and operation of plants for the production of energy from renewable sources, their connection to the grid and the related grid itself are considered as being in the overriding public interest and in the interest of public safety and qualify for the most favourable procedure available in their planning and permitting procedures". This is a strong warning no longer to delay the rapid transformation towards domestically available and increasingly cheap renewable energy.

A more detailed *REPowerEU plan* (European Commission 2022 (3), 2022) together with a *Staff Working Document* (European Commission 2022 (4), 2022) was presented on 18 May 2022. These documents further elaborated on key short- and medium-term measures suggested to mitigate the looming energy crisis and supply shortages. COM proposed—as an amendment to the RED (European Commission 2022 (5), 2022), to further increase the 2030 target for the share of renewables in GFEC to at least 45%, five percentage points above the Fit-for-55 level and—as an amendment to the Energy Efficiency Directive—to increase the energy savings target to 13% instead of nine.

Another key proposal suggested to significantly accelerate permitting procedures for renewable energy projects (European Commission 2022 (6), 2022). This proposal is already part of the *Emergency Regulation* (European Council, 2022) and also of the still to be formally decided revised RED. It obliges Member States to establish "go-to areas" (which are now labelled renewable energy acceleration areas in the most recent version of the revised RED)—areas in which the maximum time for permit granting decisions was limited and in which EIA for individual projects would widely be substituted by EIA for the whole area. This proposal also repeated the suggestion to legally enshrine renewable energy projects in the EU as being of overriding public and security interest.

A main element of moving away from fossil gas-driven heating and cooling of buildings is the target of *tripling heat pump* deployment, which translates into installing ten million new heat pumps *in the next 5 years.*

The REPowerEU plan (European Commission 2022 (3), 2022) also includes a *Biomethane Action Plan* aiming at producing at least 35 billion cubic metres of biomethane in the EU until 2030. In addition, 10,000,000 tons of *domestically produced green hydrogen* should be available by 2030 plus the same amount of imported green hydrogen.

Significantly higher shares of renewable energy and in particular renewable electricity are highlighted as major building blocks of a sustainable and secure energy supply. Solar PV capacity is poised to rapidly grow and become a key component of the EU's power supply. An ambitious *Solar Energy Strategy* (European Commission, 2022c), which includes launching a European *Solar PV Industry Alliance* and avoiding new dependencies by ensuring *supply chain resilience* and diversification, is an important part of REPowerEU. An *EU large-scale skills partnership* is initiated to grow the available qualified workforce needed for the accelerated energy transformation and also to make sure that cheap solar energy is available for everybody including people with low incomes. The objective of the Solar Energy Strategy is accelerated deployment of PV capacity in Europe—aiming at 320 GW of installed capacity by 2025 and 600 GW by 2030, which would require 45 GW of new capacity each year.

Shorter and simpler permitting procedures shall support the accelerated and increased installation of new PV capacity. A *Solar Rooftop Initiative* is proposed to speed up widespread rooftop PV installations. It is suggested to legally limit the maximum duration of permitting for rooftop PV to 3 months and to make the process simpler, particularly focussing on online applications instead of paperwork.

Installing rooftop PV is suggested to become mandatory for all buildings with roofs of more than 250 sqm—in the first step for all *new* public and commercial buildings by *2026* and for *all* other public and commercial buildings by *2027*. For new residential buildings, the obligation would kick-in by 2029. All new buildings should be built as *solar ready*.

In addition, special attention should be given to innovative solar PV solutions, such as building integrated PV (BIPV), or new and more efficient module designs. Floating PV and agrivoltaics are also suggested to receive special support by MS policies and in terms of private and public investment decisions.

17.4 Status of the Green Deal and REPowerEU

Legislative procedures on most of the legislative proposals for implementing the European Green Deal as laid out in the European Climate Law and the Fit-for-55 package and amended by the REPowerEU plan were concluded or are already published in the *Official Journal of the EU (OJ)* when this paper was written so that they are formally entering into force.

As of 31 July 2023, eight of the directives and regulations of the Fit-for-55 package had been published in the *OJ*: Energy Efficiency Directive; Emissions Trading for Road Transport and Buildings; Emissions Trading for Power, Industry,

Maritime and Aviation; Emissions Standards for Cars and Vans; Effort Sharing Regulation; Carbon Border Adjustment Mechanism; Land Use, Land Use Change and Forestry; and the Climate Social Fund. The revision of the Energy Taxation Directive is and will remain pending because the necessary unanimity for this decision cannot be reached. A provisional agreement, which most likely only needs to be formally adopted by EP and EC, was reached on the ReFuelEU Initiative, the FuelEU Maritime Initiative, and the Alternative Fuels Infrastructure Regulation. These dossiers will enter into force once they have been formally adopted and published. This will have to be finalised by early 2024. Otherwise, the legislative process would have to restart after the Elections to the European Parliament to be held in June 2024 and the resulting appointment of a new European Commission.

The need for final approval by EP and EC also applies to the provisional agreement that was reached in July 2023 on the revised RED. The agreement—which is scheduled to be adopted by EP on 13 September 2023 and foreseen by EC for October 2023—will be a major step forwards towards net zero emissions as required by the European Climate Law. The revised RED legally enshrines a 2030 target of 45% (consisting of 42.5% to be implemented by MS and additional 2.5% to be achieved through a still to be installed EU mechanism). The revised directive also enshrines renewable energy acceleration areas and other rules for accelerating permitting procedures for renewable energy projects. When this paper was finalised, the formal adoption of the revised RED and subsequent publication in the *OJ* was still pending, although it was considered very likely to happen before the end of the year.

17.5 Outlook

This climate crisis and global warming are accelerating. Already today, in September 2023, it is obvious that 2023 will be the hottest year on record again—by far. The impact of more and longer heat waves was felt across the globe, just as extreme droughts and devastating storms and wildfires were more frequent and more detrimental than before. The need for immediate and urgent climate action is obvious. Renewable energy—together with energy efficiency, energy savings and phasing out fossil fuels—is key to reduce emissions to zero as soon as possible. The transformation towards renewable energies is already happening—in many countries worldwide. But it is not happening fast enough—due to strong and continuous lobbying of short-sighted incumbents of the fossil and nuclear-driven energy world and due to decision makers in parliaments and governments who dare not take the necessary decisions to phase out unsustainable energy and to accelerate the transformation towards renewables.

Despite the lobbyists of the incumbent polluting energy system and their supporting politicians, significant progress is already under way: The drastic cost reductions of renewable energy technologies have made renewables the dominant sources of new power capacities in the last decade. For more than 10 years, more than half of global annual new power capacities were renewable energy installations—with solar PV alone providing 50% of new capacity in 2021 (43% in 2020). Wind power added 25% (29% in 2020) and hydropower contributed another 7% (8% in 2020). Fossil capacities, particularly coal and gas, made up 4% and 11% in 2021 (9% and 8% in 2020) (Bloomberg NEF, 2022). New nuclear was irrelevant in the last decade.

Renewables also attracted the major share of global investment in new power capacity. Out of 640 billion USD invested in 2022, 74% were for renewables and 11% were spent for oil and gas. Coal and nuclear had a share of 8% each (REN21, 2023a). These data show that in the power sector, the transformation to renewables and respective shift of investment is already happening and becoming stronger, although this tendency is regionally very different and even inexistent in some countries (but this would be the content of another paper). It is also true that the shift towards renewables in other end-uses, such as heating, cooling, transport, and industry, is significantly lagging behind. In 2022 nearly two thirds (953 billion USD) of global investment in the energy sector went to fossil fuels and related infrastructure (IRENA and CPI, 2023).

There is light, and there is shadow. Renewables are mature and available everywhere, and they are the lowest cost options in most markets and for most end-uses. A mix of renewable energy sources is available nearly everywhere. Shifting energy supply and demand towards renewables significantly reduces cost and supply security risks resulting from energy import dependencies—as people in Europe and several other countries could see clearly when Russia invaded Ukraine and was no longer available as a reliable exporter of oil and gas, coal and uranium. This is why many countries around the globe immediately began diversifying their energy supply by trying to buy elsewhere and to save energy wherever possible and to accelerate the deployment of domestic renewables to make supply chains more resilient and at the same time comply with the requirements of GHG reduction.

The EU Green Deal—as detailed by the Fit-for-55 package and enhanced by the REPowerEU plan—is well suited to accelerate the diversification of energy sources and reduce import dependencies to a major extent by speeding up the deployment of renewable energy and increasing energy efficiency in Europe.

If the EU successfully finalises the legislation to implement the Green Deal, and if it succeeds in reaching or even overachieving the climate and energy targets, this could serve an example to help accelerate the global energy transformation.

According to the European Climate Law, the EU has just started—in July 2023—to set out new climate and energy targets aiming at the next decade up to 2040, the decisive timespan to set up and implement policies to achieve zero emissions and climate neutrality. But already this year, given the undeniable acceleration of global warming, the EU should update and increase the bloc's climate and energy targets for 2030. This would also provide a strong signal to the international community for COP28, the next global climate conference to be held in Dubai, UAE, in November/December this year. The *EU should aim at minus 65% GHG reduction by 2030 and at minus 90–95% by 2040*. This would allow for reaching climate neutrality well before 2050, by 2045 or earlier—which would be necessary in terms of a fair global effort sharing. Such increased ambition of the EU could serve as a lever for other countries, particularly from the Global North, to step up their own ambitions and commitments. And it could help countries of the Global South to avoid some of the mistakes and deviations of the old industrialised world.

Around the globe, cities, regions and countries, but also enterprises and all kinds of businesses, are striving for and committing to 100% renewable energy in their supply and demand. This highlights a deeper understanding of the need to fully phase out subsidies for unsustainable and polluting energies, such as coal, oil and gas, and also nuclear power, which is not compatible with a flexible, modern, renewable-based energy system, and the waste of which will have to be safely stored for hundreds of thousands of year. *Aiming at 100% renewable energy*, often starting with the power sector, but reaching out to heating, cooling, transport and industrial application is the clearest signal that the age of fossil fuels and GHG emissions needs to be over—*the sooner the better.*

A global agreement to phase fossil fuels and related direct, indirect and hidden subsidies would be very much needed to accelerate this decision. Unfortunately, as of now, neither the G20, nor the G7, nor any of the 27 COPs so far could agree on this target. Instead, the latest G20 summit agreed on a "New Delhi Leaders' Declaration" (G20 New Delhi Leaders' Declaration, 2023), in which they emphasised their commitment to "*pursue and encourage efforts to triple renewable energy capacity globally* through existing targets and policies, as well as demonstrate similar ambition with respect to other zero and low-emission technologies, including abatement and removal technologies, in line with national circumstances by 2030". This is still far away from what would be needed to keep the window open for limiting global warming to a maximum of 1.5 °C above pre-industrial levels, but it would at least translate into just *over 11,000 GW of installed renewable energy capacities by 2030* as suggested by a recent IRENA report (IRENA, 2023).

It seems that demanding an agreement on a *Global Renewable Energy Target* is gaining traction in the run-up towards COP28 in Dubai. A newly founded Global Renewables Alliance (Global Renewables Alliance, 2023) together with the international multi-stakeholder network REN21 (REN21, 2023b) has launched a global campaign requesting that COP28 should agree on a Global Renewables Target of at least tripling the installed capacity to 11,000 GW by 2030, which would imply about 1000 GW of new renewable power capacity each year.

Agreeing on a Global Renewable Energy Target at COP28 would be an important breakthrough and a meaningful step to keep the window open for achieving the objectives of the Paris Agreement. But much more ambition, commitment and actual implementation will be needed: *phasing out polluting energies and related subsidies* is the logical next step—and a global agreement on this could certainly accelerate implementation. The ultimate objective—and the only unambiguous one—would be to *strive for 100% renewable energy for all end-uses, everywhere*. Technically and economically, this could be achieved within the next two decades. Real commitment beyond lip services and timely decisions are needed to make it happen—for the sake of keeping a liveable planet for future generations.

References

Bloomberg NEF. (2022). *Power transition trends*, 21 Sept 2022, https://assets.bbhub.io/professional/sites/24/BNEF-Power-Transition-Trends-2022_FINAL.pdf.

European Commission. (2019). *Communication from the Commission to the European Parliament, the European Council, the Council, European Economic and Social Committee and the Committee of the Regions*. The European Green Deal. 11 December 2019, https://eur-lex.europa.eu/resource.html?uri=cellar:b828d165-1c22-11ea-8c1f-01aa75ed71a1.0002.02/DOC_1&format=PDF

European Commission. (2020). *Proposal for a Regulation of the European Parliament and of the Council establishing the framework for achieving climate neutrality and amending Regulation (EU) 2018/1999* (European Climate Law), COM (2020) 80 final, 4 March 2020.

European Commission. (2021). Communication from the Commission to the European Parliament, the European Council, the Council, the European Economic and Social Committee and the Committee of Regions, 'Fit for 55': delivering the EU's 2030 Climate Target on the way to climate neutrality, 14 July 2021, https://eur-lex.europa.eu/legal-content/EN/TXT/PDF/?uri=CELEX:52021DC0550.

European Commission. (2022a). *Communication from the Commission to the European Parliament, the European Council, the Council, the European Economic and Social Committee and the Committee of the Regions*. REPowerEU: Joint European Action for More Affordable, Secure and Sustainable Energy, COM. 108 final, Strasbourg 8 March 2022, https://eur-lex.europa.eu/resource.html?uri=cellar:71767319-9f0a-11ec-83e1-01aa75ed71a1.0001.02/DOC_1&format=PDF

European Commission. (2022b). *Communication from the Commission to the European Parliament, the European Council, the Council, the European Economic and Social Committee and the Committee of Regions*. REPowerEU Plan. 18 May 2022, https://eur-lex.europa.eu/resource.html?uri=cellar:fc930f14-d7ae-11ec-a95f-01aa75ed71a1.0001.02/DOC_1&format=PDF

European Commission. (2022c). *Communication from the Commission to the European Parliament, the European Council, the Council, the European Economic and Social Committee and the Committee of the Regions*. EU Solar Energy Strategy, COM. 221 final, Brussels 18 May 2022, https://eur-lex.europa.eu/resource.html?uri=cellar:516a902d-d7a0-11ec-a95f-01aa75ed71a1.0001.02/DOC_1&format=PDF

European Commission 2021 (2), *Proposal for a Directive of the European Parliament and of the Council amending Directive (EU) 2018/2001 of the European Parliament and of the Council, Regulation (EU) 2018/1999 of the European Parliament and of the Council and Directive 98/70/EC* of the European Parliament and of the Council as regards the promotion of energy from renewable sources, and repealing Council Directive (EU) 2015/652, Brussels 14 July 2021., https://eur-lex.europa.eu/resource.html?uri=cellar:dbb7eb9c-e575-11eb-a1a5-01aa75ed71a1.0001.02/DOC_1&format=PDF

European Commission 2022 (3), *Communication from the Commission to the European Parliament, the European Council, the Council, the European Economic and Social Committee and the Committee of the Regions*. REPowerEU Plan, COM (2022) 230 final, Brussels 18 May 2022., https://eur-lex.europa.eu/resource.html?uri=cellar:fc930f14-d7ae-11ec-a95f-01aa75ed71a1.0001.02/DOC_1&format=PDF

European Commission 2022 (4), *Commission Staff Working Document accompanying the document Communication from the Commission to the European Parliament, the European Council, the Council, the European Economic and Social Committee and the Committee of the Regions*. REPowerEU Plan, SWD (2022) 230 final, Brussels 18 May 2022, https://eur-lex.europa.eu/legal-content/EN/TXT/?uri=SWD%3A2022%3A230%3AFIN&qid=1653033922121

European Commission 2022 (5), *Proposal for a Directive of the European Parliament and of the Council amending Directive (EU) 2018/2001 on the promotion of the use of energy from renewable sources, Directive 2010/31/EU on the energy performance of buildings and*

Directive 2012/27/EU on energy efficiency, COM (2022) 222 final, Brussels 18 May 2022., https://eur-lex.europa.eu/legal-content/EN/TXT/PDF/?uri=CELEX:52022PC0222&from=EN

European Commission 2022 (6), *Commission Recommendations of 18.5.2022 on speeding up permit-granting procedures for renewable energy projects and facilitating Power Purchase Agreements*, C (2022) 3219 final, Brussels 18 May 2022, https://eur-lex.europa.eu/legal-content/EN/TXT/PDF/?uri=PI_COM:C(2022)3219&from=EN

European Council. (2022). *Council regulation (EU) 2022/2577 of 22 December 2022 laying down a framework to accelerate the deployment of renewable energy*, 29 Dec 2022, https://eur-lex.europa.eu/legal-content/EN/TXT/PDF/?uri=CELEX:32022R2577

Eurostat, *EU's energy import dependency decreased in 2020—Products Eurostat News—Eurostat (europa.eu)*, last seen 2 Sept 2023.

G20 New Delhi Leaders' Declaration, New Delhi, India 9–10 Sept 2023., https://www.g20.org/content/dam/gtwenty/gtwenty_new/document/G20-New-Delhi-Leaders-Declaration.pdf

Global Renewables Alliance, website (last seen 11 Sept 2023), https://globalrenewablesalliance.org/

Guterres, A., 'Era of global boiling has arrived,' says UN chief as July set to be hottest month on record | Climate science | *The Guardian*, July 2023, last seen on 8 Sept 2023.

Intergovernmental Panel on Climate Change. (2023). *Synthesis report.* Climate Change 2023, https://www.ipcc.ch/report/sixth-assessment-report-cycle/, last seen on 8 Sept 2023.

IRENA, *Renewable power generation costs in 2022*, Abu Dhabi 2022., https://www.irena.org/Publications/2023/Aug/Renewable-Power-Generation-Costs-in-2022

IRENA. (2023). *World Energy Transitions Outlook 2023*: 1.5 °C Pathway, Volume 1, International Renewable Energy Agency, https://mc-cd8320d4-36a1-40ac-83cc-3389-cdn-endpoint.azureedge.net/-/media/Files/IRENA/Agency/Publication/2023/Jun/IRENA_World_energy_transitions_outlook_v1_2023.pdf?rev=cc4522ff897a4e26a47906447c74bca6

IRENA and CPI. (2023). *Global landscape of renewable energy finance*. International Renewable Energy Agency. https://mc-cd8320d4-36a1-40ac-83cc-3389-cdn-endpoint.azureedge.net/-/media/Files/IRENA/Agency/Publication/2023/Feb/IRENA_CPI_Global_RE_finance_2023.pdf?rev=8668440314f34e588647d3994d94a785

NASA, *Earth observatory* 2022., https://earthobservatory.nasa.gov/world-of-change/decadaltemp.php, last seen on 2 Sept 2023.

Official Journal of the European Union, Regulation (EU) 2021/1119 of the European Parliament and of the Council of 30 June 2021 establishing the framework for achieving climate neutrality and amending Regulations (EC) No 401/2009 and (EU) 2018/1999 ('European Climate Law') (2021), OJ L 243/1, 09 Sept 2021, https://eur-lex.europa.eu/legal-content/EN/TXT/PDF/?uri=CELEX:32021R1119&from=FR

REN21, *Renewables 2023 Global Status Report collection, Renewables in Energy Supply*, REN21 Secretariat, 2023a., https://www.ren21.net/wp-content/uploads/2019/05/GSR-2023_Energy-Supply-Module.pdf

REN21—*Renewables now*, website (last seen 11 Sept 2023b), https://www.ren21.net/

United Nations. (2015). *The Paris Agreement.* https://unfccc.int/files/essential_background/convention/application/pdf/english_paris_agreement.pdf, last seen on 8 Sept 2023

Chapter 18
Energy-Efficient Design for Care Home Buildings: A Case Study

Linyuan Ouyang, Shan Zhou, Emmanuel Essah, and Runming Yao

18.1 Introduction

Aging population has become an increasing challenge for the world (Hughes et al., 2019). According to the prediction of the United Nations Department of Economic and Social Affairs, the percentage of the elderly (more than 60 years old) in China will increase to 30.4% and 38.9% by 2035 and 2050, respectively (United Nations, 2022). The aging society has become a major challenge for the economic growth and social care because of the lack of support for the aged due to their inability to work and sustain themselves (Hughes et al., 2019). Nowadays, Chinese older people do not always live with their children, and their children have less time to take care of them (World Health Organization, 2016). In this situation, the elderly tend to live in nursing homes or care homes as substitute options where professional care and assistance are offered (Rijnaard et al., 2016). As of current global trends, the International Energy Agency (IEA) has reported that residential buildings consume a huge amount of energy accounting for 22% of global final energy (IEA, 2019). Building energy consumption has become topical, and the significant rise of residential building energy consumption needs serious attention. What is more, care homes, which are in context also residential, provide service and shelter to the

L. Ouyang · S. Zhou
Joint International Research Laboratory of Green Buildings and Built Environments (Ministry of Education), Chongqing University, Chongqing, China

E. Essah
School of the Built Environment, University of Reading, Reading, UK

R. Yao (✉)
Joint International Research Laboratory of Green Buildings and Built Environments (Ministry of Education), Chongqing University, Chongqing, China

School of the Built Environment, University of Reading, Reading, UK
e-mail: r.yao@cqu.edu.cn; r.yao@reading.ac.uk

A. Sayigh (ed.), *Transition Towards a Carbon Free Future*, Innovative Renewable Energy, https://doi.org/10.1007/978-3-031-61660-0_18

elderly, and the elderly spend a lot of time indoors (Van Hoof et al., 2017), thereby impacting on the energy use that in recent years has also increased.

Many studies (Brounen et al., 2012; Huang, 2015; Harold et al., 2015; Liao & Chang, 2002; Yamasaki & Tominaga, 1997; Pachauri, 2004) have focused on energy consumption of elderly building energy consumption. In a study in the Netherlands (Brounen et al., 2012), energy consumption of the families with the elderly is higher than other family types, and natural gas consumption increased with age for winter heating. Besides, the apartment dwelling type is the most energy conserving type compared with other types of houses due to the smaller exposed surface of apartments. Another research in Taiwan showed that households with more elderly members consumed more electricity (Huang, 2015). The same results are found in studies of Ireland (Harold et al., 2015), the USA (Liao & Chang, 2002), Japan (Yamasaki & Tominaga, 1997), and India (Pachauri, 2004). Therefore, It has been widely acknowledged that residential building energy consumption has a significant correlation with age.

It is universally considered that buildings for the elderly consume higher energy to meet their living demand. Thus, there is a potential for energy saving in care homes. However, present studies of care home energy consumption are still in its infancy. As a result, there is a need for research into energy consumption of care home with different occupancy and operation scenarios considering the research project that is located in Chengdu, China. This study aims to explore the energy consumption of care home, the local weather of Chengdu city, and whether the solar photovoltaic system in Chengdu can meet the energy needs of care/nursing homes.

18.2 Methodology

The method used in this research is based on numerical simulation. Based on the research gaps identified in the introduction, the research gaps were identified. To fill in the gap and achieve corresponding research aims, this section discusses the simulation research methods and simulation scenarios.

18.2.1 Case Study Building

To begin with, a model was obtained from existing simulation research developed in China investigating care homes. The care home is located in Pidu District, Chengdu, with a total area of 7897.7 m^2 and a building height of 10.8 m. The building is T-shaped, about 114 m long from east to west, and about 50 m wide from south to north. The care home has 3 storeys, of which the ground floor is designed into a part of a hospital and the first and second floors are a care home. The model (Fig. 18.1) was developed using the IES VE software (Table 18.1).

Fig. 18.1 The model established by IES VE

Table 18.1 Information of ground floor layout

Ground floor	Total area	Open space	Medical room	Auxiliary room (stairwell, toilet)
Area (m^2)	2634.5	1754.1	699.3	181.1

18.2.2 Energy Consumption Model by IES VE Software

To represent the layout and structure of the building under real conditions, IES VE was used to create the model. Three energy cases of the building were modeled energy consumption comparison in different scenarios.

(a) Base case

The building was partitioned into different room functions. The bedrooms were designed to accommodate between one and eight people. Also, the large windows were substituted by smaller windows with a typical size of 2.4 m × 1.8 m and a base height of 0.9 m.

A VRF HVAC system was introduced within the settings in IES VE software. The system type was a fan coil system, and both heating and cooling sources were electric air source heat pumps. The overall COPs for heating and cooling were also 1.9 and 2.3, respectively. There were no auxiliary ventilation system or domestic hot water system. As a result, in this base case, the whole building was assigned with HVAC system to reflect real thermal demand.

In the bedrooms, occupancy and equipment profiles were developed to be the same. On every weekend and holiday, the profile modulating values were always 100%. Cooling set point and heating set point were 26 and 20 °C, respectively, and the cooling and heating periods were from 31 May to 30 September and 30 November to 28 February, respectively. During the corresponding periods, the HVAC system was always on. Lighting was on from 6 pm to 12 pm every day. In comparison, office templates have some differences. Occupancy, lighting and equipment modulating value are high from 7 am to 7 pm only on weekdays. With the same heating and cooling points and periods, HVAC system is on only on weekdays assuming staff are only in the office on weekdays.

(b) Scenario 1: Modified base case

This scenario was a modification of the base case to reflect reality. The changes are as follows. For office, changed the duty rooms and main medical rooms to 24/7 occupancy and usage, so the occupancy and electrical appliance internal gains were set to constant values. Lighting was on at night from 6 pm to 8 am, and HVAC set points were constant: 26 and 20 °C for cooling and heating, respectively. For effective analyses, 18 rooms with total area of 251.14 m^2 were selected, and the profiles were set to be 24/7. For bedroom, changed the uniform occupant density (35 m^2/person) to real bed number for every room. Changing everyday occupancy rate and electrical facility usage is more practical. In the daytime (from 9 am to 8 pm), the modulating values were set from 0 in base case to 0.9 in this case.

(c) Scenario 2: Modified base case plus natural ventilation

In the as-built case, occupants and managers have to always open the windows for natural ventilation and fresh air. In addition, opening doors is even more frequent when the rooms are in use. In this regard, Microflo application was used to simulate the effect of natural ventilation because of operation of openings. For all the doors, it was assumed the maximum infiltration flow was 0.25 air changes per hour (ACH). The door openings were designed to simulate real room use and operation. For example, the bedroom door is set to 0.1 (from a possible value of 1) from 8 am to 9 pm. The canteen sets a value of 0.3 at mealtime. Similarly, the opening rate of windows needs to be adjusted accordingly. In summer and autumn, bedroom windows need to be set to 0. 3 from 7 am to 07:30. The windows of other rooms are also changed to 0.3.

18.2.3 Solar Energy Model by PVSyst Software

In order to investigate the energy potential of solar energy in the Chengdu area, PVSyst, a design software for PV systems, was used to simulate PV modules on the roof under scenario 2. The electricity production in this case is simulated. The array uses a fixed plane, an inclination angle of 16 degrees, a positive southward direction, a case roof area of 1500 m^2, a graphene photovoltaic array, and a lithium battery for energy storage. The user voltage is set to 205 V, the constant heat loss of the system is set to 20 W/m^2K, the line loss is set to 1.50%, and the voltage drop of the diode is set to 0.7 V.

18.3 Results and Discussion

18.3.1 Solar SunCast

Because building thermal performances are driven by real weather data (IES, 2020), building solar thermal gain was calculated by SunCast application of IES software. The SunCast application enables visualization of solar intensity of the external solar thermal gain on the building facade. The solar energy results are shown in Fig. 18.2.

As can be seen in Fig. 18.2, the roof of the building has the highest simulated annual solar gain with 884.9 kWh/m^2. The south facade annual solar energy is about 500 kWh/m^2 with blue color, while the north façades are lower than 400 kWh/m^2. This is because Chengdu is in the northern hemisphere and the sun is usually south. In this case, solar gains of south, east, and west facades are all between 400 and 500 kWh/m^2.

18.3.2 Energy Consumption

Annual total energy consumption is the sum of heating and cooling energy consumption, which reflects the building thermal performance and energy demand in the perspective of the whole year. Figure 18.3 shows total electricity intensity results of the three cases. The benchmark result regulates the total heating and cooling electricity consumption for green buildings. Furthermore, the results of the three cases are all higher than the benchmark value of 39.5 kWh/m^2. Especially, the result of scenario 2, which is the closest to reality, is twice the value of the national benchmark. One of the reasons is the heating set point is lower than the thermal requirements of the standard, and in winter they may prefer a higher indoor environment (Tartarini et al., 2017). In this regard, energy demand per unit area is higher in care homes than common residential buildings. Furthermore, the heating design for the elderly should be 22–24 °C in design standard (MOHURD, 2018). If this temperature range is adopted, the heating energy consumption would be even higher than the results in this study. As a result, green residential building energy

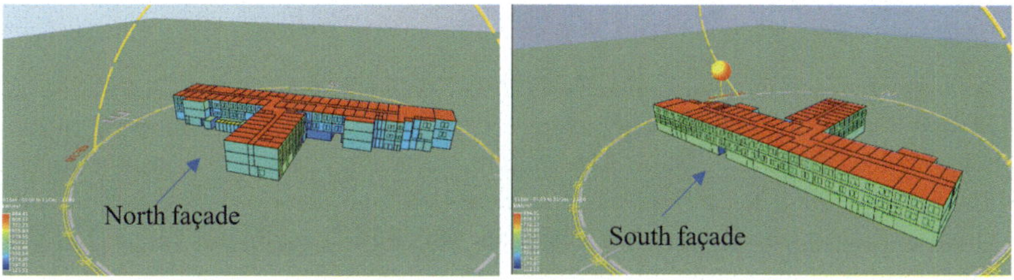

Fig. 18.2 Building façade solar energy results

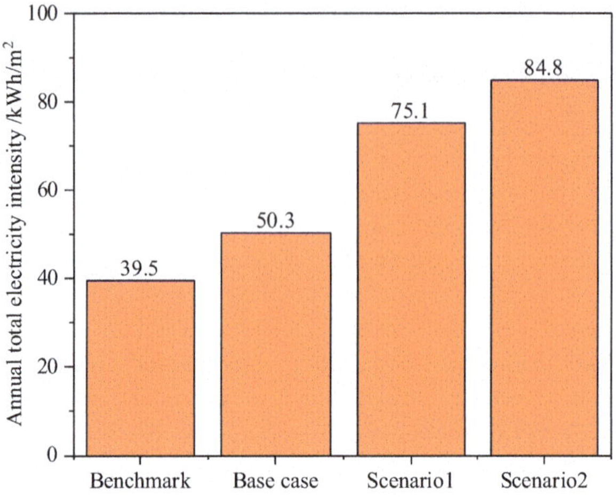

Fig. 18.3 Annual total electricity intensity results of benchmark, IES base case, scenario 1, and scenario 2

consumption standard may not be suitable for care homes, because the care home energy consumption is intrinsically higher than common residential buildings.

There is another factor that can influence energy consumption with regard to indoor environment: air exchange settings. In the three cases, different ventilation strategies are used (e.g., infiltration, natural ventilation, and no air exchange). It can be seen that the total energy intensity of scenario 2 is 12.9% higher than that of scenario 1. However, these methods are simplified and are used as a demonstration of air exchange method to calculate building energy consumption. As a result, all these assumptions cannot reflect reality. One of the reasons is that occupants can adapt to indoor environments actively by opening doors or opening windows. These activities can be observed because of complex adaptive methods. Accordingly, the simulation results may not be accurate enough. The second reason is the lack of indoor air quality monitoring. Care home rooms have high occupancy intensity. Insufficient fresh air can reduce energy consumption but may conversely be harmful to the health. Without indoor air quality monitoring, there is a risk of poor indoor air quality, which requires further research.

18.3.3 Solar Energy

The simulated annual solar capacity is 200 MWh, the unit power generation is 824 k Wh/kWp, the solar energy accounts for 28.44%, and the efficiency of the photovoltaic system is 78.36%. The monthly available electricity consumption is shown in Fig. 18.4. It can be seen that there is a high level of solar energy from March to August, mainly because Chengdu, which is in southwest China, has the highest sunshine rate during this period.

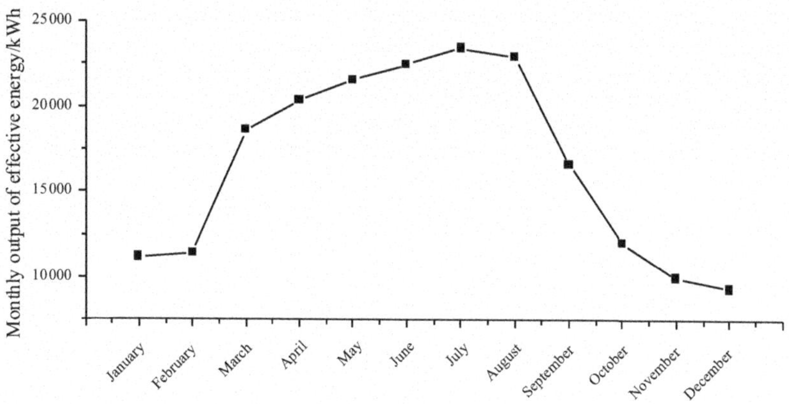

Fig. 18.4 Monthly effective energy obtained by the photovoltaic system

18.4 Conclusions

In order to reduce energy consumption in a care home of Chengdu, China, a simulation study is conducted. IES VE software is used to simulate energy consumption results. PVSyst is used to simulate the collection of solar energy. Based on the case information, the base case model is established by IES VE software, and then scenario 1 and scenario 2 are developed. Besides, the solar photovoltaic power generation system of this case is simulated by PVsyst software. Modified base case (scenario 1) is more detailed than the base case with room functions and respective profiles, while ventilated modified base case (scenario 2) can best reflect the reality with window and door operations. Care home energy consumption is much higher than national green residential building energy benchmark. The scenario 1 case considering the actual occupancy rate consumes more energy than the base case, and the ventilation model scenario 2 developed on the basis of scenario 1 has the highest energy consumption. The use of the solar photovoltaic system in Chengdu can meet 28.4% of the energy needs of buildings. This study demonstrated the feasibility of combining PV to reduce energy consumption and provided reference for energy-saving design and operation of care home buildings..

References

Brounen, D., Kok, N., & Quigley, J. M. (2012). Residential energy use and conservation: Economics and demographics. *European Economic Review, 56*(5), 931–945.

Harold, J., Lyons, S., & Cullinan, J. (2015). The determinants of residential gas demand in Ireland. *Energy Economics, 51*, 475–483.

Huang, W.-H. (2015). The determinants of household electricity consumption in Taiwan: Evidence from quantile regression. *Energy, 87*, 120–133.

Hughes, C., Natarajan, S., Liu, C., Chung, W. J., & Herrera, M. (2019). Winter thermal comfort and health in the elderly. *Energy Policy, 134*, 110954.

IEA. (2019). *Global Status Report for Buildings and Construction 2019*. IEA.

IES. (2020). *VE Application ApacheSim*. https://www.iesve.com/software/virtual-environment/applications/whole-building-energy-simulation/apachesim. Accessed 1 Aug 2020

MOHURD. (2018). *JGJ 450–2018: Standard for design of care facilities for the aged*. China Architecture Publishing and Media, Ltd.. [in Chinese].

Pachauri, S. (2004). An analysis of cross-sectional variations in total household energy requirements in India using micro survey data. *Energy Policy, 32*(15), 1723–1735.

Rijnaard, M. D., van Hoof, J., Janssen, B. M., Verbeek, H., Pocornie, W., Eijkelenboom, A., Beerens, H. C., Molony, S. L., & Wouters, E. J. M. (2016). The factors influencing the sense of home in nursing homes: a systematic review from the perspective of residents. *Journal of Aging Research, 2016*, 6143645.

Tartarini, F., Cooper, P., & Fleming, R. (2017). Thermal environment and thermal sensations of occupants of nursing homes: a field study. *Procedia Engineering, 180*, 373–382.

Chapter 19
Numerical Modelling of a Novel PV/T Collector Incorporated with PCM and CFM as a Cooling Method Compared with the Conventional PV/T Collector

Mojtaba Dayer and Kamaruzzaman Sopian

19.1 Introduction

The PV/T collectors are classified based on two most important heat exchanging features, the working fluid and collector configurations. While some collectors use water, air or a combination of both, the most recent one may include nanofluids. The collectors may have various configurations ranging from a tube to a sheet or heat pipe, etc. (Daghigh et al., 2011; Farshchimonfared et al., 2016; Meng et al., 2016; Yazdanifard et al., 2016). However, the PV/T collectors may be classified differently when glazing as cell cover has shown to exert tremendous overall effects (Tripanagnostopoulos et al., 2005). The most common type of PV/T collector uses air to reduce complexity and initial cost. However, implementing liquids, such as water, as working fluids improves the thermal capacitance of systems by yielding better cooling (Hegazy, 2000). Kazanci et al. investigated the use of PCM in a photovoltaic/thermal (PV/T) module for a plus-energy house to sustain its heating and cooling. The results obtained showed that the PV/T efficiency was around 42.8%, whereas that of the conventional PV was 13.59% (Kazanci et al., 2014). Qiu et al. theoretically investigated a new PV/T design that employed a microencapsulated phase change material slurry (MPCM), for its energy performance. Their results revealed that the maximum achieved electrical efficiency was 8.7%, while the maximum achieved thermal efficiency was 59% (Qiu et al., 2015).

M. Dayer
Solar Energy Research Institute (SERI), Universiti Kebangsaan Malaysia, Bangi, Selangor, Malaysia

K. Sopian (✉)
Department of Mechanical Engineering, Universiti Teknologi PETRONAS, Seri Iskandar, Perak Darul Ridzuan, Malaysia
e-mail: ksopian@utp.edu.my

A. Sayigh (ed.), *Transition Towards a Carbon Free Future*, Innovative Renewable Energy, https://doi.org/10.1007/978-3-031-61660-0_19

Hassan et al. conducted an electrical and thermal energy efficiency analysis for a solid-liquid PCM integrated into PV panels in two different climates. Their results indicated that the average achieved efficiencies for Ireland and Pakistan are 14.6% and 20.3%, respectively (Hasan et al., 2014). Fiorentini et al. studied a new air-based photovoltaic/thermal (PV/T) collector with the addition of a PCM storage unit integrated with a reverse cycle heat pump for an HVAC application. The maximum generated electrical efficiency was 9%, while the maximum thermal efficiency was 45% (Fiorentini et al., 2015).

Many researches have attempted to improve the low thermal conductivity of PCMs as a main disadvantage. Tao et al. studied the heat transfer of a porous medium filled with phase change materials by implementing the lattice Boltzmann method (Tao et al., 2016). Utilizing the phase field method, Zhou and Zhao analysed the heat transfer of phase change material in a metal foam (Zhou & Zhao, 2011). Others presented a two-temperature model to evaluate the performance of a metallic foam, filled with PCM. They used Darcy-Brinkman equations to shed light on the effect of metal foam (Daghigh et al., 2011). Du and Ding set a numerical modelling on the heat transfer to improve charging and discharging of the phase change material. To this end, they used porous aluminium (0.8) and paraffin as the porous medium and phase change material, respectively (Du & Ding, 2016).

In this study, both water and metal-foam phase change material (PCM) have been incorporated to further enhance the total efficiency of the PV/T module. The study aimed to increase the overall performance of the photovoltaic/thermal (PV/T) in a bid to introduce a novel PV/T design with higher efficiency and better performance for the demanding market.

19.2 Methodology

19.2.1 Physical Model

The conceptual design of the proposed PV/T collector is provided in Fig. 19.1, showcasing a cross-sectional drawing that illustrates the different layers of the PV/T collector.

To understand the performance of the proposed system, it is important to compare it to a reference system. The systems that are employed are (i) conventional PV/T with PCM only including water flowing to extract the heat and (ii) novel proposed

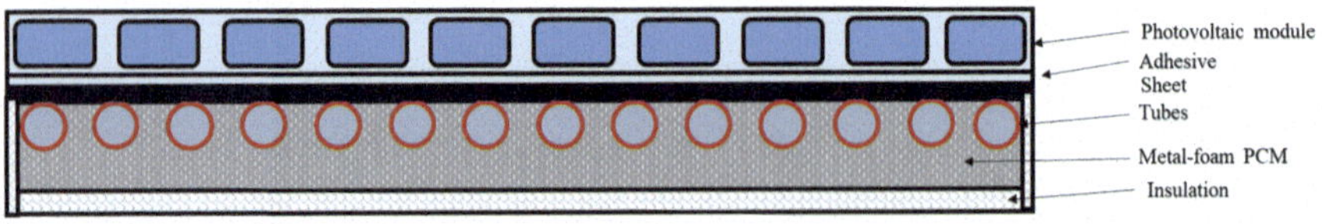

Fig. 19.1 Cross-sectional drawing of the metal-foam PCM and water fluid-based PV/T collector

system PCM embedded in copper foam with water flowing. Such comparison would reflect the performance of the collector when employing copper foam with PCM.

19.2.2 Assumptions

The problem has been solved under the assumption of local thermal non-equilibrium (LTNE) among copper foam and phase change material (PCM). PCM velocity solved with Brinkman-Forchheimer equation extended the Darcy equation under the presumption of incompressibility, Boussinesq assumption and neglecting micro-inertial effects. The flowing water is assumed to be laminar and incompressible. For the boundary condition, it is assumed that all walls present a velocity of equal to zero. The heat transfer input was applied perpendicular to the upper surface of the panel. In this study, the heat flux is split into two-parallel PCM and copper foam on the basis of porosity weight:

$$q_{\mathrm{fe}} = \varepsilon q_{\mathrm{in}} \tag{19.1}$$

$$q_{\mathrm{se}} = (1 - \varepsilon) q_{\mathrm{in}} \tag{19.2}$$

Incorporating this assumption is a reliable cause effective thermal conductivity of two phases, which are more or less of the same order of magnitude. As a result, the heat flux can be partitioned based on porosity and boundary condition scaling, as previously demonstrated by Hwang et al. (1995).

19.2.3 Mathematical Model

For heat transfer analysis, the applied governing equations for the water fluid, encompassing mass, momentum and energy are Eqs. (19.3)–(19.5):

$$\frac{\partial \rho_{\mathrm{w}}}{\partial t} + \nabla.(\rho_{\mathrm{w}} V_{\mathrm{w}}) = 0 \tag{19.3}$$

$$\rho_{\mathrm{w}} \left[\frac{\partial V_{\mathrm{w}}}{\partial t} + (\partial V_{\mathrm{w}}.\nabla) V_{\mathrm{w}} \right] = \nabla P + \nabla.(\mu_{\mathrm{w}} \nabla V_{\mathrm{w}}) \tag{19.4}$$

$$\rho_{\mathrm{w}} C_{\mathrm{p,w}} \frac{\partial T_{\mathrm{w}}}{\partial t} + \rho_{\mathrm{w}} C_{\mathrm{p,w}} V_{\mathrm{w}}.\nabla T_{\mathrm{w}} = \nabla.(k_{\mathrm{w}} \nabla T_{\mathrm{w}}) \tag{19.5}$$

where the subscript f denotes heat transfer in fluid and P, V, ρ and μ are the fluid pressure, velocity vector, density and dynamic viscosity, respectively. In this 3D modelling with coordinates (x, y, z) and unit vectors of axes $\{\vec{e_x}, \vec{e_y}, \vec{e_z}\}$, ∇ is defined as:

$$\nabla = \vec{e_x}\frac{\partial}{\partial x} + \vec{e_y}\frac{\partial}{\partial y} + \vec{e_z}\frac{\partial}{\partial z} = \left(\frac{\partial}{\partial x}, \frac{\partial}{\partial y}, \frac{\partial}{\partial z}\right) \tag{19.6}$$

However, there are two approaches to solve heat transfer between copper foam and PCM, namely, local thermal equilibrium and non-equilibrium; in this case, the latter is implemented. Local thermal non-equilibrium (LTNE) is a more accurate method to simulate the heat transfer since thermal conductivity difference among two regions is significant. The continuity, momentum and energy equations for the PCM embedded in the copper foam domain are derived as mentioned below:

$$\frac{\partial \rho_\mathrm{f}}{\partial t} + \frac{\partial(\rho_\mathrm{f} u_\mathrm{f})}{\partial x} + \frac{\partial(\rho_\mathrm{f} v_\mathrm{f})}{\partial y} + \frac{\partial(\rho_\mathrm{f} w_\mathrm{f})}{\partial z} = 0 \tag{19.7}$$

$$\frac{\rho_\mathrm{f}}{\varepsilon}\frac{\partial u_\mathrm{f}}{\partial t} + \frac{\rho_\mathrm{f}}{\varepsilon^2}\left(\vec{V}\vec{\nabla}\right).u_\mathrm{f} = -\frac{\partial P_\mathrm{f}}{\partial x} + \frac{\mu_\mathrm{f}}{\varepsilon}\nabla^2 u_\mathrm{f} - \left(\frac{\mu_\mathrm{f}}{K} + \frac{\rho_\mathrm{f} C_\mathrm{p,f}}{\sqrt{K}}|u_\mathrm{f}|\right)u_\mathrm{f} + S_u \tag{19.8}$$

$$\frac{\rho_\mathrm{f}}{\varepsilon}\frac{\partial v_\mathrm{f}}{\partial t} + \frac{\rho_\mathrm{f}}{\varepsilon^2}\left(\vec{V}\vec{\nabla}\right).v_\mathrm{f} = -\frac{\partial P_\mathrm{f}}{\partial y} + \frac{\mu_\mathrm{f}}{\varepsilon}\nabla^2 v_\mathrm{f} - \left(\frac{\mu_\mathrm{f}}{K} + \frac{\rho_\mathrm{f} C_\mathrm{p,f}}{\sqrt{K}}|v_\mathrm{f}|\right)v_\mathrm{f} + S_v \tag{19.9}$$

$$\frac{\rho_\mathrm{f}}{\varepsilon}\frac{\partial w_\mathrm{f}}{\partial t} + \frac{\rho_\mathrm{f}}{\varepsilon^2}\left(\vec{V}\vec{\nabla}\right).w_\mathrm{f} = -\frac{\partial P_\mathrm{f}}{\partial z} + \frac{\mu_\mathrm{f}}{\varepsilon}\nabla^2 w_\mathrm{f} - \left(\frac{\mu_\mathrm{f}}{K} + \frac{\rho_\mathrm{f} C_\mathrm{p,f}}{\sqrt{K}}|w_\mathrm{f}|\right)w_\mathrm{f}$$
$$+ S_w + \rho_\mathrm{f} g\beta[T_\mathrm{f} - (T_\mathrm{m} - \Delta T_\mathrm{m})] \tag{19.10}$$

$$\rho_\mathrm{f} c_\mathrm{p,f}\left(\varepsilon\frac{\partial T_\mathrm{f}}{\partial t} + U.\nabla T_\mathrm{f}\right) = k_\mathrm{fe}\nabla^2 T_\mathrm{f} + h_\mathrm{sf} a_\mathrm{sf}(T_\mathrm{s} - T_\mathrm{f}) \tag{19.11}$$

$$(1-\varepsilon)\rho_\mathrm{s} c_\mathrm{p,s}\frac{\partial T_\mathrm{s}}{\partial t} = k_\mathrm{se}\nabla^2 T_\mathrm{s} + h_\mathrm{sf} a_\mathrm{sf}(T_\mathrm{f} - T_\mathrm{s}) \tag{19.12}$$

The last term in the z-coordinate of the momentum equation is the expression of Forchheimer-extended Darcy model, which demonstrates non-Darcy and inertia effect incurred by copper foam. The physical parameters of the copper foam structure are expressed below:

$$K = \frac{\varepsilon^2}{36\chi(\chi-1)}d_\mathrm{k}^2 \tag{19.13}$$

$$\chi = 2 + 2\cos\left[\frac{4\pi}{3} + \frac{1}{3\cos(2\varepsilon-1)}\right] \tag{19.14}$$

$$d_\mathrm{k} = \frac{\chi d_\mathrm{p}}{(3-\chi)} \tag{19.15}$$

$$d_\mathrm{p} = \frac{22.4\times10^{-3}}{\omega} \tag{19.16}$$

$$C = 0.00212(1-\varepsilon)^{-0.132}\left(\frac{d_f}{d_p}\right)^{-1.63} \tag{19.17}$$

$$d_f = 1.18\sqrt{\frac{1-\varepsilon}{3\pi}}d_p \tag{19.18}$$

where K, χ, C, d_f, d_p, d_k, ε and ω are the permeability, tortuosity coefficient, inertial coefficient, fibre diameter, pore size, characteristic length, porosity and pores per inch of copper foam, respectively.

Many models have been introduced to investigate the effective thermal conductivity of copper foam and PCM. Nevertheless, in the present study, the tetrakaidecahedron cell model proposed by Boomsma and Poulikakos was adopted to describe the effective thermal conductivity (Boomsma & Poulikakos, 2011):

$$k_{fe} = k_{eff}|k_s = 0 \tag{19.19}$$

$$k_{se} = k_{eff}|k_f = 0 \tag{19.20}$$

$$k_{eff} = \frac{\sqrt{2}}{2(R_A + R_B + R_C + R_D)} \tag{19.21}$$

$$R_A = \frac{4d}{(2e^2\pi d(1-e))k_s + (4 - 2e^2 - \pi d(1-e))k_f} \tag{19.22}$$

$$R_B = \frac{(e-2d)^2}{(e-2d)e^2k_s + (2e - 4d - (e-2d)e^2)k_f} \tag{19.23}$$

$$R_C = \frac{\left(\sqrt{2} - 2e\right)^2}{2\pi d^2\left(1 - 2e\sqrt{2}\right)k_s + 2\left(\sqrt{2} - 2e - \pi d^2\left(1 - e\sqrt{2}\right)\right)k_f} \tag{19.24}$$

$$R_D = \frac{2e}{e^2k_s + (4 - e^2)k_f} \tag{19.25}$$

$$d = \sqrt{\frac{\sqrt{2}\left(2 - (5/8)e^3\sqrt{2} - 2\varepsilon\right)}{\pi\left(3 - 4e\sqrt{2} - e\right)}},\ e = 0.339 \tag{19.26}$$

The term β is employed to make sure that when PCM is solid, the velocity remains equal to the initial value which is zero. β and S are defined as below respectively:

$$\beta = \begin{cases} 0 & T < \left(T_m - \Delta T_m/2\right) \\ \dfrac{T - \left(T_m - \dfrac{\Delta T_m}{2}\right)}{\Delta T_m} & \left(T_m - \Delta T_m/2\right) \le T < \left(T_m + \Delta T_m/2\right) \\ 1 & T > \left(T_m + \Delta T_m/2\right) \end{cases} \tag{19.27}$$

$$S_u = A_{\text{mush}} \frac{(1-\beta)^2}{b_{\text{mush}} + \beta^3} u_f, S_v = A_{\text{mush}} \frac{(1-\beta)^2}{b_{\text{mush}} + \beta^3} v_f, S_w = A_{\text{mush}} \frac{(1-\beta)^2}{b_{\text{mush}} + \beta^3} w_f \quad (19.28)$$

The two terms A_{mush} which denotes geometrical constant and b_{mush} which is used to avoid division by zero are subtracted from the Carman-Kozeny equation. In this study, we assumed $A_{\text{mush}} = 10^4$, while $b_{\text{mush}} = 10^{-3}$.

The interfacial heat transfer coefficient h_{sf} and interfacial surface area a_{sf} are calculated based on Zhukauskas' correlations:

$$a_{sf} = \frac{3\pi d_f}{d_p^2} \quad (19.29)$$

$$h_{sf} = Nu \frac{k_f}{d_f} \quad (19.30)$$

The Nusselt number Nu can be estimated by the following correlation:

$$Nu = \begin{cases} 0.76\,Re^{0.4} Pr^{0.37}, & 0 < Re \le 40 \\ 0.52\,Re^{0.5} Pr^{0.37}, & 40 < Re \le 1000 \\ 0.26\,Re^{0.6} Pr^{0.37}, & 1000 < Re \le 2 \times 10^5 \end{cases} \quad (19.31)$$

Re and Pr, which are non-dimensional variables, are calculated as follows:

$$Re = \frac{u_m \rho_f d_k}{\mu_f}, u_m = \sqrt{u_f^2 + v_f^2 + w_f^2} \quad (19.32)$$

$$Pr = \frac{\upsilon_f}{\alpha_f}, \upsilon_f = \frac{\mu_f}{\rho_f} \quad (19.33)$$

19.3 Results

19.3.1 Phase Change Materials and Metal Foam

Phase change materials are used to retain thermal energy, allowing temperature stabilization in a PV/T collector. PCM experiences a change in their physical state during the melting and freezing cycle during which the energy could be stored and discharged later as "latent" heat. However, as the structure of paraffin as an organic hydrocarbon has a low thermal conductivity, a copper foam was used to retain the temperature gradients within the system. Table 19.1 illustrates the thermophysical

Table 19.1 Thermophysical properties of PCM and the copper foam matrix

Property	PCM	Copper foam matrix
Solid state density (ρ, kg/m^3)	930	8960
Liquid state density (ρ, kg/m^3)	830	–
Thermal conductivity (k, W/m°K)	0.21	400
Latent heat of fusion (L_f, kJ/kg)	196	–
Dynamic viscosity (μ, Pa.s)	7.1×10^{-3}	–
Thermal expansion coefficient (α, 1/°K)	7.78×10^{-4}	–
Solid state specific heat (c_p, J/kg°K)	2100	385
Liquid state specific heat (c_p, J/kg°K)	2100	–
Melting point temperature (T_m, °C)	49	–
Melting range (ΔT_m, °C)	2	–
Porosity (ε, 1)	–	90%
Pores per inch (PPI) (ω, 1)	–	40

properties and other specifications of the utilized PCMs including paraffin wax and copper foam.

19.3.2 Verification

We have verified our results of simulation, by comparing them with the results of an experimental study by Browne et al. (2016) and a simulation study by Mousavi et al. (2018). To this end, the mass flow rate, the phase change material and the porous metal foam of both studies were considered as the input variables for the present simulation. The effects of mass flow rates on thermal efficiency in both studies were compared with the present study in order to determine the accuracy of our numerical modelling.

19.3.3 Efficiency Analysis

For the purpose of evaluating the performance of the proposed PV/T collector, the thermal and electrical efficiencies are as below. The thermal efficiency was calculated based on the Duffie and Beckman equation as follows (Duffie & Beckman, 1980):

$$\eta_{th} = \frac{Q_u}{G_{STC} \times A_c} \tag{19.34}$$

where η_{th} is the thermal efficiency, G_{STC} is the solar irradiance at the standard test conditions, A_c is the area of the PV panel and Q_u is the efficient collected heat which can be extracted from the following equation:

$$Q_u = \dot{m} \times C_p \times (T_{out} - T_{in}) \tag{19.35}$$

Besides, the electrical efficiency of the PV/T collectors can be written as:

$$\eta_{el} = \eta_{el,n} \times [1 - \beta(T_{panel} - 25°)] \tag{19.36}$$

19.3.4 Exergy Analysis

With respect to the second law of thermodynamics, the maximum available work can be excluded from analysing the exergy of the system (Saidur et al., 2012). Therefore, the thermal exergy was described according to the following equation:

$$\dot{E}_{x,thermal} = \dot{m}_{fluid} \times Cp_{fluid} \times (T_{out} - T_{in}) \times \left(1 - \frac{T_0}{T_m}\right) \tag{19.37}$$

Furthermore, solar and electrical exergy are given as the following equations:

$$\dot{E}_{x,solar} = G_t \times A_p \times \left[1 - \frac{4}{3}\frac{T_0}{T_{sun}} + \frac{1}{3}\left(\frac{T_0}{T_{sun}}\right)^4\right] \tag{19.38}$$

$$\dot{E}_{x,electrical} = \eta_{pv} \times G_t \times A_p \tag{19.39}$$

As it was earlier mentioned, the second law efficiency can be obtained as below:

$$\eta_{II} = \frac{\dot{E}_{x,useful}}{\dot{E}_{x,solar}} = \frac{\dot{E}_{x,electrical} + \dot{E}_{x,thermal}}{\dot{E}_{x,solar}} \tag{19.40}$$

By replacing those exergy terms in the second law efficiency equation, the equation can be written as follows:

$$\eta_{II} = \frac{1}{\left[1 - \frac{4}{3}\frac{T_0}{T_{sun}} + \frac{1}{3}\left(\frac{T_0}{T_{sun}}\right)^4\right]}\left[\eta_{I,thermal}\left(1 - \frac{T_0}{T_m}\right) + \eta_{pv}\right] \tag{19.41}$$

The dead state conditions, for calculating the exergy terms, are considered as 298.15 and 100 kPa. The sun's temperature is considered to be 5774 K.

19.3.5 The Effect of Mass Flow Rate

The mass flow rate is an important determinant of the thermal efficiency of PV/T collectors. However, the optimum mass flow rate must be determined as both low and high rates results in reduced efficiency and increased energy cost. Figure 19.2 shows that as the mass flow rate increases, the thermal efficiency improves. Our results indicate that the optimum mass flow rate is about 0.01 kg s^{-1}. Figure 19.3 presents the effects of increasing mass flow rates of the electrical efficiency of the PV/T collector when PCM was both used alone and incorporated with porous media.

As shown, the incorporation of PCM reduces the PV/T temperature by acting as a thermal storage. During the phase change process, a fraction of solar radiation will be absorbed by melting PCM, hence dissipating the PV/T temperature. During the melting process, the heat transfer rate from the PV/T to the PCM is reduced because of the formation of a thermal-resistant interface. This is why implementation of a metal foam will overcome the thermal resistance by lending higher conductivity. While using PCM caused higher thermal efficiency of the PV/T collector, using PCM incorporated with a porous medium resulted in higher electrical efficiency as shown in Figs. 19.2 and 19.3.

Figure 19.4 indicates that as the average PV temperature increases, the electrical efficiency decreases drastically. In fact, reduction of PV/T temperature will increase the electrical efficiency of the system at 0.06% per 1 °C decrease in PV/T temperature.

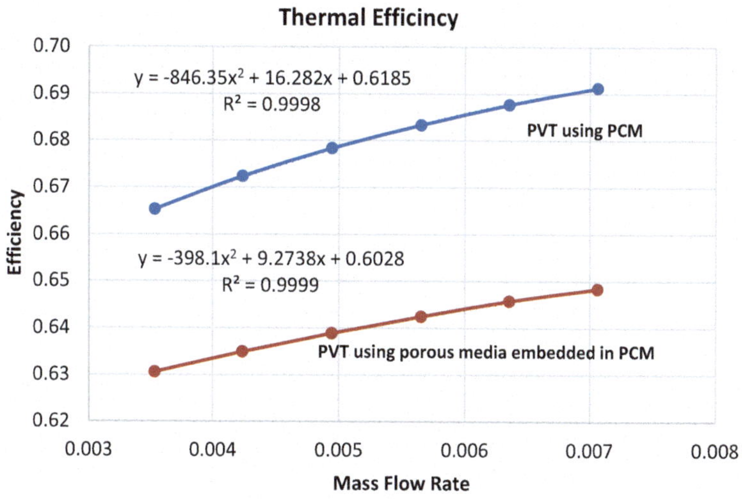

Fig. 19.2 Thermal efficiency of PV/T using PCM and PCM plus copper foam for various mass flow rates

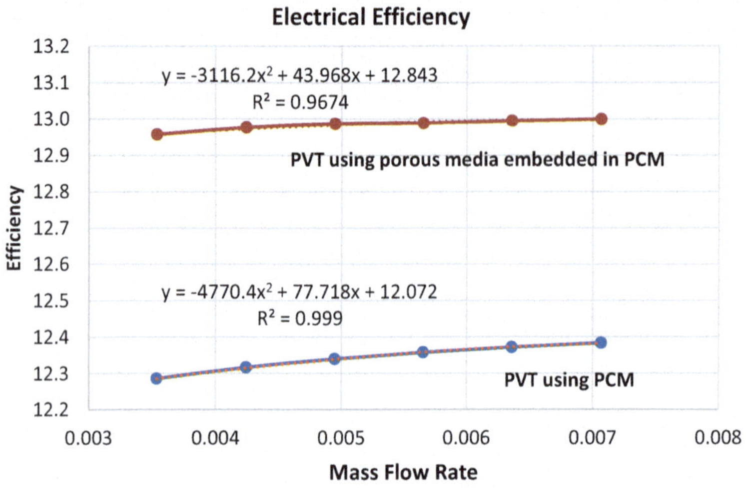

Fig. 19.3 Electrical efficiency of PV/T using PCM and PCM plus copper foam for various mass flow rates

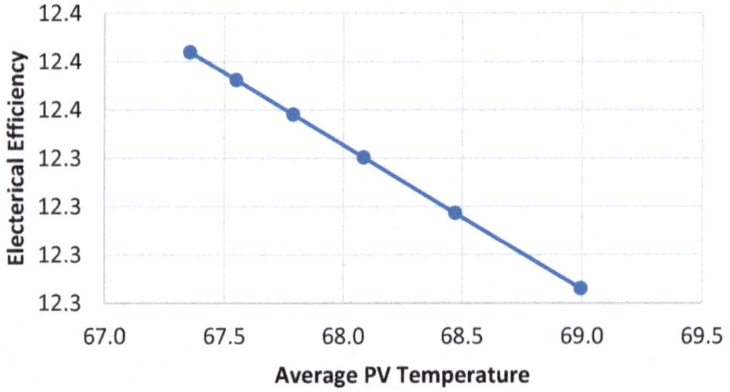

Fig. 19.4 Effects of average PV temperature on electrical efficiency

19.3.6 Exergy Efficiency Analysis

Figure 19.5 shows that the exergy of PV/T used with PCM is slightly lower than the PV/T using PCM plus porous copper foam. This means that the electrical exergy of the PCM plus CFM-incorporated PV/T has compensated the reduced thermal exergy in comparison to PCM-incorporated PV/T. Moreover, the exergy of the PV/T using porous media filled with PCM is considerably higher than PCM-incorporated PV/T with an exergy efficiency equal to 0.144 considering the dead state condition.

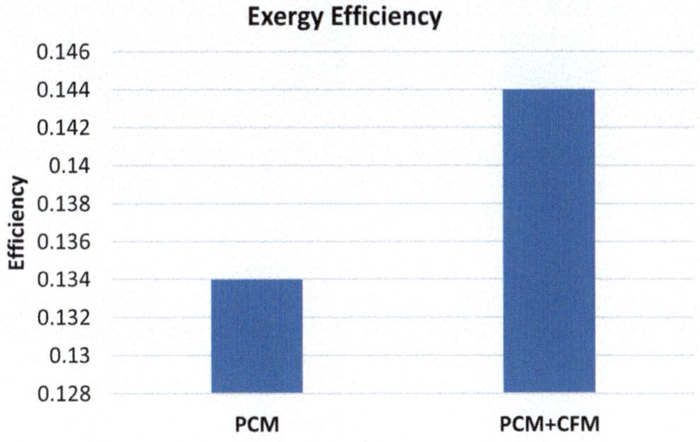

Fig. 19.5 Exergy efficiency influenced by implementation of stand-alone PCM or combined with CFM

19.4 Conclusions

In summary, a numerical CFD modelling was utilized to investigate the effects of incorporating phase change materials (PCMs) embedded in copper foam matrix (CFM) on the enhancement of overall efficiencies of a PV/T collector. Our results revealed that incorporation of CFM caused 12% increase in heat storage through phase changing in PCM due to improved heat conductivity. This means that heat stored in PCM may be utilized later at night when the sun is absent. Moreover, the total exergy efficiencies of both the newly designed PV/T and conventional systems were calculated, and the former showed higher efficiency than the latter. However, the numerical analysis may sometimes indicate a trivial higher thermal efficiency for the conventional system over the newly designed one. This could be due to the increment of temperature distribution as well as thermal energy stored in PCM. The most challenging factor for the newly proposed system is the high price of CFM. For this reason, further investigation of using porous media in solar energy would be an interesting topic for future work. Further research should be done towards improving the total efficiency of PV/T as well as discovering ways to reduce the cost of fabrication.

References

Boomsma, K., & Poulikakos, D. (2011). Corrigendum for the paper: K. Boomsma, D. Poulikakos, "On the effective thermal conductivity of a three-dimensionally structured fluid-saturated metal foam" [International Journal of Heat and Mass Transfer, 44 (2001) 827–836]. *International Journal of Heat and Mass Transfer, 54*(1), 746–748. https://doi.org/10.1016/j.ijheatmasstransfer.2010.08.023

Browne, M. C., Norton, B., & McCormack, S. J. (2016). Heat retention of a photovoltaic/thermal collector with PCM. *Solar Energy, 133*, 533–548. https://doi.org/10.1016/j.solener.2016.04.024

Daghigh, R., Ruslan, M. H., Zaharim, A., & Sopian, K. (2011). Monthly performance of a photovoltaic thermal (PV/T) water heating system. In *Recent researches in energy & environment*. WSEAS Press.

Du, Y., & Ding, Y. (2016). Towards improving charge/discharge rate of latent heat thermal energy storage (LHTES) by embedding metal foams in phase change materials (PCMs). *Chemical Engineering and Processing: Process Intensification, 108*, 181–188. https://doi.org/10.1016/j.cep.2016.08.003

Duffie, J. A., & Beckman, W. A. (1980). *Solar engineering of thermal processes*. Wiley. https://books.google.com.my/books?id=HudSAAAAMAAJ

Farshchimonfared, M., Bilbao, J. I., & Sproul, A. B. (2016). Full optimisation and sensitivity analysis of a photovoltaic–thermal (PV/T) air system linked to a typical residential building. *Solar Energy, 136*, 15–22. https://doi.org/10.1016/j.solener.2016.06.048

Fiorentini, M., Cooper, P., & Ma, Z. (2015). Development and optimization of an innovative HVAC system with integrated PVT and PCM thermal storage for a net-zero energy retrofitted house. *Energy and Buildings, 94*, 21–32. https://doi.org/10.1016/j.enbuild.2015.02.018

Hasan, A., McCormack, S. J., Huang, M. J., & Norton, B. (2014). Energy and cost saving of a photovoltaic-phase change materials (PV-PCM) system through temperature regulation and performance enhancement of photovoltaics. *Energies, 7*(3), 1318–1331. https://www.mdpi.com/1996-1073/7/3/1318

Hegazy, A. (2000). Comparative study of performance of four photovoltaic/thermal solar air collectors. *Energy Conversion and Management, 41*, 861–881. https://doi.org/10.1016/S0196-8904(99)00136-3

Hwang, G., Wu, C., & Chao, C.-H. (1995). Investigation of non-Darcian forced convection in an asymmetrically heated sintered porous channel. *Journal of Heat Transfer, 117*, 725–732. https://doi.org/10.1115/1.2822636

Kazanci, O. B., Skrupskelis, M., Sevela, P., Pavlov, G. K., & Olesen, B. W. (2014). Sustainable heating, cooling and ventilation of a plus-energy house via photovoltaic/thermal panels. *Energy and Buildings, 83*, 122–129. https://doi.org/10.1016/j.enbuild.2013.12.064

Meng, X.-L., Xia, X.-L., Sun, C., Li, Y., & Li, X.-L. (2016). A novel free-form Cassegrain concentrator for PV/T combining utilization. *Solar Energy, 135*, 864–873. https://doi.org/10.1016/j.solener.2016.06.034

Mousavi, S., Kasaeian, A., Shafii, M. B., & Jahangir, M. H. (2018). Numerical investigation of the effects of a copper foam filled with phase change materials in a water-cooled photovoltaic/thermal system. *Energy Conversion and Management, 163*, 187–195. https://doi.org/10.1016/j.enconman.2018.02.039

Qiu, Z., Zhao, X., Li, P., Zhang, X., Ali, S., & Tan, J. (2015). Theoretical investigation of the energy performance of a novel MPCM (Microencapsulated Phase Change Material) slurry based PV/T module. *Energy, 87*, 686–698. https://doi.org/10.1016/j.energy.2015.05.040

Saidur, R., BoroumandJazi, G., Mekhlif, S., & Jameel, M. (2012). Exergy analysis of solar energy applications. *Renewable and Sustainable Energy Reviews, 16*(1), 350–356. https://doi.org/10.1016/j.rser.2011.07.162

Tao, Y. B., You, Y., & He, Y. L. (2016). Lattice Boltzmann simulation on phase change heat transfer in metal foams/paraffin composite phase change material. *Applied Thermal Engineering, 93*, 476–485. https://doi.org/10.1016/j.applthermaleng.2015.10.016

Tripanagnostopoulos, Y., Souliotis, M., Battisti, R., & Corrado, A. (2005). Energy, cost and LCA results of PV and hybrid PV/T solar systems. *Progress in Photovoltaics: Research and Applications, 13*, 235–250. https://doi.org/10.1002/pip.590

Yazdanifard, F., Ebrahimnia-Bajestan, E., & Ameri, M. (2016). Investigating the performance of a water-based photovoltaic/thermal (PV/T) collector in laminar and turbulent flow regime. *Renewable Energy, 99*, 295–306. https://doi.org/10.1016/j.renene.2016.07.004

Zhou, D., & Zhao, C. Y. (2011). Experimental investigations on heat transfer in phase change materials (PCMs) embedded in porous materials. *Applied Thermal Engineering, 31*(5), 970–977. https://doi.org/10.1016/j.applthermaleng.2010.11.022

Chapter 20
Optimal Sizing Techniques for Hybrid Photovoltaic Systems Using Artificial Neural Networks (ANNs): A Review Paper

Ali H. A. Al-Waeli, Anita Askander, Hussein A. Kazem, and Miqdam T. Chaichan

20.1 Introduction

The popularization of solar energy technology is attributed to many factors that include its technical utility, cost viability, and environmental effects (Lorenz et al., 2008; Apostoleris et al., 2019). Solar energy technology has gone through different breakthroughs that allowed it to be a competitive, and potential energy alternative, source that is widely utilized in residential, commercial, and industrial settings (Zahedi, 2006; Agnew & Dargusch, 2015). Thus, it is acceptable to assume that improving the techno-economic and environmental features of solar energy devices would further position this technology toward being an alternative energy source. Solar energy systems usually fall into three categories which are photovoltaic modules (PV), solar thermal collectors (ST), and hybrid photovoltaic-thermal (PVT) collectors (Al-Waeli et al., 2017). PV modules capitalize on the light component, photons, of the solar energy, solar thermal collectors utilize the heat component, and PVT collectors utilize both. However, out of the three devices, it is apparent that PV modules form most solar energy devices in the industry (IEA Solar PV. https://www.iea.org/reports/solar-pv). As of 2022, PV technology has a total capacity in operation of 1153 GW electrical, while the solar thermal was at around 542 GW thermal (IRENA, 2022; Solar Heat World Wide. https://www.iea-shc.org/

A. H. A. Al-Waeli (✉) · A. Askander
Engineering Department, American University of Iraq, Kurdistan Region, Sulaimani, Iraq
e-mail: ali.alwaeli@auis.edu.krd

H. A. Kazem
Faculty of Engineering, Sohar University, Sohar, Oman

M. T. Chaichan
Energy and Renewable Energies Technology Center, University of Technology – Iraq, Baghdad, Iraq

A. Sayigh (ed.), *Transition Towards a Carbon Free Future*, Innovative Renewable Energy, https://doi.org/10.1007/978-3-031-61660-0_20

Data/Sites/1/publications/Solar-Heat-Worldwide-2023.pdf). Certainly, investment in PV systems is capital intensive in many parts of the world but with improvements in performance and implementation along with the governmental subsidies plays a pivotal role in the growth of this technology. In 2021, around 3% of the global electricity generation came from solar photovoltaic systems with annual PV capacity additions reaching 150 GW. Around 95% of the existing capacity is crystalline silicon-based modules and only 5% comes from thin-film PV technology (IRENA, 2022). The economic viability of PV modules comes from the fact that its prices have dropped by 80% over the last decade which is mainly a result of economies of scale and the improvements made to the supply chain.

The techno-economic and environmental aspects of PV systems are often evaluated depending on (i) the technology or device, for instance PV, ST, or PVT; (ii) the performance of the device in terms of its power and efficiency; (iii) the setting such as stand-alone or on-grid; (iv) the application such as illumination, drying, or remote power generation; (v) the system components such as inverters, power conditioners, and mounting structures; and (vi) the location that impacts the cost of equipment and cost of energy as well as the amount of solar energy received by the system (Dellosa et al., 2021; Sommerfeldt & Madani, 2017). Research in this discipline can target any of the mentioned categories or other categories as well. The improvement in the design, manufacturing, construction, and employment of this technology would lead to further adoption of solar PV modules.

The practice of sizing a PV system involves (i) determining the appropriate system capacity to meet electricity demands by estimating energy consumption and load profile; (ii) assessing the availability of ground space, considering operational and environmental factors such as tilt angle, orientation, and shading; (iii) determining system efficiency; (iv) calculating system capacity; (v) considering system losses; (vi) considering other factors such as cloudy days; and (vii) examining the size of the remaining components in the system (Prasetyaningsari et al., 2013; Salam et al., 2013; Mirzahosseini & Taheri, 2012). Traditionally, sizing PV systems is often done using numerical methods, and their economic viability is often compared with conventional energy source. Meanwhile, optimal sizing is the process of determining the most suitable configuration or capacity for a PV system based on a set of specific criteria. The optimization is made to maximize system performance, efficiency, and economic viability. Various factors and constraints are used in this process such as the solar resource availability, system components, costs, energy demand, and technical limitations (Sharma et al., 2014; Khare & Rangnekar, 2014; Lupangu & Bansal, 2017). Figure 20.1 shows typical PV system configurations in block diagram forms.

Artificial neural networks (ANNs) have emerged as a valuable tool in the field of renewable energy, including the optimal sizing of photovoltaic (PV) systems. ANNs are computational models inspired by the structure and functioning of biological neural networks in the human brain. They excel at learning complex patterns and relationships from data, making them well-suited for solving optimization problems in various domains (Moayedi et al., 2020). In the context of PV system sizing,

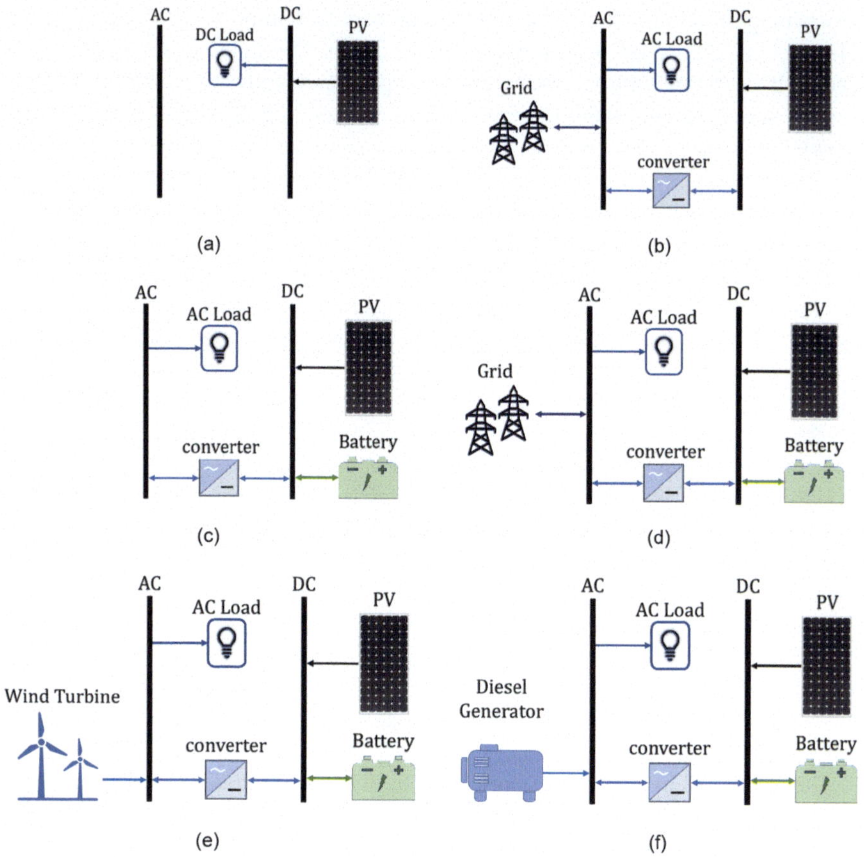

Fig. 20.1 Block diagram drawing of different PV systems: (**a**) stand-alone PV system, (**b**) grid-connected PV system, (**c**) stand-alone PV system with inverter and battery storage, (**d**) grid-connected PV system with battery storage, (**e**) hybrid PV-wind turbine system with battery storage, and (**f**) hybrid PV-diesel generator system with battery storage. The converter represents both the inverter and rectifier. The lines refer to DC and AC busbars, while the arrows represent the possible directions of energy flow

ANNs offer several advantages. They can learn from historical data on energy consumption, solar radiation, system configurations, and other relevant parameters to develop accurate models (Sivaneasan et al., 2017). These models can then be used to predict the optimal size or configuration of a PV system based on input parameters.

ANNs can capture nonlinear relationships and intricate dependencies between input variables and the optimal system size. This is particularly valuable in PV system sizing, as multiple factors such as energy demand, solar resource availability, system losses, and cost considerations interact in a complex manner. ANNs can handle this complexity and provide reliable predictions (Bermejo et al., 2019).

By training ANNs on large datasets, researchers can improve the accuracy and robustness of their models (Rumbayan et al., 2012). The training process involves feeding the network with known input-output pairs and adjusting its internal parameters to minimize the prediction errors. Once trained, the ANN can generalize its knowledge to new, unseen data, enabling it to estimate the optimal sizing of PV systems for various scenarios (Mellit et al., 2005a). The use of ANNs in PV system sizing research has shown promising results. It allows for more comprehensive analyses and better understanding of the factors influencing system sizing decisions. Moreover, ANNs can account for dynamic and uncertain conditions, such as varying energy demand patterns or changing solar radiation levels, enhancing the adaptability and flexibility of PV system designs. Overall, the application of artificial neural networks in the optimal sizing of PV systems offers a powerful and data-driven approach to inform decision-making processes. It enables researchers and practitioners to optimize system configurations, improve energy efficiency, and enhance the economic and environmental viability of PV installations.

Hybrid photovoltaic (PV) systems have gained significant attention in recent years as a promising solution for enhancing the efficiency, reliability, and sustainability of renewable energy generation. These systems combine PV technology with other complementary energy sources, such as wind, battery storage, or diesel generators, to overcome the limitations of stand-alone PV systems and optimize overall system performance. By integrating multiple energy sources, hybrid PV systems offer the potential to meet diverse energy demands, improve energy generation and consumption balance, and provide reliable power supply in various applications and locations.

The concept of hybrid PV systems stems from the recognition that PV generation alone may not always be sufficient to meet the energy requirements throughout the day or in areas with intermittent sunlight. For instance, during periods of low solar radiation or at night, the power output of PV systems diminishes (Chrifi-Alaoui et al., 2023). By incorporating additional energy sources, hybrid PV systems can address these limitations and provide a more consistent and reliable energy supply. Moreover, hybridization enables better utilization of renewable energy resources and minimizes reliance on conventional fossil fuel-based generation, leading to reduced carbon emissions and environmental impact. Figure 20.2 shows global solar and wind resource maps to showcase the availability of these resources worldwide.

The design and operation of hybrid PV systems require careful consideration of various factors, including load profiles, renewable resource availability, storage capacity, system sizing, control strategies, and economic feasibility (Erdinc & Uzunoglu, 2012). Advances in technology, such as smart control algorithms, efficient energy storage systems, and hybrid system optimization techniques, have facilitated the integration and management of different energy sources in a coordinated manner, ensuring efficient utilization of available resources. Hybrid PV

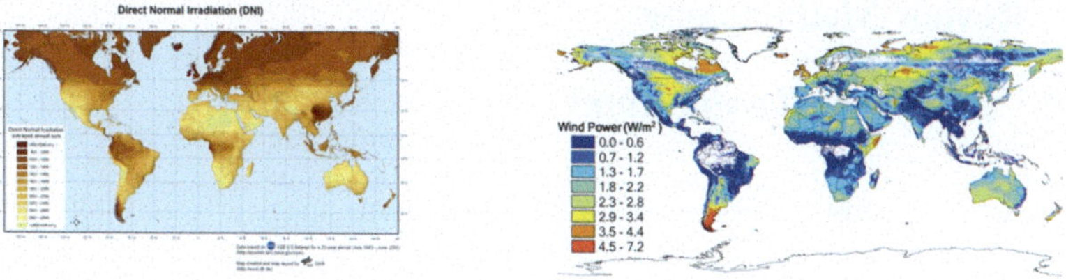

(a)
Source: https://www.solarpaces.org/csp-technologies/csp-potential-solar-thermal-energy-by-member-nation/ (accessed on 1st Jul 2023)

(b)
Source: https://www.pnas.org/doi/full/10.1073/pnas.0904101106 (accessed on 1st Jul 2023)

Fig. 20.2 Maps of global (**a**) annual solar irradiation and (**b**) wind power density (Solarspaces, 2023; Pnas, 2023)

systems find applications in diverse sectors, including residential, commercial, industrial, and remote off-grid areas. They offer numerous benefits, such as increased self-consumption of renewable energy, improved grid stability and reliability, reduced dependence on fossil fuels, and potential cost savings. Hybrid PV systems also enable the integration of intermittent renewable energy sources into existing power grids, promoting the transition toward a sustainable and decentralized energy landscape (Muselli et al., 1999).

There is much interest in this area of research which is evident by the diversity of ideas published recently—within the past 5 years—on this topic. However, there is a lack of a comprehensive review that consolidates the existing knowledge on ANN-based optimal sizing techniques for PV systems. Moreover, the identification of gaps and limitations in this topic is needed to guide future investigations and development in specific aspects such as system modeling, training algorithms, performance evaluation, and application domains. Additionally, a comparative analysis can provide a more holistic and accurate perspective on these algorithms. Thus, examining the results reported in previous studies will help in informing the efforts toward the performance evaluation of such techniques—assessing prediction accuracy, computational efficiency, robustness, and applicability, in order to provide insights into the reliability and limitations of these approaches.

This paper aims to provide an overview of hybrid PV systems, exploring their key components, design considerations, system configurations, control strategies, and performance evaluation. It will review the latest advancements and research in this field, highlighting the benefits, challenges, and potential applications of hybrid PV systems. By understanding the principles and possibilities of hybrid PV systems, researchers, policymakers, and industry professionals can make informed decisions and contribute to the development and deployment of sustainable energy solutions.

20.2 Review Methodology

20.2.1 Research Classifications

The research concerning optimal sizing of PV systems spans a large multitude of topics; thus, a classification of the research topics is needed to produce a meaningful evaluation of the reviewed literature. The selected classifications are based on (i) research focus, (ii) methodology, (iii) ANN architecture, (iv) system configuration, and (v) performance metrics. Figure 20.3 shows the classification of studies in this area based on different criteria.

20.2.2 Review Demographics

The distributions of published research on this topic throughout the years has also been tracked and is provided in Fig. 20.4. According to the figure, the number of published documents throughout the years in relation to this topic has increased which highlights the novelty and the shift toward ANN-based optimization for PV system configuration sizing. The findings show that most of the studies on this topic have been made in 2022, while the least amount was published in 2008.

20.2.3 Review Protocol

The main steps of this review were created to answer the following questions:

(i) What are the implemented nonlinear techniques for optimal sizing of hybrid PV systems?
(ii) What are the most effective algorithms used for sizing of hybrid PV systems?
(iii) What are the research trends surrounding this topic?

The first step was identifying the keywords and search terms that included the following terms: "hybrid photovoltaic systems," "optimal sizing," "artificial neural networks," "ANN," and "hybrid PV." Moreover, the databases selected for the search were IEEE Xplore, ScienceDirect, and Google Scholar. The selected search query for the research was the following:

Search Query

ScienceDirect ("Optimal sizing") AND ("ANN" OR "Artificial Neural Networks" OR "Predictive Algorithms") AND ("PV-wind" or "hybrid PV").

IEEE Xplore ("Document Title":"Optimal sizing" OR "Document Title":"Sizing") AND ("Document Title":"ANN" OR "Document Title":"Artificial Neural Networks" OR "Document Title":"Predictive Algorithms") AND ("Document Title":"PV-wind" OR "Document Title":"hybrid PV" OR "Document Title":"PV" OR "Document Title":"Photovoltaic").

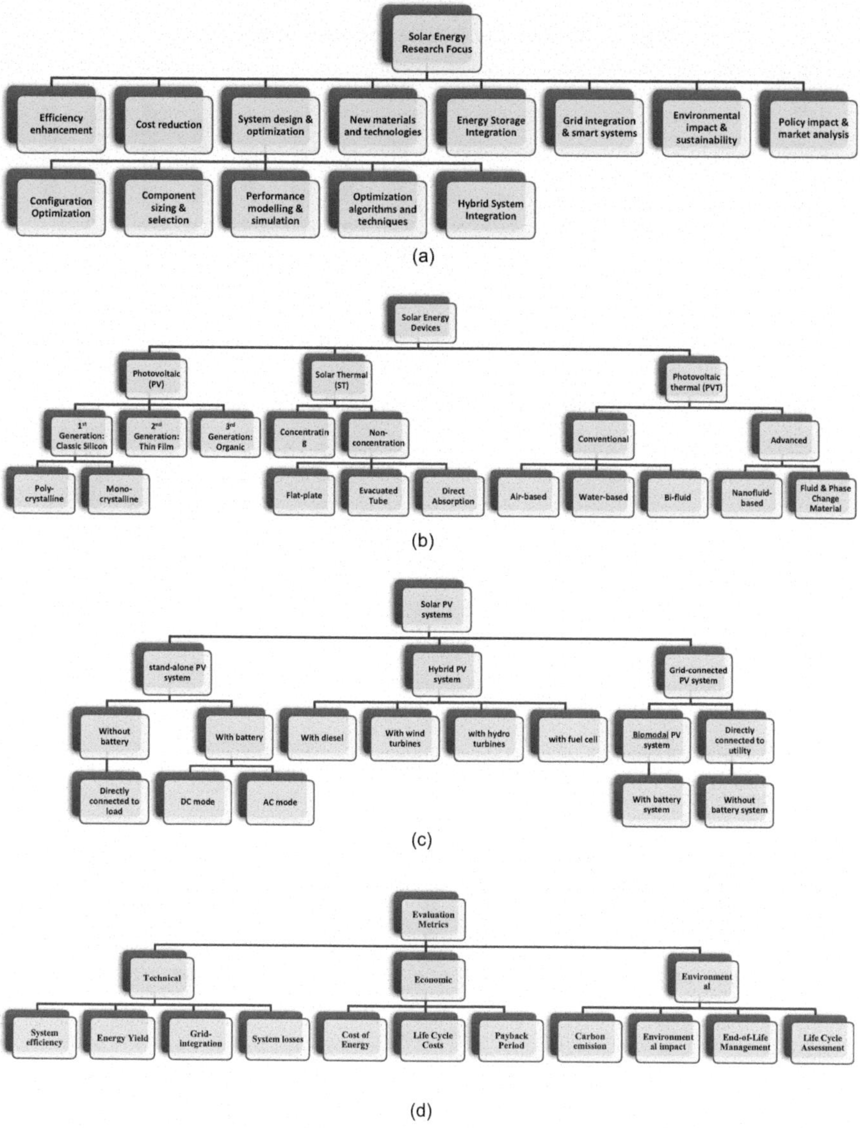

Fig. 20.3 Classification of hybrid PV systems based on (**a**) research focus, (**b**) device/technology, (**c**) system type, and (**d**) evaluation metrics

Google Scholar intitle:("Optimal sizing") AND ("ANN" OR "Artificial Neural Networks" OR "Predictive Algorithms") AND ("PV-wind" OR "hybrid PV").

The query syntax was modified for each database accordingly. Furthermore, the inclusion and exclusion criteria were identified as follows:

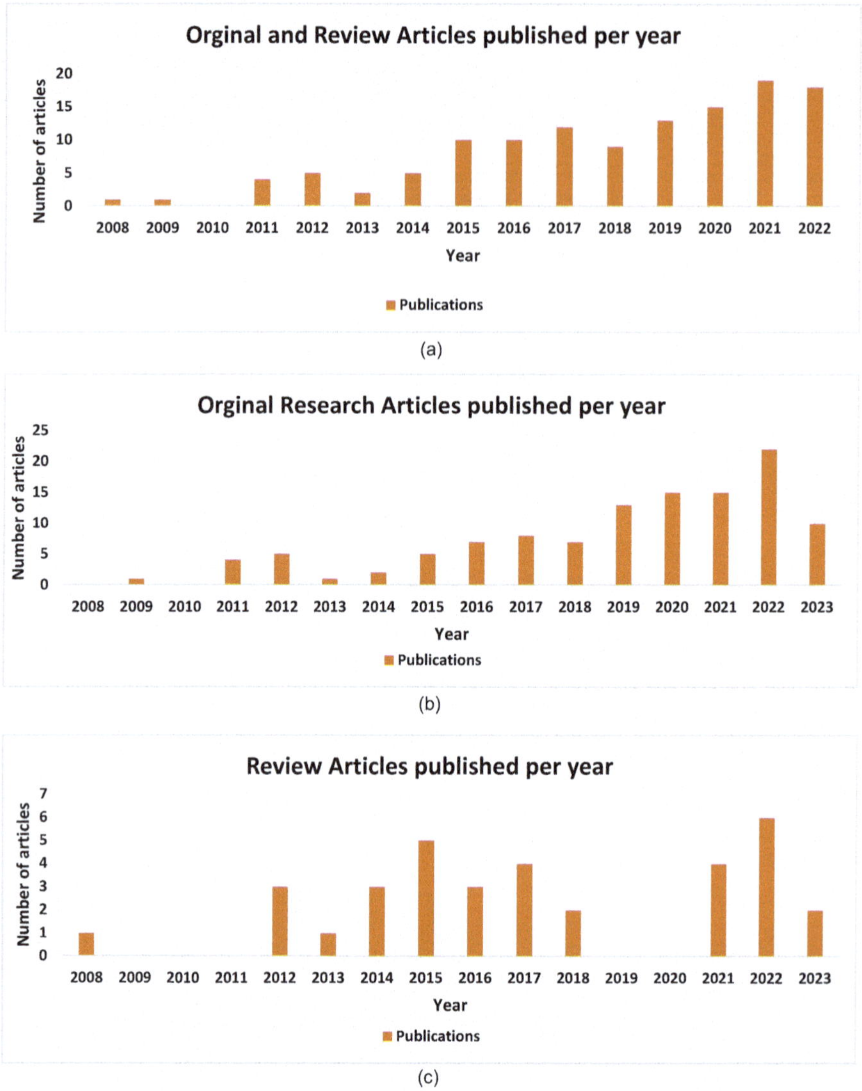

Fig. 20.4 Number of articles published by year: (**a**) combined original and review articles, (**b**) original articles, and (**c**) review articles

Inclusion Criteria

1. Articles that present an analysis of optimization techniques for hybrid PV systems.
2. Articles that review existing nonlinear techniques for optimal sizing applications.
3. Articles that focus on hybrid PV systems that include PV-wind, PV-diesel, and PV-ST.
4. Articles that focus on the integration of ANN models in PV optimal sizing.

Exclusion Criteria

1. Articles that use languages other than English.
2. Articles that focus on linear techniques or analytical calculations.
3. Articles that target regular PV systems.
4. Duplicates.

The search was conducted across the databases using the query mentioned above and utilizing the appropriate filters. Papers that were irrelevant or out of scope were excluded as well as research papers that were omitted. The inclusion criteria were as follows: (1) articles that present an analysis of optimization techniques for hybrid PV systems, (2) articles that review existing nonlinear techniques for optimal sizing applications, and (3) articles that focus on hybrid PV systems that include PV-wind, PV-diesel, and PV-ST. Research articles that did not mean the suitability criteria were eliminated. The exclusion criteria included the following: (1) articles that use languages other than English, (2) articles that focus on linear techniques or analytical calculations, and (3) articles that target regular PV systems. Only research articles were considered in this study, while patents and short communications were not targeted. The screening process went through three levels that include (1) screening of the titles, (2) abstract screening, and (3) full text screening. The inclusion and exclusion criteria were applied when screening the articles. The results from the search execution are as follows: initial results were $N = 151$, 1, and 142 papers in ScienceDirect, IEEE Xplore, and Google Scholar, respectively. After removing duplicates and screening the papers, the final subset was $N = 80$, 1, and 70, respectively. Moreover, other studies were added through citation tracking (Koutroulis et al., 2006a; Dufo-Lopez et al., 2007; Hernández et al., 2007; El-Hefnawi, 1998; Xu et al., 2005).

20.3 Hybrid Renewable Energy System Sizing

Hybrid renewable energy system sizing involves determining the optimal configuration and capacity of renewable energy sources, such as solar, wind, hydro, or biomass, in combination with energy storage and backup systems. Figure 20.5 provides a diagram of hybrid energy systems (HESs) adopted from Upadhyay and Sharma (Upadhyay & Sharma, 2014). The sizing process aims to design a system that meets the energy demands while maximizing the utilization of renewable resources and minimizing reliance on nonrenewable sources.

Factors considered during sizing include the energy consumption profile, renewable resource availability, load balancing requirements, storage capacity, and economic viability. By appropriately sizing the components of a hybrid renewable energy system, it is possible to achieve a reliable and sustainable energy supply, reduce greenhouse gas emissions, and enhance energy independence. The sizing process requires comprehensive analysis, simulation, and optimization techniques to determine the optimal combination and capacity of renewable energy sources and

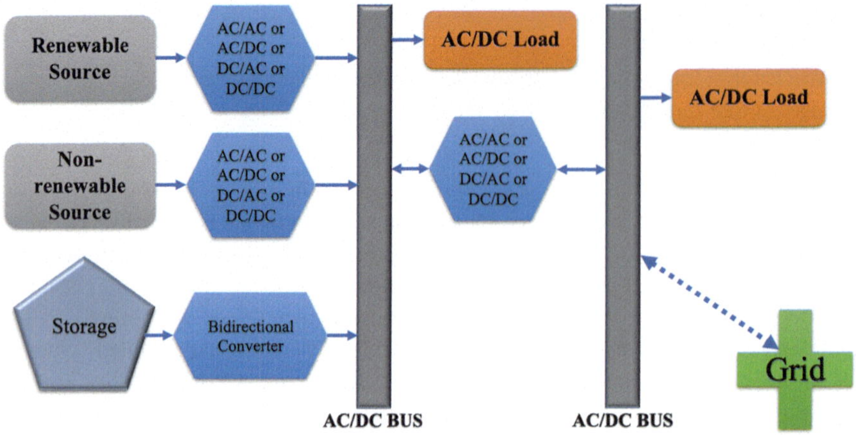

Fig. 20.5 Hybrid energy system (HES) diagram

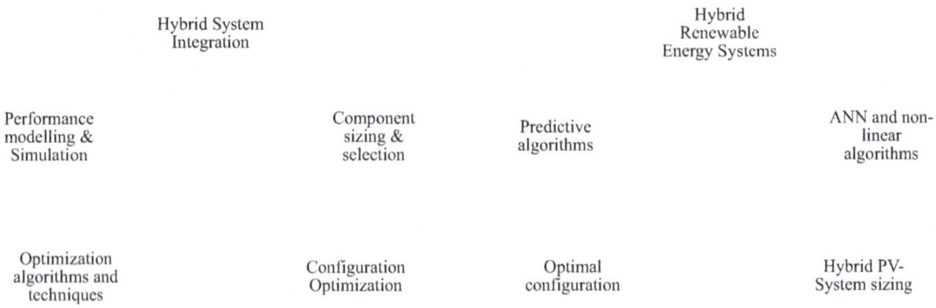

Fig. 20.6 Basic Venn diagrams of the topics related to the present review

their integration with storage and backup systems. Figure 20.6 is provided to further elucidate the targeted research area of this review.

Figure 20.7 shows a diagram describing the general steps for sizing a hybrid renewable energy system. The first two steps can be switched depending on the situation. If the user requests for a system to meet a specific load, then the process initiates with the analysis of the load profile and then an assessment of the meteorological conditions. However, if the load could be placed at any location, then it is more suitable to first select a location with preferrable energy resources.

Mellit (2010) developed an artificial neural network (ANN)-based genetic algorithm (GA) to generate sizing curve for photovoltaic systems in stand-alone setting. The GA was used to mitigate the issue of the high computation time required by ANNs to generate such curves. The GA is used to obtain the optimal ANN architecture as in finding the suitable number of hidden layers and neurons within these layers. The loss-of-load probability (LLP) corresponding to 40 different sites in Algeria was used to numerically generate the sizing curves. The sizing curves require massive solar irradiation data, but the issue for isolated locations is their

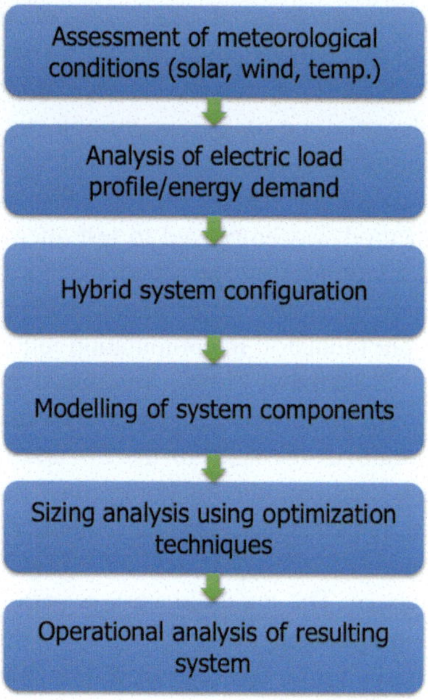

Fig. 20.7 Hybrid energy system (HES) diagram

lack of data. Thus, synthetic approaches were utilized to generate the data. The data used were divided into two sets for training and testing. The training dataset for the ANN-GA model included 36 sites, leaving four sites for the testing phase. LLP and the corresponding geographical coordinates were inputs to the model, while the generated sizing curves were the output—making it a multi-input single-output network.

In a study conducted by Koutroulis et al. (2006a), a methodology was proposed based on genetic algorithms (GAs) and compared to linear programming. Through simulation results, it was confirmed that hybrid PV/wind generator systems exhibit reduced system costs compared to scenarios where either only wind generator or only PV sources are utilized.

Dufo-López et al. (2007) introduced a novel hybrid optimization strategy termed hybrid optimization by genetic algorithms (HOGA) for designing PV-diesel systems. HOGA employs a genetic algorithm to determine the optimal sizing and operational control of a PV-diesel system. The software for this approach was created using the C++ programming language. A direct comparison was drawn between a PV-diesel system optimized by the HOGA approach and a stand-alone PV-only system devised through a conventional technique grounded in available energy under the most unfavorable conditions. Both systems were configured to meet the same energy demand and solar irradiation levels.

The computational outcomes vividly highlighted the economic advantages of the PV-hybrid system. Furthermore, HOGA was assessed against a commercial optimization software designed for hybrid systems, offering additional insights into its efficiency and capabilities.

Hernández et al. (2007) introduced a systematic algorithm designed to determine the optimal placement and sizing of photovoltaic grid-connected systems (PVGCSs) in feeders, aiming to achieve the maximum overall impact on the feeder. The algorithm employs a multi-objective optimization approach to obtain the optimal solution. The authors claim that the proposed methodology's applicability is evident through the results obtained for various feeders discussed in the literature. The method was employed to test alternative solutions for PVGCS allocation. Simulations conducted on actual feeders demonstrate that allocating PVGCSs based on stability voltage distribution yields the best outcomes. This finding enables a significant reduction in computation required for future analyses.

El-Hefnawi (1998) introduced a method for designing PV-diesel systems. The optimization process begins with defining a model for the diesel generator and then optimizes the sizes of the PV array and battery. The objective is to determine the minimum number of storage days and the minimum area of the PV array. In another study (Xu et al., 2005), the authors explored the use of a genetic algorithm (GA) with an elitist strategy to achieve optimal sizing of a stand-alone hybrid wind/PV power system. The main objective was to minimize the total capital cost while ensuring a constraint on the loss of power supply probability (LPSP). The LPSP for each individual in the GA population was calculated through simulating all 8760 h of a year. The authors reported that the GA quickly converges, and the proposed methodology proves feasible for optimal sizing of stand-alone hybrid wind/PV power systems.

Raj Kumar et al. (2011) introduced an optimization methodology that utilizes the adaptive neuro-fuzzy inference system (ANFIS) for modeling PV and wind sources. The developed algorithm is compared to HOMER and HOGA software, revealing a high accuracy of 96% for both PV and wind. The optimized system is simulated using PSCAD/EMTDC, and the results indicate minimal excess energy. Additionally, the optimized system demonstrates the capability to supply power to the load for an extended duration even in the absence of renewable sources while meeting the desired loss of power supply probability (LPSP) requirements.

Furthermore, another stochastic optimization technique used for sizing hybrid energy systems is known as particle swarm optimization (PSO), which was introduced by Kennedy and Eberhart in 1995 (Kennedy, 1997). This method takes inspiration from certain social behaviors. To provide a brief overview of PSO, imagine a swarm consisting of p particles, with each particle's position representing a potential solution within the design problem space D. Each particle is characterized by its position and speed, and throughout an iterative process, it continually tracks the best solution found during its trajectory.

In energy system sizing, artificial neural networks (ANNs) have been utilized to enhance the analytical method of loss of load probability (LOLP) curves, as demonstrated by the works of Hontoria et al. (2005) and Mellit et al. (2005b, 2007). These studies aimed to improve upon the analytical method of LOLP curves originally proposed by Egido and Lorenzo (1992). ANN has been combined with genetic algorithms (GAs) to enhance the design and calibration process of the neural networks. This combination allows for the generation of sizing curves for the power system based on the LOLP concept.

Hakimi and Tafreshi (2007) employed particle swarm optimization (PSO) to determine the optimal unit sizing of a stand-alone hybrid power system. Their study also focused on optimizing the sizing of a stand-alone wind fuel cell hybrid power system specifically for the Kahnouj area in South East Iran (Hakimi & Tafreshi, 2009). The primary objective of this research was to minimize the total costs associated with the hybrid power system while ensuring that the residential area's energy demand was met. A noteworthy aspect of this work is the utilization of biomass as a source of hydrogen. The model considered in the study includes various components such as the Bergey Wind Power's BWC Excel-R148 turbine, Ballard fuel cell, Avalence electrolyzer, MAHLER reformer, hydrogen tank, and anaerobic reactor for methane production. The optimization procedure incorporates inputs such as capital costs, operation and maintenance costs, efficiency, replacement costs, component lifespan, project lifespan, component specifications, information about the area's population, and waste production. Based on simulation results, the recommended specifications include a nominal power of 7.5 kW for each wind turbine and 1 kW for each fuel cell and electrolyzer, a hydrogen tank size of 1 kg, and a project lifespan of 20 years. The study yielded optimal sizing values for various components: wind turbine (687), electrolyzer (2504), hydrogen tank (2810), and fuel cell (1222). The corresponding optimal cost amounted to $35.5 million. The sizes of the reactor, reformer, and compressor were determined to be fixed at 750 kg/day, 31.2 kg H2/day, and 50 kW, respectively. Notably, the longer lifespan of wind turbines contributes to reducing project costs. This hybrid system is particularly suitable in areas characterized by high wind speeds, the presence of abundant agricultural waste, high fuel transmission costs, and pollution levels.

In some optimization problems, the standard particle swarm optimization (PSO) suffers from premature convergence. To address this issue, modified PSO algorithms such as hierarchical PSO and CPSO have been implemented. Airashidi and Naggar (2010) proposed the use of PSO to minimize the errors associated with model parameter estimation for annual peak load forecasting in electrical systems. Actual recorded data from Kuwaiti and Egyptian networks were employed as input data. The results were compared with well-known least square techniques and exhibited improved estimates. Hence, PSO technique shows great promise and is utilized as a valuable tool for parameter estimation.

In this section, the utilization of artificial intelligence (AI) techniques to optimize hybrid PV/wind systems is discussed. Some researchers, including those in reference (Koutroulis et al., 2006b), combine AI methods with numerical methods. In Koutroulis et al. (2006b), the authors propose a method for determining the appropriate sizing of PV/wind systems within a particular geographical area. This methodology involves three main stages. Firstly, they establish a mathematical model of the PV/wind system. Then, they engage in iterative simulations using hourly meteorological data. These simulations aim to create multiple system configurations that can fulfill a specific load requirement. Lastly, an optimization challenge is formulated, incorporating system cost and a predetermined level of reliability. To address this challenge, the authors employ a genetic algorithm for optimization. However, it is important to mention that in this research, the authors express the PV model in terms of the fill factor (FF) but do not elaborate on how they acquired the maximum voltage and current values for the PV module. Additionally, the precise reliability level set for the optimized system is not clearly defined; the authors merely mention that the energy generated must exceed the load demand.

Furthermore, some researchers employ AI techniques for modeling hybrid systems. In the context of (Khatib et al., 2012), an optimization investigation is conducted on a hybrid photovoltaic (PV)/wind system designed for an island in Malaysia. The study employs hourly datasets containing load demand, wind speed, and solar radiation information. It is worth highlighting that the research title might imply the use of the neuro-fuzzy technique for optimization, although the neuro-fuzzy technique is not inherently an optimization tool. In actuality, this study applies the neuro-fuzzy technique for modeling the PV module and wind turbine. The optimization of the system, however, is achieved through an iterative approach based on the limited load point (LLP) concept.

In terms of the modeling procedure utilizing the neuro-fuzzy technique, the author employs established wind turbine and PV module mathematical models to create an extensive dataset comprising inputs (such as solar radiation, temperature, and wind speed) and corresponding outputs (power generated by the PV module and wind turbine). This dataset is utilized to train the neuro-fuzzy model, enabling it to predict output power based on provided inputs. It is important to note that a limitation of this modeling approach is that the developed model is accurate primarily for PV modules and wind turbines resembling the commercial brands used during the training phase of the model.

20.3.1 Reviews on Optimal Sizing of Hybrid PV Systems

As mentioned in Sect. 20.2.3, much review research works have been conducted in relation to this topic. In some cases, the focus is on general optimization algorithms for sizing PV-wind, while in others the focus is on finding the most appropriate system to meet the load demand, regardless of the source of energy. Table 20.1 offers a summary of the review articles on sizing of hybrid PV systems (Upadhyay & Sharma, 2014; Mellit & Kalogirou, 2008; Luna-Rubio et al., 2012; Bajpai & Dash, 2012; Sharma, 2012; Khatib et al., 2013; Karabacak & Cetin, 2014; Iqbal et al., 2014; Mahesh & Sandhu, 2015; Mohammed et al., 2015; Sinha & Chandel, 2015; Siddaiah & Saini, 2016; Zahraee et al., 2016; Khare et al., 2016; Al Busaidi et al., 2016; Nadjemi et al., 2017; Jha et al., 2017; Al-Falahi et al., 2017; Anoune et al., 2018; Twaha & Ramli, 2018; Bukar & Tan, 2019; Memon & Patel, 2021; Ridha et al., 2021; Al-Shahri et al., 2021; Mazzeo et al., 2021; Liu et al., 2022; Alabi et al., 2022; Khan et al., 2022; Arsad et al., 2022; Gusain et al., 2023; Islam & Roy, 2023; Maghami & Mutambara, 2023; Basnet et al., 2023; Thirunavukkarasu et al., 2023).

Table 20.1 A comprehensive overview of review articles focusing on the sizing of hybrid photovoltaic (PV) systems, encompassing references ranging from (Mellit & Kalogirou, 2008) to (Thirunavukkarasu et al., 2023)

Study	Year	Features	Study focus	Emphasis on hybrid PV system sizing using AI/ANN
Mellit and Kalogirou (2008)	2008	(i) Forecasting and modeling of meteorological data (ii) Sizing of photovoltaic systems and modeling (iii) Simulation and control of photovoltaic systems	Overall review of existing AI-based approaches related to solar PV systems including stand-alone, grid-connected, and hybrid, as well as metrological data and PV system control	Low
Luna-Rubio et al. (2012)	2012	(i) Hybrid energy system architecture (ii) Hybrid energy metrics (iii) Sizing methods in general	General sizing methodologies including probabilistic, analytical, iterative, and numerical	Low
Bajpai and Dash (2012)	2012	(i) Modeling hybrid energy resources (ii) Unit sizing and optimization (iii) Energy flow management techniques	Overall review of unit sizing, hybrid energy system component modeling including electrical and thermal models, and modeling of remaining components	Medium
Sharma (2012)	2012	(i) Solar thermal systems (ii) Simulation of solar systems (iii) Optimization of solar systems	Overview of optimization techniques for solar systems with focus on solar thermal. Covers the simulation and optimization techniques	Low
Khatib et al. (2013)	2013	(i) Sizing of system components (ii) Various sizing methods (iii) Challenges for PV size optimization	Review of size optimization techniques used for stand-alone PV systems, hybrid PV/diesel and PV/wind, grid-connected PV, and inverter sizing	Low
Karabacak and Cetin (2014)	2014	(i) ANN applications for PV systems (ii) ANN applications for wind energy conversion systems (iii) ANN applications for hybrid systems	Review on the use of ANN for wind, PV, and hybrid power systems for solar resource prediction, performance simulation, and maximum PowerPoint tracking	Low

(continued)

Table 20.1 (continued)

Study	Year	Features	Study focus	Emphasis on hybrid PV system sizing using AI/ANN
Upadhyay and Sharma (2014)	2014	(i) HES sizing methods and evaluation criteria (ii) AI and hybrid sizing methods (iii) Control and energy management	Emphasis on the sizing methods of hybrid energy systems (HESs) including intuitive, analytical, numerical, AI-based, and hybrid methods	Low
Iqbal et al. (2014)	2014	(i) Analysis of the literature (ii) Types of optimizations in renewable energy (iii) Optimization techniques for renewable energy sources for small units, home users, and the industry	General review of renewable energy sources with consideration for different modes of operations, types of objective functions for optimization, and geographical areas	Low
Mahesh and Sandhu (2015)	2015	(i) Unit sizing (ii) Converter and controller design (iii) Performance assessment	Emphasis on hybrid PV/wind energy systems with battery storage and overview of various applications of AI-based techniques	Medium
Mohammed et al. (2015)	2015	(i) Modeling of hybrid PV/diesel systems (ii) Comparison of hybrid PV/diesel with stand-alone PV (iii) Process and operational system control procedures	Emphasis on hybrid photovoltaic/diesel generator system control strategies	Low
Sinha and Chandel (2015)	2015	(i) Requirements of hybrid PV/wind system optimization (ii) Optimization criteria for PV/wind hybrid systems (iii) Modeling of hybrid PV/wind systems	Emphasis on hybrid PV/wind energy system sizing methodologies	High
Siddaiah and Saini (2016)	2016	(i) HRES configuration and planning (ii) Modeling and optimization techniques (iii) Economic and reliability-based modeling schemes	Overview of planning, configurations, modeling, and optimization techniques of hybrid renewable energy systems (HRESs) for off-grid applications	Medium

Zahraee et al. (2016)	2016	(i) AI methods for hybrid system energy optimization	Analysis of AI-based sizing of hybrid energy systems	High
		(ii) Summary of hybrid models and software tools for optimization		
		(iii) Promising approaches for system sizing		
Khare et al. (2016)	2016	(i) Sizing, modeling, and control of HRES	Emphasis on solar-wind hybrid renewable energy system sizing, modeling, and control through different techniques	High
		(ii) Evolutionary techniques in HRES		
		(iii) Game theory in HRES		
Al Busaidi et al. (2016)	2016	(i) Overview of sizing techniques	Emphasis on PV-wind renewable energy hybrid systems with consideration for sizing of different system components and preferable logistical environmental and economical considerations	Low
		(ii) Overview of optimization techniques		
		(iii) Discussion of wind-solar energies in Oman		
Nadjemi et al. (2017)	2017	(i) Grid-connected hybrid energy systems	Hybrid PV/wind-battery energy systems for sizing and energy management	Medium
		(ii) Energy management strategies		
		(iii) Cuckoo search (CS) algorithm is proposed for sizing optimization		
Jha et al. (2017)	2017	(i) Overview of renewable energy systems	Emphasis on the integration of both hybrid and single AI approaches in the process of research and development for designing and optimizing renewable energy systems	Medium
		(ii) AI utilization in different renewable energies		
		(iii) Hybrid AI methods in renewable energy systems		
Al-Falahi et al. (2017)	2017	(i) Combinations of solar-wind HES systems	Review of hybrid renewable energy systems (HRESs)	High
		(ii) Developments in sizing algorithms		
		(iii) Comparison of single and hybrid algorithms as well as sizing tools		
Anoune et al. (2018)	2018	(i) Discussion of PWHS systems and coupling topology schemes	Emphasis on PV-wind hybrid systems (PWHS) for isolated areas with focus on techniques/methods of obtaining the best compromise between system costs and power reliability	Medium
		(ii) Mathematical models of hybrid system components		
		(iii) Analysis of algorithms and software tools used in optimization		

(continued)

Table 20.1 (continued)

Study	Year	Features	Study focus	Emphasis on hybrid PV system sizing using AI/ANN
Twaha and Ramli (2018)	2018	(i) Overview of optimization methods (ii) Mathematical/deterministic models (iii) Optimization techniques for on-grid and off-grid systems	Focus on optimization approaches, analytical and AI-based, hybrid distributed energy generation (DEG) systems for stand-alone and grid-connected systems	High
Bukar and Tan (2019)	2019	(i) Renewable energy architecture (ii) Optimization formulation framework (iii) Energy management strategies	The scope is the "optimization and energy management strategies (EMSs)" of wind energy and stand-alone PV hybrid systems integrated with fuel cells	High
Memon and Patel (2021)	2021	(i) Conventional techniques for HRES sizing (ii) Modern optimization techniques for HRES sizing	Evolutionary algorithms and hybrid optimization techniques for HRES sizing and optimization	High
Ridha et al. (2021)	2021	(i) Review of multi-objective optimization methodologies (ii) Review of multi-criteria decision-making methodologies (iii) Mathematical models for power output estimation	Comprehensive review of stand-alone PV system design based on "multi-objective optimization (MOO)" and "multi-criteria decision-making (MCDM) methodologies"	Low
Al-Shahri et al. (2021)	2021	(i) Issues and challenges related to PV system optimization are presented and evaluated (ii) The review of the optimization methods includes aspects such as cells type, temperature variation, and efficiency (iii) A suggested proposal to further advance optimization of such systems is provided	Review of solar energy systems' optimization in terms of issues and challenges and a breakdown of the existing methods in terms of their function, research gaps, and constraints	Medium

Mazzeo et al. (2021)	2021	(i) A description and review of indicators and optimization algorithms	Provided a comprehensive review of PV-wind hybrid systems from 550 research articles in terms of study methodology, combined analysis (of studies), and statistical analysis	High
		(ii) A review of system configurations and components		
		(iii) Technical analysis of the reviewed literature		
Liu et al. (2022)	2022	(i) A description of common AI-related approaches for large-scale hybrid renewable energy integration	Emphasizes the role of artificial intelligence (AI) techniques in the large-scale integration of renewable energy for achieving carbon neutrality	High
		(ii) A discussion of advanced applications of AI techniques for those systems		
		(iii) Describes optimization and the system from generator and distribution network sides		
Alabi et al. (2022)	2022	(i) Overview of integrated energy systems (IES)	Provides a comprehensive overview of IES systems which includes technologies, modeling methods, configurations, along with a critical review of optimization methods and machine-learning (ML) techniques used in IES research.	High
		(ii) Applications of optimization techniques in IES		
		(iii) Applications of machine-learning and deep-learning applications in IES		
Khan et al. (2022)	2022	(i) Introduction of optimization framework	Offers a comprehensive literature survey on various optimization techniques used in the research and development of hybrid renewable energy system with hydrogen (HRES-H2)	Low
		(ii) Review of literature covering optimization techniques such as classical, modern, hybrid meta-heuristic, and software tools		
		(iii) Overview of optimization techniques used for HRES		
Arsad et al. (2022)	2022	(i) Analytical evaluation of the highly cited articles discussing hybrid energy storage (HydESS) systems	Focuses on hydrogen energy storage systems and their integration with other renewable energy systems from 120 publications	Low
		(ii) Detailed analysis of the literature related to HydESS		
		(iii) Future directions for HydESS		

(continued)

Table 20.1 (continued)

Study	Year	Features	Study focus	Emphasis on hybrid PV system sizing using AI/ANN
Gusain et al. (2023)	2023	(i) Optimization methodologies for PV-wind systems (ii) Discusses the recent optimization techniques in the literature (iii) Provides future scope in this area	The review provides a critical comparative analysis of HRES optimization methodologies including objective functions, decision variables, and evaluation indicators	High
Islam and Roy (2023)	2024	(i) A critical analysis of different AI techniques for renewables integration is provided (ii) Applications of ANN, CNN, and PSO in solar energy (iii) Integration of AI techniques in solar, wind, hybrid renewable, and energy storage systems	Overview of AI techniques in different renewable energy systems such as solar, wind, hybrid renewables, and energy storage systems. In addition to reviewing techno-economic feasibility studies	High
Maghami and Mutambara (2023)	2024	(i) Discussion on energy management (ii) Review of energy unit sizing techniques (iii) Review of AI solutions for demand side management (DSM)	Emphasizes the integration of AI techniques in hybrid energy systems (HESs) as a solution to the common issues related to costs, complexity, and reliability	High
Basnet et al. (2023)	2024	(i) Overview of HRES architecture (ii) Optimization methods and solvers of HRES (iii) Discussion on energy management strategy (EMS)	Reviews the recent developments in the area of energy management strategies of HRES with consideration of locations, applications, and technical, social, and environmental factors	High
Thirunavukkarasu et al. (2023)	2024	(i) Classical optimization methods for sizing HRES (ii) AI-based optimization techniques (iii) Hybrid and recently developed AI algorithms	Offers a concise overview of the recent advancements in optimizing various hybrid renewable energy systems (HRESs) through the utilization of different optimization techniques	High

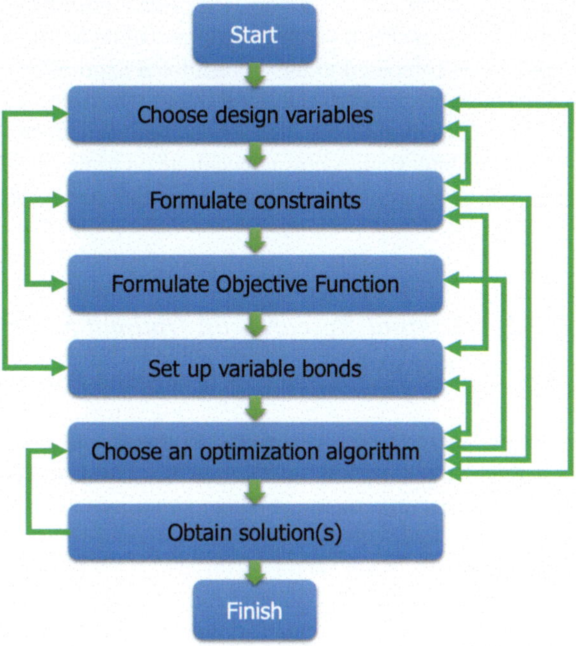

Fig. 20.8 Flowchart of the procedure used for optimal sizing of HRES (Bukar & Tan, 2019)

From the existing literature, it is evident that researchers initially relied on graphical and iterative methods to address optimal sizing problems. However, as the complexity of constraints and objectives grew, the focus shifted toward employing multi-objective optimization algorithms to tackle these challenges effectively. The adoption of such algorithms enabled researchers to handle the increasing complexity and achieve more comprehensive solutions in the optimization process. The general methodology that can be considered the trend for HRES optimization is shown in Fig. 20.8.

The main points to summarize the review of other literature review studies on this topic are provided below:

1. Artificial algorithms have attracted considerable attention in optimizing the sizes of hybrid energy systems (HESs) owing to their capacity to handle intricate, nonlinear, and multi-objective challenges. This has led to a shift from classical methods to artificial approaches in HES size optimization.
2. While the utilization of the simulated annealing (SA) algorithm for HES size optimization is not widespread, it holds the advantage of evading local optima. Therefore, combining SA with other evolutionary algorithms and/or search methods can enhance the accuracy of optimal solutions and improve global search capabilities.
3. Particle swarm optimization (PSO) has found extensive application in HES sizing. However, the conventional PSO algorithm often encounters premature convergence. As a result, recent studies have observed a decrease in the exclusive

use of traditional PSO for HES size optimization. Instead, researchers are employing modified versions and improved variants of PSO, which yield superior outcomes compared to the conventional approach.

4. Given the complexity of optimizing hybrid energy systems, it is essential to account for multiple objective functions and constraints to achieve high-quality results. Many studies now concentrate on triple-objective optimization problems, as they offer more realistic solutions and, consequently, more accurate optimal outcomes.

20.3.2 Original Research on Optimal Sizing of Hybrid PV Systems

Many studies have been carried out to investigate novel algorithms and techniques for optimal sizing of hybrid PV systems. These studies span from optimizing the input parameters, predicting weather conditions, and optimizing the system design based on the load requirements and other constraints.

In their study, Maleki and Pourfayaz (2015a) focused on analyzing a photovoltaic (PV)/wind generator (WG)/diesel hybrid system in a remote region of Rafsanjan, Iran. They created mathematical models for each component and developed a probabilistic solo algorithm (HS) to optimize sizing and cost-effectively meet load requirements. A comprehensive analysis was conducted, comparing harmony search (HS), simulated annealing (SA), and the HOMER software to gauge their accuracy. The investigation also encompassed the assessment of multiple hybrid configurations, such as PV/WG/diesel/fuel cell (FC), PV/diesel/FC, WG/diesel/FC, PV/diesel/battery, WG/diesel/battery, diesel-only, and PV/WG/diesel/battery systems. This assessment considered both cost-effectiveness and the sources of air pollution (CO_2, SO_2, and NO_2). Figure 20.9 showcases a schematic of the PV/WG/diesel/fuel cell (FC) system. The methodology was implemented using MATLAB, where the discrete harmony search (DHS) and discrete simulated annealing (DSA) algorithms were employed to optimize PV panels, wind generators, and hydrogen tanks within the PV/WG/diesel/FC system. Additionally, PV panels, wind generators, and batteries within the PV/WG/diesel/battery system were optimized using the same algorithms. Notably, DHS exhibited the most precise and cost-effective outcomes, followed by DSA and HOMER software.

Jakhrani et al. (2012) presented an innovative analytical framework to efficiently determine the dimensions of independent photovoltaic (PV) systems while adhering to desired cost-effectiveness and reliability criteria. They deduced straightforward algebraic expressions for the optimal area of the PV array (A_{opt}), the optimal utility of battery storage capacity (Cu,opt), and the integration constant (k_{opt}). Comparisons were made with other methods such as Aguiar's, Graham's Markvart, Bartoll's, and Barra's methods at LLP 0, 0.01, and 1.00. The proposed method showed smaller PV array capacity and larger battery storage capacity, outperforming Barra's and Bartoli's methods. It also demonstrated better results compared to Markvart's

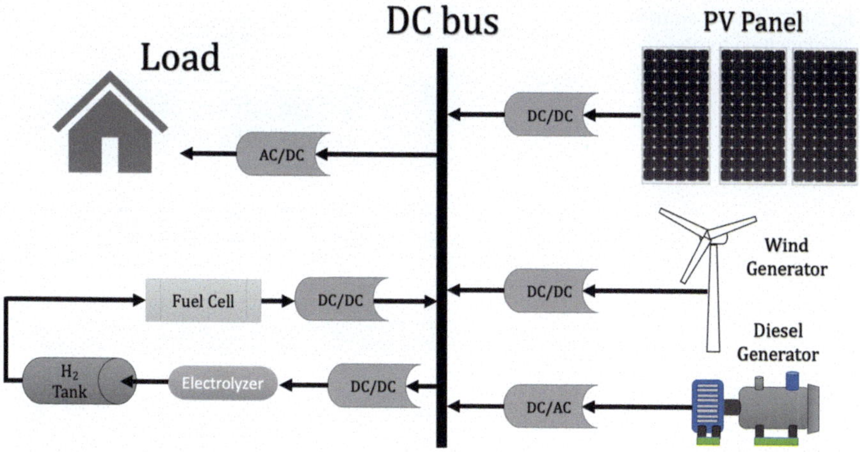

Fig. 20.9 A schematic redrawing of the PV/WG/diesel/fuel cell (FC) system proposed by Maleki and Pourfayaz (2015a)

method. However, achieving zero LLP required an oversized system with a significant budget. In a case study in Kuching, Malaysia, the proposed method effectively determined the optimal PV array area and battery storage capacity, considering energy demand, insolation, meteorological conditions, requested LLP values, and element costs. Overall, the proposed model proved to be more effective and practical than existing models. Figures 20.10 and 20.11 showcase the methodology flowcharts for sizing stand-alone PV systems and hybrid PV systems, respectively.

Maleki and Askarzadeh (2014) evaluated and compared four heuristic algorithms (PSO, TS, SA, HS) for optimizing PV/wind/FC and PV/wind/battery systems in terms of accuracy, computational cost, and load satisfaction. PSO showed the most promising results, outperforming TS, SA, and HS in total annual cost (TAC). It is considered one of the most popular and efficient optimization algorithms due to its simplicity and ease of implementation. MATLAB was used to execute the algorithms and determine the optimal number of PV panels, wind turbines, and hydrogen tanks for the PV/wind/FC system and PV panels, wind turbines, and batteries for the PV/wind/battery system. The study found that while hydrogen storage systems are not economically competitive with batteries, they have other benefits such as being environmentally friendly and having a smaller footprint.

Maleki and Pourfayaz (2015b) evaluated the performance of seven heuristic algorithms (PSO, TS, SA, IPSO, IHS, IHSBSA, ABSO) for optimizing the sizing of PV/WT/battery, WT/battery, and PV/battery hybrid systems. The objective was to continuously satisfy load demand with minimal total annual cost (TAC), considering capital and maintenance costs. Heuristic algorithms are superior for solving non-convex and nonlinear optimization problems such as hybrid system sizing. Mathematical models were introduced for each component, and MATLAB was used to execute the algorithms and find the optimal number of PV panels, wind turbines, and batteries. PV/WT/battery system was economically better

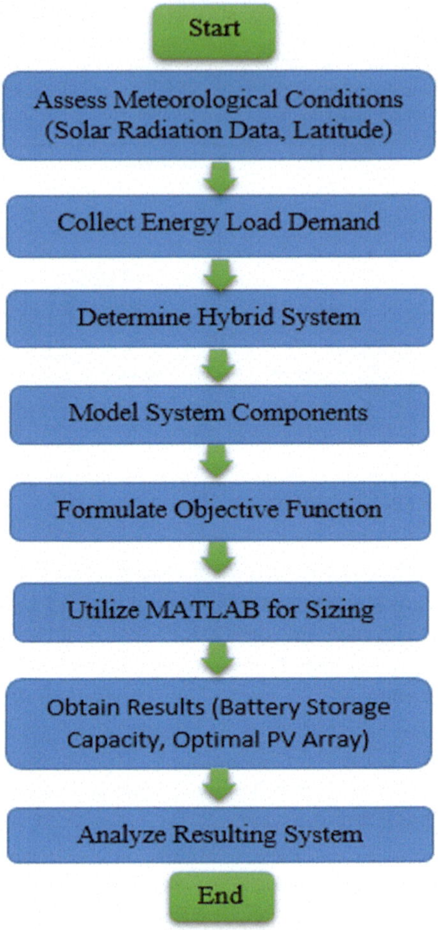

Fig. 20.10 Flowchart of the procedure used for the optimal sizing of off-grid PV systems

for LPSPmax = (0%, 0.3%, 1%, 2%), while PV/battery system was preferable at LPSPmax = 5%. Among the algorithms, ABSO performed the best and quickly found optimal sizing, with the average ranks as ABSO > IHSBSA > IHS > IPSO > PSO > TS > SA.

Hadidian Moghaddam et al. (2019) aimed to optimize the sizing and energy management of a stand-alone hybrid photovoltaic/wind system based on hydrogen storage (PWFHS) while considering reliability indices (LOEE and LOLEs). They used the flower pollination algorithm (FPA) to minimize the total net present cost of PV panels, wind turbines, and fuel cells based on hydrogen storage. The algorithm's performance was compared with teaching-learning-based optimization (TLBO) and particle swarm optimization (PSO). The proposed FPA quickly found optimal sizing with faster convergence in fewer iterations compared to TLBO and PSO. The hybrid PV/WT/FC system was the most cost-effective for various reliability index values, while the PV/FC system had lower cost and better reliability than the WT/FC

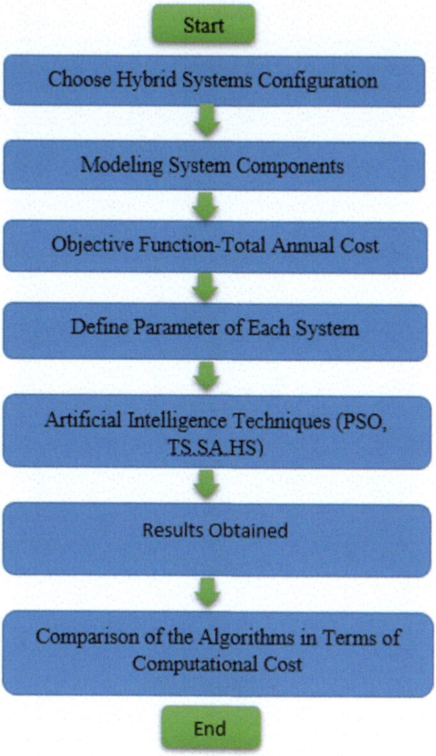

Fig. 20.11 Flowchart of the procedure used for optimal sizing of the (**a**) hybrid PV, wind and fuel cell, and (**b**) hybrid PV, wind, and battery systems

system. Tables 20.2 and 20.3 provide a deconstruction of the reviewed studies with hybrid PV systems, while Table 20.4 provides a summary of the different system configurations studied in the literature.

Table 20.4 showcases a sample of the studies in this research area and the type of system configurations that are generally considered for hybrid PV systems. It is clear that solar thermal collectors are not usually considered for such systems even though they provide useful heat for various thermal applications. Moreover, biomass is rarely considered as well. The most considered types are PV-wind systems, followed by PV-wind-diesel generator systems.

20.4 Artificial Neural Networks and Predictive Algorithms

Artificial neural networks (ANNs) and predictive algorithms have become instrumental in the field of renewable energy, providing valuable insights and solutions for various applications. With the rapid growth and integration of renewable energy sources, such as solar, wind, and hydroelectric power, there is a need for accurate

Table 20.2 Summary of reviewed studies on sizing of hybrid PV systems

Reference	Year published	Location, country	System configuration	Sizing technique/software	Load (type and size)
Maleki and Pourfayaz (2015a)	2015	Rafsanjan, Iran	Stand-alone photovoltaic/wind/diesel system with battery and fuel cell storage devices	HOMER and MATLAB software (DHS and DSA)	Electric (4.97 kW/h at peak during a year)
Jakhrani et al. (2012)	2012	Kuching, Sarawak, Malaysia	Stand-alone photovoltaic (PV)/battery storage	Novel analytical method	Electric (5000.00 kWh/day)
					For domestic purpose
Maleki and Askarzadeh (2014)	2014	Rafsanjan, Iran	Hydrogen-based stand-alone photovoltaic/wind hybrid system	MATLAB software (PSO, TS, HS, SA)	Electric (2.90 Kw/h at peak during a day)
					Referring to Fig. 20.4
Maleki and Pourfayaz (2015b)	2015	Rafsanjan, South of Iran	Autonomous hybrid photovoltaic/wind/battery/power system	MATLAB software (ABSO, IHSBSA, IHS, IPSO, PSO, TS, SA)	Electric (7.3 kWh/h at peak during a year)
					Referring to Fig. 20.5
Hadidian Moghaddam et al. (2019)	2019	Northwest region of Iran	Stand-alone hybrid photovoltaic/wind system based on hydrogen storage	MATLAB software (FPA, TLBO, PSO)	Electric (50 kW/h at peak during a year)

Table 20.3 Summary of methodology and outcomes of the reviewed studies on sizing of hybrid PV systems

Reference	Methodology	Outcomes
Maleki and Pourfayaz (2015a)	The study introduces and applies a discrete version of the discrete harmony search (DHS) algorithm to achieve the optimal sizing of system components, ensuring load satisfaction in the most economically efficient manner. Furthermore, the DHS algorithm is utilized to assess the volume of CO_2 emissions generated through the utilization of a diesel generator in both PV/WG/diesel/fuel cell and PV/WG/diesel/battery systems. The outcomes obtained through the harmony search (HS) algorithm are juxtaposed with the results derived from the discrete simulated annealing (DSA) algorithm and the HOMER software	The results from HOMER and DSA are less accurate than those produced by DHS
		The hybrid system containing PV, wind turbine, diesel generator, and a battery system is more cost-effective than the PV, wind turbine, diesel generator, and fuel cell system. However, the latter can be more reliable with enhanced fuel cell and electrolyzer efficiency
		The best option for providing power is the PV with diesel generator and fuel cells, compared to the remaining systems
		The system with the least pollutions due to emissions is the hybrid PV, wind generator, and diesel generator system with battery storage
Jakhrani et al. (2012)	Creating an innovative analytical approach to ascertain the optimal dimensions of an independent photovoltaic (PV) system. This method relies on both reliability and cost considerations. This is achieved through the utilization of monthly average daily data concerning solar radiation and various meteorological factors	It is more practical in terms of reliability and cost and simple to implement for the size optimization of stand-alone PV systems as compared to conventional methods
		The output of the system is provided directly without the need for calculating the battery storage capacity (C_b) and the PV array capacity (C_a)
		At LLP 0.1 and LLP 0.01, the proposed method leads to small size of PV array capacity and larger battery storage capacity
Maleki and Askarzadeh (2014)	Evaluating and comparing the performance of four different heuristic algorithm techniques for optimum sizing of a hydrogen-based stand-alone photovoltaic/wind hybrid system in terms of accuracy, computational cost, and loads required satisfaction	PSO leads to smaller total annual cost
		Based on the average ranks, the order of the search power of the algorithms is as follows: PSO > TS > HS > SA
		Hydrogen energy storage systems do not offer cost competitiveness when compared to battery storage systems which is indicated through the total annual cost

(continued)

Table 20.3 (continued)

Reference	Methodology	Outcomes
Maleki and Pourfayaz (2015b)	Seven distinct heuristic methods are employed to determine the optimal quantities of each component, with the objective of minimizing the total annual cost (TAC). Additionally, the optimization process takes into account the maximum permissible probability of power supply loss (LPSPmax) to ensure the establishment of a dependable system	ABSO is more promising than those of the other algorithms
		Based on the average ranks, the order of the search power of the algorithms is as follows: ABSO > IHSBSA > IHS > IPSO > PSO > TS > SA
		At LPSPmax = (0%, 0.3% and 1%, 2%) among the battery-based hybrid system, PV/WT/battery system is, economically, a better choice for power generation
		At LPSPmax = 5%, economically the PV/battery system is a better choice
Hadidian Moghaddam et al. (2019)	Employing a novel and intelligent algorithm named the flower pollination algorithm (FPA). The study aims to achieve the optimal configuration and effective control of an independent hybrid photovoltaic/wind system dependent on hydrogen storage. This endeavor involves the consideration of three distinct reliability indices, namely, loss of expected energy (LOEE) and loss of expected load (LOLE), quantified at reliability levels of 1%, 3%, and 5%	FPA reaches lower TNPC than TLBO and PSO algorithms and also lower LOLE, LOEE, and EENS in different hybrid system combinations and different reliability indices
		(FPA) quickly finds the optimum sizing of the different systems with more convergence speed in less iteration
		The hybrid PV/WT/FC system is the most cost-effective hybrid system for different LOLEmax (%) and LOEEmax(%)
		The hybrid PV/FC system has lower cost and better reliability than hybrid WT/FC system

Table 20.4 Summary of system configurations used in the literature

Reference	System components						
	PV module	Wind turbine	Fuel cell	Diesel generator	Solar thermal collector	Battery system	Other
Maleki and Pourfayaz (2015a)	✓	✓	✓	✓		✓	
Fares et al. (2022)	✓	✓				✓	
Abd El-Sattar et al. (2021)	✓	✓				✓	Biomass
Boussetta et al. (2017)	✓	✓					
Rajkumar et al. (2011)	✓	✓				✓	
Al Garni et al. (2022)	✓	✓		✓		✓	
Abba et al. (2021)	✓	✓				✓	
Altun and Kilic (2020)	✓	✓		✓		✓	
Jedou et al. (2021)	✓			✓		✓	
Zeljković et al. (2022)	✓	✓		✓		✓	
Afrouzi et al. (2021)	✓	✓					
Halabi et al. (2017)	✓			✓		✓	
Esan et al. (2019)	✓			✓		✓	
Elmaadawy et al. (2020)	✓	✓		✓		✓	
Fodhil et al. (2019)	✓			✓		✓	
Babatunde et al. (2022)	✓	✓	✓			✓	
Sagani et al. (2017)	✓	✓				✓	
Ashourian et al. (2013)	✓	✓		✓		✓	
Zhang et al. (2019)	✓	✓				✓	
Dufo-López et al. (2017)	✓	✓				✓	

prediction models to optimize their utilization and integration into existing energy systems (Rahman et al., 2021). ANNs and predictive algorithms offer powerful tools for forecasting energy generation, optimizing system performance, and enabling effective decision-making in the renewable energy sector.

Artificial neural networks are computational models inspired by the structure and functioning of biological neural networks in the human brain. ANNs consist of interconnected nodes, known as neurons, which process and transmit information (Jain et al., 1996). By simulating the behavior of neural networks, ANNs can learn from data, recognize patterns, and make predictions or decisions based on the learned knowledge. ANNs excel at handling complex, nonlinear relationships in data and have proven to be highly effective in a wide range of applications, including renewable energy. Meanwhile, predictive algorithms are mathematical models or techniques that analyze historical data to make predictions about future outcomes. These algorithms use patterns and relationships in the available data to forecast future events or behaviors. In the context of renewable energy, predictive algorithms can leverage historical weather data, energy production data, demand patterns, and other relevant variables to estimate future energy generation, consumption, or system performance (Ghritlahre & Prasad, 2018). These algorithms are crucial for optimizing the integration of renewable energy sources into the grid, planning energy storage systems, and managing energy demand (Hassan et al., 2014). Figure 20.12 shows an illustration of a NN used to predict available PV power based on weather parameters.

The use of ANNs and predictive algorithms in renewable energy has numerous applications. Some key areas where these technologies are applied include the following:

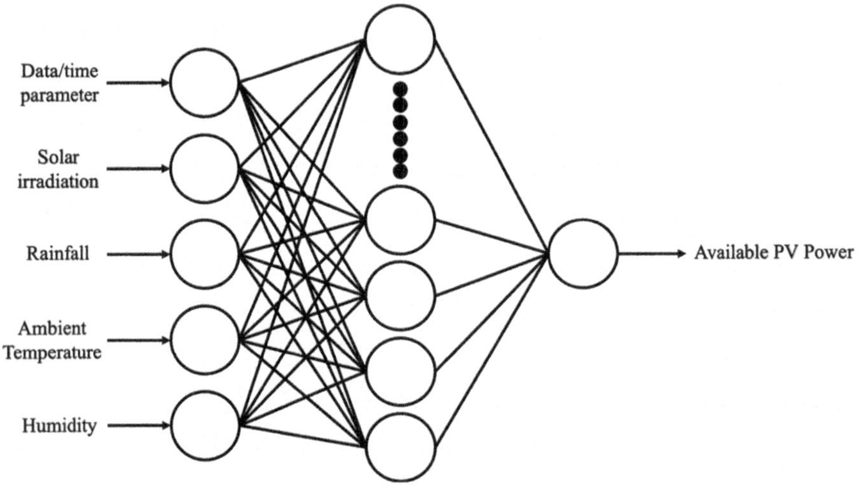

Fig. 20.12 Illustration of an ANN developed to estimate power yield of PV modules based on environmental conditions for a more accurate system sizing

(i) Energy generation forecasting: ANNs and predictive algorithms can analyze historical weather data, solar radiation levels, wind speed, and other parameters to forecast energy generation from renewable sources. Accurate predictions enable efficient grid management, improved energy trading, and optimal utilization of renewable resources.
(ii) Load forecasting: Predictive algorithms can analyze historical energy consumption data to forecast future demand patterns. This information helps utilities and grid operators plan and optimize energy generation and distribution, leading to improved energy management and cost reduction.
(iii) System optimization: ANNs and predictive algorithms can optimize the configuration and control of renewable energy systems, such as photovoltaic arrays, wind farms, and energy storage systems. By considering various factors, including weather conditions, system parameters, and energy demand, these technologies can determine the optimal operating settings for maximizing energy production or minimizing costs.
(iv) Energy trading and market analysis: ANNs and predictive algorithms can analyze market data, price fluctuations, and historical energy production/consumption patterns to support energy trading decisions. These technologies enable traders and market participants to make informed decisions, mitigate risks, and identify profitable opportunities.

20.4.1 Research Methodologies for ANN-Based Optimization of Hybrid PV Systems

The researchers utilized a combined approach of the flower pollination algorithm (FPA) and simulated annealing (SA), integrating computational fluid dynamics (CFD) simulation. Their aim was to optimize a stand-alone PV/wind/battery system (Tahani et al., 2015). By employing a hybrid FPA/SA algorithm, the researchers sought to minimize costs and enhance reliability of the system. Furthermore, they investigated the impact of the tilt angle of PV panels on wind speed through CFD simulation. The decision variables were defined as the tilt angle of the PV panel, the number of batteries, and the number of PV panels, while the objective functions included the payback time and loss of power supply probability.

In reference (Alsayed et al., 2013), a new optimization method using "multi-criteria decision analysis (MCDA)" is introduced, aiming to determine the optimal system size by ranking different available alternatives. This MCDA approach provides the flexibility to incorporate additional constraints in order to observe their impact on system sizing. Apart from the conventional analysis of the levelized cost of electricity (LCOE) and system performance (LPSP), the authors have also incorporated parameters to consider the synergistic effects of wind and solar characteristics, power fluctuations when connected to the grid, and the overall cost of the system.

20.4.2 Evaluation Metrics for ANN-Based Algorithms

According to the literature, the evaluation metrics of AI-based algorithms for sizing hybrid PV systems are helpful in assessing the performance of those algorithms/models. The metrics found in the literature include root mean squared error (RMSE), coefficient of determination (R^2), mean absolute error (MAE), mean square error (MSE), and mean absolute percentage error (MAPE).

RMSE is a widely used metric for measuring the accuracy of a model's predictions. It calculates the average difference between the predicted values and the actual values (ground truth) and then takes the square root of the mean of the squared differences. The formula for RMSE is as follows:

$$\mathrm{RMSE} = \sqrt{\frac{\sum(\mathrm{predicted} - \mathrm{actual})^2}{N}} \tag{20.1}$$

where Σ denotes the sum over all data points, *predicted* is the predicted value from the model, *actual* is the actual value (ground truth), and N is the total number of data points. A lower RMSE indicates better performance, as it represents a smaller average error between predicted and actual values.

The coefficient of determination (R^2) is a statistical measure that represents the proportion of the variance in the dependent variable (target) that is predictable from the independent variables (features) in the model. It provides a measure of how well the model fits the data compared to a simple mean. The formula of R^2 is provided below:

$$R^2 = 1 - \frac{\sum(\mathrm{predicted} - \mathrm{actual})^2}{\sum(\mathrm{actual} - \mathrm{mean})^2} \tag{20.2}$$

where Σ denotes the sum over all data points, *predicted* is the predicted value from the model, *actual* is the actual value (ground truth), and *mean* is the mean of the actual values. R^2 ranges from 0 to 1. A higher R-squared value indicates a better fit of the model to the data.

The mean absolute error (MAE) measures the average absolute difference between the predicted and actual values. Unlike RMSE, it does not involve squaring the errors, which can be beneficial if you want to avoid giving more weight to large errors. The formula for MAE is as follows:

$$\mathrm{MAE} = \frac{\sum|\mathrm{predicted} - \mathrm{actual}|}{N} \tag{20.3}$$

where Σ denotes the sum over all data points, *predicted* is the predicted value from the model, *actual* is the actual value (ground truth), and N is the total number of data points. Similar to RMSE, a lower MAE indicates better performance.

Similar to RMSE, the mean squared error (MSE) is a metric used to measure the average squared difference between predicted and actual values. The main difference is that MSE is not rooted, so it is more sensitive to large errors:

$$\mathrm{MSE} = \frac{\sum (\mathrm{predicted} - \mathrm{actual})^2}{N} \tag{20.4}$$

where Σ denotes the sum over all data points, *predicted* is the predicted value from the model, *actual* is the actual value (ground truth), and *N* is the total number of data points. MSE is closely related to RMSE (root mean squared error), which is simply the square root of MSE. Both MSE and RMSE are used to evaluate the accuracy of regression models, with RMSE being more interpretable as it is in the same units as the target variable. A lower MSE (or RMSE) value indicates better model performance as it signifies smaller errors between the predicted and actual values.

The mean absolute percentage error (MAPE) is a percentage-based metric that measures the average absolute percentage difference between predicted and actual values. It is often used in time series forecasting and is helpful for interpreting errors relative to the magnitude of the target variable. It provides a percentage-based measure of the model's accuracy, making it easy to interpret errors relative to the magnitude of the target variable. The formula for MAPE is as follows:

$$\mathrm{MAPE} = \sum \left(\left| \frac{(\mathrm{predicted} - \mathrm{actual})}{\mathrm{actual}} \right| \right) * (100/N) \tag{20.5}$$

where Σ denotes the sum over all data points, *predicted* is the predicted value from the model, *actual* is the actual value (ground truth), and *N* is the total number of data points. MAPE calculates the absolute percentage error for each data point, sums them up, and then takes the average by dividing by the number of data points. Finally, the result is multiplied by 100 to express the error as a percentage. It is important to note that MAPE is more suitable for situations where the target variable has a nonzero mean and does not contain any zero or near-zero values (as division by small values could lead to instability). If the target variable can have zero values, alternative metrics such as symmetric mean absolute percentage error (SMAPE) or mean absolute scaled error (MASE) might be considered.

These metrics provide a quantitative way to compare different AI techniques and assess how well they perform in sizing hybrid PV systems. Researchers often use a combination of these metrics to get a comprehensive understanding of the model's accuracy and predictive capabilities. Table 20.5 provides a summary of evaluation metrics used for different sizing techniques in the literature.

Table 20.5 Summary of evaluation metrics for different AI techniques

Reference	System configuration	Optimal sizing algorithm	Evaluation metric			
			R^2	Coefficient of correlation (R)	MSE	RMSE
Yap and Karri (2015)	A PV/diesel hybrid system		0.9919	–	–	-
			0.9925			
Sadeghi et al. (2020)	PV/wind/battery/EVs	Multi-objective particle swarm optimization	–	–	–	–
Charrouf et al. (2020)	ANN power management	Hybrid PV-wind desalination system	–	0.99995	–	–
Gang et al. (2014)	Ground source heat pump (GSHP) systems	ANN-backpropagation (BP) network	–	–	0.0833	–
					0.1462	
					0.1580	
					0.1574	
					0.1439	
					0.1558	
Jani et al. (2016)	Solid desiccant-vapor compression hybrid air-conditioning system		0.958	0.99934	N/A	–
Roseline et al. (2022)	Hybrid PV/wind energy systems		0.79	–	–	–
			0.74			
			0.80			
Amadou et al. (2023)	Photovoltaic system		–	–	0.8523	0.029
Abba et al. (2021)	Stand-alone photovoltaic (PV)/wind/battery hybrid		0.5081	0.7128	0.0589	0.2427
			0.6789	0.8240	0.0384	0.1961
			0.9569	0.9782	0.0052	0.0718
			0.9628	0.9812	0.0044	0.0667
			0.9872	0.9936	0.0015	0.0392
			0.2739	0.5234	0.0599	0.2448
			0.3209	0.5665	0.0560	0.2367
			0.6186	0.7865	0.0315	0.1774
			0.7188	0.8478	0.0232	0.1523
			0.9755	0.9877	0.0020	0.0450

20.5 Critical Review and Challenges

1. The off-grid application of PV, wind, and battery systems holds significant promise. By improving battery lifespan and power converter efficiency, this combination can gain wider acceptance, driven by its environmentally beneficial aspect of producing zero emissions.
2. Including the tilt angle of the PV array as a constraint in the size optimization problem is crucial, as it directly affects the accuracy of the optimal solutions.
3. The optimization outcomes are significantly affected by the height of the wind turbine (WT) and the area it sweeps. Therefore, it is essential to incorporate these constraints in the optimization problem to achieve meaningful results (Al-Falahi et al., 2017).
4. The researchers have predominantly chosen artificial intelligent algorithms for optimization strategies, as evident from Table 20.1. Recent studies on optimal sizing heavily rely on algorithms such as genetic algorithms (GAs), modified GA such as nondominated sorting differential evolution algorithm II (NDSA II), particle swarm optimization (PSO) and its variants, ant colony optimization (ACO), artificial bee colony (ABC), and improved bat algorithm (IBA).
5. The reason behind this preference lies in the heuristic/metaheuristic nature of these algorithms. They possess the ability to efficiently find near-optimal solutions, which is advantageous for size optimization problems. These problems involve a vast number of potential solutions, making it unnecessary to pinpoint the exact optimal solution, as a near-optimal solution that satisfies all constraints is sufficient.
6. In addition to size optimization, another challenge tackled by researchers is the optimal allocation of various components in the microgrid. This allocation aims to minimize the overall system loss.
7. Regarding the vulnerability of the system, most researchers correctly identify the battery as the most susceptible component. Issues related to battery lifespan and cost are significant concerns, as batteries typically need replacement around three to four times within the typical plant life cycle, and they also demand frequent maintenance.
8. Numerous investigations have integrated limitations related to battery lifespan into the optimization procedure. These constraints encompass factors such as the count of charge or discharge cycles and the extent of discharge. By enforcing these restrictions, batteries can be preserved in an optimal charge state and experience a diminished cycle count, leading to an enhancement in their longevity. Although these strategies offer a certain degree of extension to battery life, there exists an urgent requirement for extensive research in battery technology. This research aims to not only continue extending lifespan but also to substantially curtail costs and the need for replacements within hybrid renewable energy systems (HRESs).

9. To enhance the dependability of the system, certain studies have examined whether a battery or a diesel generator serves as a more economically viable third source within HRES. Although the diesel generator emerges as a cost-effective choice, it brings about environmental risks stemming from emissions.

A variety of storage systems are accessible, including ultra-capacitors, flywheel storage, hydrogen storage, superconducting magnetic energy storage (SMES), compressed air energy storage (CAES), and pumped hydro storage. CAES and pumped hydro offer larger storage capacities, but their applicability is restricted to particular locations. SMES stands out in its capability to store substantial power with an efficiency of roughly 98%, yet it is hindered by limited storage duration and high cost. Hydrogen storage mandates high-pressure hydrogen tanks and operates at an efficiency of approximately 40–60%. Ultra-capacitors boast a lengthy life cycle but are suitable for shorter-term power storage. Each storage system carries its own merits and limitations, underscoring the importance of selecting the most fitting system according to the specific demands and constraints of the HRES.

In the existing literature, various energy management strategies for hybrid renewable energy systems (HRESs) have been explored. Many of these strategies have addressed issues related to power flow, power control, and efficient battery utilization to enhance battery lifespan. However, the analysis of the literature highlights the need for a comprehensive control strategy that addresses all aspects of energy management in the field. Such a strategy would be beneficial to tackle the diverse challenges faced in HRES effectively.

One significant issue in HRES is the intermittent nature of power generated by renewable sources. This power fluctuation poses a serious concern for future grid integration, particularly as the share of renewable energy is expected to increase. While only a few researchers have identified this problem of power fluctuations injected into the grid, it demands more attention and research to find effective solutions.

Converters have a vital function in shifting power among distinct elements of HRES. Various sorts of converters have been devised and utilized for this objective. These encompass dedicated converters for photovoltaic and wind systems, along with unified multi-input converters. In order to tackle concerns related to power quality and the seamless integration with power grids, proficient converters and their corresponding control approaches assume a pivotal role.

Implementing local dc-micro grids can help address power quality and stability issues at the user end while reducing the number of converters required in the system.

The expense of generating electricity through HRES continues to exceed that of traditional methods. It is urgent to make substantial enhancements in the effectiveness and affordability of HRES elements. Governments bear the crucial duty of fostering research in these domains and advocating for power generation through HRES, given that renewable sources are the sole sustainable and environmentally friendly energy sources for a more reliable and efficient energy infrastructure.

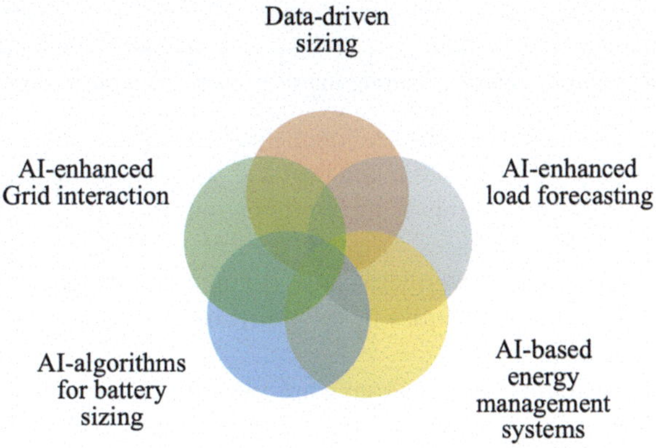

Fig. 20.13 Venn diagram of the research directions related to this topic utilizing AI techniques

The research in this topic is headed in the direction of improving the predictions for energy generation and consumption patterns by using more machine-learning algorithms and utilizing data-driven sizing that caters to specific environmental conditions and energy needs (this area is generally done through case study research), in addition to enhancing the accuracy of the load forecasting by using AI techniques that will help in determining the appropriate system size to meet the energy demands efficiently.

The development of an energy management system that is cost-effective and environmentally friendly is also a major direction to improve the reliability of these systems and allow for dynamic real-time decisions to maximize the efficiency. Thus, the optimization of battery selection and sizing is also an area of interest and can be done by also considering the weather forecasts, energy prices, and load profiles. AI techniques are used, in this context, to optimize the capacity of the storage system to balance between the demand and the supply.

These techniques can also be invoked to enable the system to have a more intelligent interaction with the grid, i.e., optimized grid interaction that is based on grid stability, electricity tariffs, and renewable energy availability. The AI would also be very beneficial in the process of monitoring the performance of the hybrid PV system and detecting the faults or anomalies in real time, which in turns helps in preventative and reliability maintenance schemes. Figure 20.13 provides a basic Venn diagram of the research directions related to this area of research.

20.6 Conclusions and Recommendations

In conclusion, the application of artificial neural networks (ANNs) in PV system sizing research has demonstrated promising outcomes. ANNs enable more comprehensive analyses and a better understanding of the factors that influence system sizing decisions. The utilization of ANNs allows for the incorporation of dynamic and uncertain conditions, such as varying energy demand patterns or changing solar radiation levels and enhancing the adaptability and flexibility of PV system designs:

1. The fusion of optimization methods involves merging two or three algorithms for the purpose of achieving the most effective sizing and positioning of solar energy systems. By employing hybrid optimization, one can explore numerous variables and uncertainties to attain the most favorable outcomes, thereby enhancing efficiency. Nevertheless, identifying the appropriate combination of hybrid optimization techniques poses a difficult and time-consuming challenge (Al-Shahri et al., 2021).
2. The conventional approaches for sizing photovoltaic (PV) systems, particularly the hybrid method, offer effective solutions. Nevertheless, these techniques prove inadequate when it comes to sizing PV systems in remote regions where the necessary data is inaccessible. Furthermore, most of the aforementioned methods rely on long-term meteorological data, including variables such as total solar irradiation, air temperature, and wind speed. As a result, these methods become unusable when such meteorological data is unavailable, particularly in isolated areas (Mellit & Kalogirou, 2008).
3. To ensure the sizing of a hybrid system, it is crucial to meet the loss of power supply probability (LPSP) requirements, which determine the system's ability to fulfill the load demands. Once the configurations that fulfill the LPSP criteria are identified, the cost of the system becomes a determining factor in selecting the solution with the lowest overall cost (Rajkumar et al., 2011).
4. When weather data is available, conventional sizing approaches are employed and classified based on the concepts of energy balance and supply reliability. However, these conventional techniques rely on long-term meteorological data for sizing PV systems, which may not be accessible in remote isolated locations. As a result, the lack of weather data in remote sites has prompted researchers to explore artificial intelligence (AI) techniques such as artificial neural networks (ANNs), fuzzy logic (FL), genetic algorithms (GAs), or hybrid combinations of these techniques (Bajpai & Dash, 2012).

Overall, the integration of ANNs in the optimal sizing of PV systems offers a powerful and data-driven approach that significantly contributes to informed decision-making processes. It empowers researchers and practitioners to optimize system configurations, improve energy efficiency, and enhance the economic and environmental viability of PV installations. By harnessing the potential of ANNs, the PV industry can drive the adoption of optimized and sustainable PV systems, leading to a cleaner and more resilient energy future.

References

Abba, S. I., Rotimi, A., Musa, B., Yimen, N., Kawu, S. J., Lawan, S. M., & Dagbasi, M. (2021). Emerging Harris Hawks Optimization based load demand forecasting and optimal sizing of stand-alone hybrid renewable energy systems – A case study of Kano and Abuja, Nigeria. *Results in Engineering, 12*, 100260.

Abd El-Sattar, H., Sultan, H. M., Kamel, S., Khurshaid, T., & Rahmann, C. (2021). Optimal design of stand-alone hybrid PV/wind/biomass/battery energy storage system in Abu-Monqar, Egypt. *Journal of Energy Storage, 44*, 103336.

Afrouzi, H. N., Hassan, A., Wimalaratna, Y. P., Ahmed, J., Mehranzamir, K., Liew, S. C., & Malek, Z. A. (2021). Sizing and economic analysis of stand-alone hybrid photovoltaic-wind system for rural electrification: A case study Lundu, Sarawak. *Cleaner Engineering and Technology, 4*, 100191.

Agnew, S., & Dargusch, P. (2015). Effect of residential solar and storage on centralized electricity supply systems. *Nature Climate Change, 5*(4), 315–318.

Ahmad, A. S., Hassan, M. Y., Abdullah, M. P., Rahman, H. A., Hussin, F., Abdullah, H., & Saidur, R. (2014). A review on applications of ANN and SVM for building electrical energy consumption forecasting. *Renewable and Sustainable Energy Reviews, 33*, 102–109.

Airashidi, M. R., & EL-Naggar KM. (2010). Load term electric load forecasting based on particle swarm optimization. *Applied Energy, 87*, 320–326.

Al Busaidi, A. S., Kazem, H. A., Al-Badi, A. H., & Khan, M. F. (2016). A review of optimum sizing of hybrid PV–Wind renewable energy systems in Oman. *Renewable and Sustainable Energy Reviews, 53*, 185–193.

Al Garni, H. Z., Mas'ud, A. A., Baseer, M. A., & Ramli, M. A. (2022). Techno-economic optimization and sensitivity analysis of a PV/Wind/diesel/battery system in Saudi Arabia using a combined dispatch strategy. *Sustainable Energy Technologies and Assessments, 53*, 102730.

Alabi, T. M., Aghimien, E. I., Agbajor, F. D., Yang, Z., Lu, L., Adeoye, A. R., & Gopaluni, B. (2022). A review on the integrated optimization techniques and machine learning approaches for modeling, prediction, and decision making on integrated energy systems. *Renewable Energy, 194*, 822–849.

Al-Falahi, M. D., Jayasinghe, S. D. G., & Enshaei, H. J. E. C. (2017). A review on recent size optimization methodologies for standalone solar and wind hybrid renewable energy system. *Energy Conversion and Management, 143*, 252–274.

Alsayed, M., Cacciato, M., Scarcella, G., & Scelba, G. (2013). Multicriteria optimal sizing of photovoltaic/wind turbine grid connected systems. *IEEE Transactions on Energy Conversion, 28*(2), 370–379.

Al-Shahri, O. A., Ismail, F. B., Hannan, M. A., Lipu, M. H., Al-Shetwi, A. Q., Begum, R. A., et al. (2021). Solar photovoltaic energy optimization methods, challenges and issues: A comprehensive review. *Journal of Cleaner Production, 284*, 125465.

Altun, A. F., & Kilic, M. (2020). Design and performance evaluation based on economics and environmental impact of a PV-wind-diesel and battery standalone power system for various climates in Turkey. *Renewable Energy, 157*, 424–443.

Al-Waeli, A. H., Sopian, K., Kazem, H. A., & Chaichan, M. T. (2017). Photovoltaic/Thermal (PV/T) systems: Status and future prospects. *Renewable and Sustainable Energy Reviews, 77*, 109–130.

Amadou, B. A., Ndiaye, A., & Mbodji, S. (2023). Power optimization of a photovoltaic system with artificial intelligence algorithms over two seasons in tropical area. *MethodsX, 10*, 101959.

Anoune, K., Bouya, M., Astito, A., & Abdellah, A. B. (2018). Sizing methods and optimization techniques for PV-wind based hybrid renewable energy system: A review. *Renewable and Sustainable Energy Reviews, 93*, 652–673.

Apostoleris, H., Sgouridis, S., Stefancich, M., & Chiesa, M. (2019). Utility solar prices will continue to drop all over the world even without subsidies. *Nature Energy, 4*(10), 833–834.

Arsad, A. Z., Hannan, M. A., Al-Shetwi, A. Q., Mansur, M., Muttaqi, K. M., Dong, Z. Y., & Blaabjerg, F. (2022). Hydrogen energy storage integrated hybrid renewable energy systems: A review analysis for future research directions. *International Journal of Hydrogen Energy, 47*(39), 17285–17312.

Ashourian, M. H., Cherati, S. M., Mohd Zin, A. A., Niknam, N., Mokhtar, A. S., & Anwari, M. (2013). Optimal green energy management for Island resorts in Malaysia. *Renewable Energy, 51*, 36–45.

Babatunde, O. M., Munda, J. L., & Hamam, Y. (2022). Hybridized off-grid fuel cell/wind/solar PV/battery for energy generation in a small household: A multi-criteria perspective. *International Journal of Hydrogen Energy, 47*(10), 6437–6452.

Bajpai, P., & Dash, V. (2012). Hybrid renewable energy systems for power generation in stand-alone applications: A review. *Renewable and Sustainable Energy Reviews, 16*(5), 2926–2939.

Basnet, S., Deschinkel, K., Le Moyne, L., & Péra, M. C. (2023). *A review on recent standalone and grid integrated hybrid renewable energy systems: System optimization and energy management strategies*. Renewable Energy Focus.

Bermejo, F., Jesús, J. F., Fernández, G., Polo, F. O., & Márquez, A. C. (2019). A review of the use of artificial neural network models for energy and reliability prediction. A study of the solar PV, hydraulic and wind energy sources. *Applied Sciences, 9*(9), 1844.

Boussetta, M., El Bachtiri, R., Khanfara, M., & El Hammoumi, K. (2017). Assessing the potential of hybrid PV–Wind systems to cover public facilities loads under different Moroccan climate conditions. *Sustainable Energy Technologies and Assessments, 22*, 74–82.

Bukar, A. L., & Tan, C. W. (2019). A review on stand-alone photovoltaic-wind energy system with fuel cell: System optimization and energy management strategy. *Journal of Cleaner Production, 221*, 73–88.

Charrouf, O., Betka, A., Abdeddaim, S., & Ghamri, A. (2020). Artificial Neural Network power manager for hybrid PV-wind desalination system. *Mathematics and Computers in Simulation, 167*, 443–460.

Chrifi-Alaoui, L., Drid, S., Ouriagli, M., & Mehdi, D. (2023). Overview of photovoltaic and wind electrical power hybrid systems. *Energies, 16*(12), 4778. https://doi.org/10.3390/en16124778

Dellosa, J. T., Panes, M. J. C., & Espina, R. U. (2021). Techno-economic analysis of a 5 MWp solar photovoltaic system in The Philippines. In *2021 IEEE international conference on environment and electrical engineering and 2021 IEEE industrial and commercial power systems Europe (EEEIC/I&CPS Europe)* (pp. 1–6). IEEE.

Dufo-Lopez, R., Bernal-Agustín, J. L., & Contreras, J. (2007). Optimization of control strategies for stand-alone renewable energy systems with hydrogen storage. *Renewable Energy, 32*(7), 1102–1126.

Dufo-López, R., Alfredo Fernández-Jiménez, L., Ramírez-Rosado, I. J., Sergio Artal-Sevil, J., Domínguez-Navarro, J. A., & Bernal-Agustín, J. L. (2017). Daily operation optimisation of hybrid stand-alone system by model predictive control considering ageing model. *Energy Conversion and Management, 134*, 167–177.

Egido, M. A., & Lorenzo, E. (1992). The sizing of stand alone PV systems: A review and proposed method. *Solar Energy Materials and Solar Cells, 26*(1–2), 51–69.

El-Hefnawi, S. H. (1998). Photovoltaic diesel-generator hybrid power system sizing. *Renewable Energy, 13*(1), 33–40.

Elmaadawy, K., Kotb, K. M., Elkadeem, M. R., Sharshir, S. W., Dán, A., Moawad, A., & Liu, B. (2020). Optimal sizing and techno-enviro-economic feasibility assessment of large-scale reverse osmosis desalination powered with hybrid renewable energy sources. *Energy Conversion and Management, 224*, 113377.

Erdinc, O., & Uzunoglu, M. (2012). Optimum design of hybrid renewable energy systems: Overview of different approaches. *Renewable and Sustainable Energy Reviews, 16*(3), 1412–1425.

Esan, A. B., Agbetuyi, A. F., Oghorada, O., Ogbeide, K., Awelewa, A. A., & Esan Afolabi, A. (2019). Reliability assessments of an islanded hybrid PV-diesel-battery system for a typical rural community in Nigeria. *Heliyon, 5*(5).

Fares, D., Fathi, M., & Mekhilef, S. (2022). Performance evaluation of metaheuristic techniques for optimal sizing of a stand-alone hybrid PV/wind/battery system. *Applied Energy, 305*, 117823.

Fodhil, F., Hamidat, A., & Nadjemi, O. (2019). Potential, optimization and sensitivity analysis of photovoltaic-diesel-battery hybrid energy system for rural electrification in Algeria. *Energy, 169*, 613–624.

Gang, W., Wang, J., & Wang, S. (2014). Performance analysis of hybrid ground source heat pump systems based on ANN predictive control. *Applied Energy, 136*, 1138–1144.

Ghritlahre, H. K., & Prasad, R. K. (2018). Application of ANN technique to predict the performance of solar collector systems – A review. *Renewable and Sustainable Energy Reviews, 84*, 75–88.

Gusain, C., Tripathi, M. M., & Nangia, U. (2023). Study of meta-heuristic optimization methodologies for design of hybrid renewable energy systems. *Thermal Science and Engineering Progress, 101711*.

Hadidian Moghaddam, M. J., Kalam, A., Nowdeh, S. A., Ahmadi, A., Babanezhad, M., & Saha, S. (2019). Optimal sizing and energy management of stand-alone hybrid photovoltaic/wind system based on hydrogen storage considering LOEE and LOLE reliability indices using flower pollination algorithm. *Renewable Energy, 135*, 1412–1434.

Hakimi, S. M., & Tafreshi, S. M. M. (2007). Unit sizing of a stand-alone hybrid power system using particle swarm optimization (PSO). *IEEE ICAL, 1*(8–21), 3107–3112.

Hakimi, S. M., & Tafreshi, S. M. M. (2009). Optimal sizing of a stand-alone hybrid power system via particle swarm optimization for Kahnouj area in the South-East of Iran. *Renewable Energy, 34*, 1855–1862.

Halabi, L. M., Mekhilef, S., Olatomiwa, L., & Hazelton, J. (2017). Performance analysis of hybrid PV/diesel/battery system using HOMER: A case study Sabah, Malaysia. *Energy Conversion and Management, 144*, 322–339.

Hernández, J. C., Medina, A., & Jurado, F. (2007). Optimal allocation and sizing for profitability and voltage enhancement of PV systems on feeders. *Renewable Energy, 32*(10), 1768–1789.

Hontoria, L., Aguilera, J., & Zufiria, P. (2005). A new approach for sizing stand alone photovoltaic systems based in neural networks. *Solar Energy, 78*(2), 313–319.

IEA Solar PV. https://www.iea.org/reports/solar-pv

Iqbal, M., Azam, M., Naeem, M., Khwaja, A. S., & Anpalagan, A. (2014). Optimization classification, algorithms and tools for renewable energy: A review. *Renewable and Sustainable Energy Reviews, 39*, 640–654.

IRENA. Renewable Energy Capacity Statistics 2022.

Islam, S., & Roy, N. K. (2023). Renewables integration into power systems through intelligent techniques: Implementation procedures, key features, and performance evaluation. *Energy Reports, 9*, 6063–6087.

Jain, A. K., Mao, J., & Moidin Mohiuddin, K. (1996). Artificial neural networks: A tutorial. *Computer, 29*(3), 31–44.

Jakhrani, A. Q., Othman, A. K., Rigit, A. R. H., Samo, S. R., & Kamboh, S. A. (2012). A novel analytical model for optimal sizing of standalone photovoltaic systems. *Energy, 46*(1), 675–682.

Jani, D. B., Mishra, M., & Sahoo, P. K. (2016). Performance prediction of solid desiccant–vapor compression hybrid air-conditioning system using artificial neural network. *Energy, 103*, 618–629.

Jedou, E., Ndongo, M., Ali, M. M., Yetilmezsoy, K., Bilal, B., Ebeya, C. C., Kébé, C. M. F., Ndiaye, P. A., Kıyan, E., & Bahramian, M. (2021). A cartographic approach coupled with optimized sizing and management of an on-grid hybrid PV-solar-battery-group based on the state of the sky: An african case study. *Solar Energy, 227*, 101–115.

Jha, S. K., Bilalovic, J., Jha, A., Patel, N., & Zhang, H. (2017). Renewable energy: Present research and future scope of Artificial Intelligence. *Renewable and Sustainable Energy Reviews, 77*, 297–317.

Karabacak, K., & Cetin, N. (2014). Artificial neural networks for controlling wind–PV power systems: A review. *Renewable and Sustainable Energy Reviews, 29*, 804–827.

Kennedy, J. (1997). The particle swarm: Social adaptation of knowledge. In *Proceedings of the IEEE international conference on evolutionary computation, Indianapolis, IN* (pp. 303–308).

Khan, T., Yu, M., & Waseem, M. (2022). Review on recent optimization strategies for hybrid renewable energy system with hydrogen technologies: State of the art, trends and future directions. *International Journal of Hydrogen Energy, 47*(60), 25155–25201.

Khare, A., & Rangnekar, S. (2014). Optimal sizing of a grid integrated solar photovoltaic system. *IET Renewable Power Generation, 8*(1), 67–75.

Khare, V., Nema, S., & Baredar, P. (2016). Solar–wind hybrid renewable energy system: A review. *Renewable and Sustainable Energy Reviews, 58*, 23–33.

Khatib, T., Mohamed, A., & Sopian, K. (2012). Optimization of a PV/wind micro-grid for rural housing electrification using a hybrid iterative/genetic algorithm: Case study of Kuala Terengganu, Malaysia. *Energy Buildings, 47*, 321–331.

Khatib, T., Mohamed, A., & Sopian, K. (2013). A review of photovoltaic systems size optimization techniques. *Renewable and Sustainable Energy Reviews, 22*, 454–465.

Koutroulis, E., Kolokotsa, D., Potirakis, A., & Kalaitzakis, K. (2006a). Methodology for optimal sizing of stand-alone photovoltaic/wind-generator systems using genetic algorithms. *Solar Energy, 80*(9), 1072–1088.

Koutroulis, E., Kolokotsa, D., Potirakis, A., & Kalaitzakis, K. (2006b). Methodology for optimal sizing of stand-alone photovoltaic/wind-generator systems using genetic algorithms. *Solar Energy, 80*, 1072–1088.

Liu, Z., Sun, Y., Xing, C., Liu, J., He, Y., Zhou, Y., & Zhang, G. (2022). Artificial intelligence powered large-scale renewable integrations in multi-energy systems for carbon neutrality transition: Challenges and future perspectives. *Energy and AI*, 100195.

Lorenz, P., Pinner, D., & Seitz, T. (2008). The economics of solar power. *The McKinsey Quarterly, 4*, 66–78.

Luna-Rubio, R., Trejo-Perea, M., Vargas-Vázquez, D., & Ríos-Moreno, G. J. (2012). Optimal sizing of renewable hybrids energy systems: A review of methodologies. *Solar Energy, 86*(4), 1077–1088.

Lupangu, C., & Bansal, R. C. (2017). A review of technical issues on the development of solar photovoltaic systems. *Renewable and Sustainable Energy Reviews, 73*, 950–965.

Maghami, M. R., & Mutambara, A. G. O. (2023). Challenges associated with hybrid energy systems: An artificial intelligence solution. *Energy Reports, 9*, 924–940.

Mahesh, A., & Sandhu, K. S. (2015). Hybrid wind/photovoltaic energy system developments: Critical review and findings. *Renewable and Sustainable Energy Reviews, 52*, 1135–1147.

Maleki, A., & Askarzadeh, A. (2014). Comparative study of artificial intelligence techniques for sizing of a hydrogen-based stand-alone photovoltaic/wind hybrid system. *International Journal of Hydrogen Energy, 39*(19), 9973–9984.

Maleki, A., & Pourfayaz, F. (2015a). Sizing of stand-alone photovoltaic/wind/diesel system with battery and fuel cell storage devices by harmony search algorithm. *Journal of Energy Storage, 2*, 30–42.

Maleki, A., & Pourfayaz, F. (2015b). Optimal sizing of autonomous hybrid photovoltaic/wind/battery power system with LPSP technology by using evolutionary algorithms. *Solar Energy, 115*, 471–483.

Mazzeo, D., Matera, N., De Luca, P., Baglivo, C., Congedo, P. M., & Oliveti, G. (2021). A literature review and statistical analysis of photovoltaic-wind hybrid renewable system research by considering the most relevant 550 articles: An upgradable matrix literature database. *Journal of Cleaner Production, 295*, 126070.

Mellit, A. (2010). ANN-based GA for generating the sizing curve of stand-alone photovoltaic systems. *Advances in Engineering Software, 41*(5), 687–693.

Mellit, A., & Kalogirou, S. A. (2008). Artificial intelligence techniques for photovoltaic applications: A review. *Progress in Energy and Combustion Science, 34*(5), 574–632.

Mellit, A., Menghanem, M., & Bendekhis, M. (2005a). Artificial neural network model for prediction solar radiation data: Application for sizing stand-alone photovoltaic power system. In *IEEE power engineering society general meeting* (pp. 40–44). IEEE.

Mellit, A., Benghanem, M., Hadj Arab, A., & Guessoum, A. (2005b). An adaptative artificial neural network model for sizing stand-alone photovoltaic systems: Application for isolated sites in Algeria. *Renewable Energy, 30*(10), 1501–1524.

Mellit, A., Benghanem, M., & Kalogirou, S. A. (2007). Modeling and simulation of a stand-alone photovoltaic system using an adaptative artificial neural network: Proposition for a new sizing procedure. *Renewable Energy, 32*(2), 285–313.

Memon, S. A., & Patel, R. N. (2021). An overview of optimization techniques used for sizing of hybrid renewable energy systems. *Renewable Energy Focus, 39*, 1–26.

Mirzahosseini, A. H., & Taheri, T. (2012). Environmental, technical and financial feasibility study of solar power plants by RETScreen, according to the targeting of energy subsidies in Iran. *Renewable and Sustainable Energy Reviews, 16*(5), 2806–2811.

Moayedi, H., Mosallanezhad, M., Rashid, A. S. A., Jusoh, W. A. W., & Muazu, M. A. (2020). A systematic review and meta-analysis of artificial neural network application in geotechnical engineering: theory and applications. *Neural Computing and Applications, 32*, 495–518.

Mohammed, A., Pasupuleti, J., Khatib, T., & Elmenreich, W. (2015). A review of process and operational system control of hybrid photovoltaic/diesel generator systems. *Renewable and Sustainable Energy Reviews, 44*, 436–446.

Muselli, M. N. G. L. A., Notton, G., & Louche, A. (1999). Design of hybrid-photovoltaic power generator, with optimization of energy management. *Solar Energy, 65*(3), 143–157.

Nadjemi, O., Nacer, T., Hamidat, A., & Salhi, H. (2017). Optimal hybrid PV/wind energy system sizing: Application of cuckoo search algorithm for Algerian dairy farms. *Renewable and Sustainable Energy Reviews, 70*, 1352–1365.

Pnas. Website. https://www.pnas.org/doi/full/10.1073/pnas.0904101106. Accessed 1 July 2023.

Prasetyaningsari, I., Setiawan, A., & Setiawan, A. A. (2013). Design optimization of solar powered aeration system for fish pond in Sleman Regency, Yogyakarta by HOMER software. *Energy Procedia, 32*, 90–98.

Rahman, M. M., Shakeri, M., Tiong, S. K., Khatun, F., Amin, N., Pasupuleti, J., & Hasan, M. K. (2021). Prospective methodologies in hybrid renewable energy systems for energy prediction using artificial neural networks. *Sustainability, 13*(4), 2393.

Rajkumar, R. K., Ramachandaramurthy, V. K., Yong, B. L., & Chia, D. B. (2011). Techno-economical optimization of hybrid pv/wind/battery system using Neuro-Fuzzy. *Energy, 36*(8), 5148–5153.

Ridha, H. M., Gomes, C., Hizam, H., Ahmadipour, M., Heidari, A. A., & Chen, H. (2021). Multi-objective optimization and multi-criteria decision-making methods for optimal design of standalone photovoltaic system: A comprehensive review. *Renewable and Sustainable Energy Reviews, 135*, 110202.

Roseline, J., Femila, D. D., Saravana Selvan, M., Yuvaraj, P., Duraipandy, S. S., Kumar, A. R., Prasad, R. S., & Mohanavel, V. (2022). Neural Network modelling for prediction of energy in hybrid renewable energy systems. *Energy Reports, 8*, 999–1008.

Rumbayan, M., Abudureyimu, A., & Nagasaka, K. (2012). Mapping of solar energy potential in Indonesia using artificial neural network and geographical information system. *Renewable and Sustainable Energy Reviews, 16*(3), 1437–1449.

Sadeghi, D., Naghshbandy, A. H., & Bahramara, S. (2020). Optimal sizing of hybrid renewable energy systems in presence of electric vehicles using multi-objective particle swarm optimization. *Energy, 209*, 118471.

Sagani, A., Vrettakos, G., & Dedoussis, V. (2017). Viability assessment of a combined hybrid electricity and heat system for remote household applications. *Solar Energy, 151*, 33–47.

Salam, M. A., Aziz, A., Alwaeli, A. H., & Kazem, H. A. (2013). Optimal sizing of photovoltaic systems using HOMER for Sohar, Oman. *International Journal of Renewable Energy Research, 3*(2), 301–307.

Sharma, N. (2012). Stochastic techniques used for optimization in solar systems: A review. *Renewable and Sustainable Energy Reviews, 16*(3), 1399–1411.

Sharma, D. K., Verma, V., & Singh, A. P. (2014). Review and analysis of solar photovoltaic softwares. *International Journal of Current Engineering and Technology, 4*(2), 725–731.

Siddaiah, R., & Saini, R. P. (2016). A review on planning, configurations, modeling and optimization techniques of hybrid renewable energy systems for off grid applications. *Renewable and Sustainable Energy Reviews, 58*, 376–396.

Sinha, S., & Chandel, S. S. (2015). Review of recent trends in optimization techniques for solar photovoltaic–wind based hybrid energy systems. *Renewable and Sustainable Energy Reviews, 50*, 755–769.

Sivaneasan, B., Yu, C. Y., & Goh, K. P. (2017). Solar forecasting using ANN with fuzzy logic pre-processing. *Energy Procedia, 143*, 727–732.

Solar Heat World Wide. https://www.iea-shc.org/Data/Sites/1/publications/Solar-Heat-Worldwide-2023.pdf

Solarspaces. Website. https://www.solarpaces.org/csp-technologies/csp-potential-solar-thermal-energy-by-member-nation/. Accessed 1 July 2023.

Sommerfeldt, N., & Madani, H. (2017). Revisiting the techno-economic analysis process for building-mounted, grid-connected solar photovoltaic systems: Part 1: Review. *Renewable and Sustainable Energy Reviews, 74*, 1379–1393.

Tahani, M., Babayan, N., & Pouyaei, A. (2015). Optimization of PV/Wind/Battery stand-alone system, using hybrid FPA/SA algorithm and CFD simulation, case study: Tehran. *Energy Conversion and Management, 106*, 644–659.

Thirunavukkarasu, M., Sawle, Y., & Lala, H. (2023). A comprehensive review on optimization of hybrid renewable energy systems using various optimization techniques. *Renewable and Sustainable Energy Reviews, 176*, 113192.

Twaha, S., & Ramli, M. A. (2018). A review of optimization approaches for hybrid distributed energy generation systems: Off-grid and grid-connected systems. *Sustainable Cities and Society, 41*, 320–331.

Upadhyay, S., & Sharma, M. P. (2014). A review on configurations, control and sizing methodologies of hybrid energy systems. *Renewable and Sustainable Energy Reviews, 38*, 47–63.

Xu, D., Kang, L., Chang, L., & Cao, B. (2005). Optimal sizing of standalone hybrid wind/PV power systems using genetic algorithms. In *Canadian conference on electrical and computer engineering, 2005* (pp. 1722–1725). IEEE.

Yap, W. K., & Karri, V. (2015). An off-grid hybrid PV/diesel model as a planning and design tool, incorporating dynamic and ANN modelling techniques. *Renewable Energy, 78*, 42–50.

Zahedi, A. (2006). Solar photovoltaic (PV) energy; latest developments in the building integrated and hybrid PV systems. *Renewable Energy, 31*(5), 711–718.

Zahraee, S. M., Assadi, M. K., & Saidur, R. (2016). Application of artificial intelligence methods for hybrid energy system optimization. *Renewable and Sustainable Energy Reviews, 66*, 617–630.

Zeljković, Č., Mršić, P., Erceg, B., Lekić, Đ., Kitić, N., & Matić, P. (2022). Optimal sizing of photovoltaic-wind-diesel-battery power supply for mobile telephony base stations. *Energy, 242*, 122545.

Zhang, W., Maleki, A., & Rosen, M. A. (2019). A heuristic-based approach for optimizing a small independent solar and wind hybrid power scheme incorporating load forecasting. *Journal of Cleaner Production, 241*, 117920.

Chapter 21
Energy Transition: Dynamics and Prospects

Muhammad Asif

21.1 Introduction

Energy is one of the most critical commodities in the modern age. The human-energy interaction was transformed by the Industrial Revolution of the eighteenth century. Since then, the efficient and extensive use of energy has been crucial to the advancement of humanity. On a variety of fronts, including technological, socio-economic, and geopolitical ones, energy is becoming into a more important commodity. All essential societal functions, including mobility, agriculture, industry, health, education, and trade and commerce, now require energy. Numerous different physical states of energy resources exist, and they can be tapped into and profited from using a variety of technologies. Availability of refined and efficient energy resources has played a decisive role in the advancement of societies (Asif, 2022a).

The human use of energy, especially of fossil fuels, contributes to global warming and climate change. The uncontrolled emissions of greenhouse gases (GHS) are leading to global warming. Climate change, as a result of global warming, is regarded as the biggest challenge facing the world. Different types of energy resources, especially fossil fuels, contribute to greenhouse gas (GHG) emissions. Fossil fuels are considered to be the primary reason for the anthropogenic emission of carbon dioxide (CO_2)—the eighteenth-century Industrial Revolution is considered to have triggered the rapid growth in the release of greenhouse gases (Asif & Muneer, 2007; Asif, 2022b). Climate change is leading to a wide range of consequences such as seasonal disorder; a pattern of intense and more frequent weather-related events such as floods, droughts, storms, heat waves, and wildfires; financial loss; and health problems (Asif, 2019a, 2021a; Qudratullah & Asif, 2020). Climate change is also adversely affecting the water and food supplies around the world.

M. Asif (✉)
King Fahd University of Petroleum and Minerals, Dhahran, Kingdom of Saudi Arabia
e-mail: asifm@kfupm.edu.sa

A. Sayigh (ed.), *Transition Towards a Carbon Free Future*, Innovative Renewable Energy, https://doi.org/10.1007/978-3-031-61660-0_21

Warmer temperatures are increasing sea level as a result of the melting of glaciers. During the twentieth century, global sea level rose by around 20 cm. The pace of rise in sea level is accelerating every year—over the last two decades, it has been almost double that of the last century (UN News, 2023). As a result of warmer temperatures, glaciers are shrinking across the world including in the Himalayas, Alps, Alaska, Rockies, and Africa. An extremely alarming dimension of climate change is that it is growing in momentum. Most of the temperature rise since the Industrial Revolution has occurred since the 1960s. Extreme weather conditions and climate abnormalities are becoming so frequent that the situation is already being widely dubbed as a climate crisis. With the recorded acceleration in the accumulation of greenhouse gases and consequent increase in atmospheric temperature, climate change-driven weather-related disasters are becoming more intense and recurrent. For the last few years, every year sets new records for natural disasters especially heat waves, wildfires, storms, and floods. Extreme weather events are now considered to be a new normal as experts predict more intense natural calamities including wildfires, storms, floods, and droughts. The UN General Secretary Antonio Guterres warns during the first Climate Ambition Summit in New York in 20 September 2023 that "Humanity has opened the gates to hell." He further warned that climate action was being "dwarfed by the scale of the challenge," with humanity heading toward a 2.8 °C temperature rise, increasing danger and instability (UN News, 2023).

The global energy scenario also faces a number of other challenges such as rapid growth in energy demand, depletion of fossil fuel reserves, volatile energy prices, and a lack of universal access to energy. A fast growth in the global energy demand—owing to factors such as surging population, economic and infrastructural development, and urbanization—is adding pressure on the energy supply chain. According to the Energy Information Administration (EIA), between 2018 and 2050, the world's energy requirements are projected to increase by 50% (EIA, 2019). Access to refined energy resources remains a major challenge for significant proportions of population in the developing countries.

The energy sector is experiencing a major transformation in order to address the energy and environmental challenges. The primary aim of this transition is to shift the global energy system away from fossil fuels. Renewable and low-carbon technologies are at the heart of this energy transition. In the recorded history, there have been two major energy transitions. The first one was a shift from wood and biomass to coal during the eighteenth-century Industrial Revolution and the second one was the twentieth-century transition from coal to oil and gas. With the beginning of the twenty-first century, the world is witnessing dawning of the third energy transition (Asif, 2022a).

This paper examines the key dimensions of the unfolding energy transition. It studies the major drivers and technological dimensions of the energy transitioning. It also explores the challenges facing the energy transition. It presents an overview of the key dynamics of the ongoing energy transition. It defines the energy transition to have four main dimensions: decarbonization, energy efficiency, decentralization, and digitalization. The paper also discusses the prospects of these four dimensions

(4Ds) of the energy transition especially in terms of relevant technological and policy advancements.

21.2 Energy Transition: Technological Dimensions

The use of energy is closely linked to the environment (Asif, 2013, 2019b; Nahiduzaman et al., 2018). It is estimated that despite the pledges and efforts by the global community to tackle climate change, CO_2 emissions from energy and industry have increased by 60% since the United Nations Framework Convention on Climate Change (UNFCCC) was signed in 1992 (IEA, 2021). Climate change is already there with its implications such as seasonal disorder, rising sea level, and a trend of more frequent and intense weather-driven disasters such as flooding, droughts, heat waves, wildfires, storms, and associated financial losses (Asif et al., 2017; Khan et al., 2017a). The situation calls for an urgent and paradigm shift across the entire energy sector. Responding to the challenges on hand, and ensuring a supply of energy compatible with the demands of a sustainable future for the planet, the global energy sector is going through a transformation. This energy transition can be defined as follows: "The energy transition is a pathway toward the transformation of the global energy sector from fossil-based to zero-carbon by the second half of this century." At the heart of the ongoing energy transition is the need to reduce energy-related CO_2 emissions to limit climate change (IRENA).

In the recent history, mankind has witnessed two major energy transitions. The first energy transition propelled the Industrial Revolution mainly attributed to coal replacing biomass and wood as a more efficient and effective fuel to drive machines. The second energy transition was a shift from coal to more refined forms of fossil fuels—oil and gas—in the later part of the twentieth century. The world is now experiencing the third energy transition. This energy transition is much more vibrant, intriguing, and impactful compared to the earlier ones. It is fundamentally a sustainability-driven energy pathway with a focus on decarbonization of the energy sector by shifting away from fossil fuels. This energy transition, therefore, can also be termed as "sustainable energy transition" or "low-carbon energy transition." Holistically, however, the ongoing energy transition is not just about going low carbon or shifting away from fossil fuels. It is rather much more diverse and comprehensive in terms of scope and impact. The twenty-first-century energy transition is being propelled by unprecedented developments on the fronts of energy resources and their consumption, technological advancements, socioeconomic and political response, and evolving policy landscape. This energy transition has four key dimensions: decarbonization, energy efficiency, decentralization, and digitalization (Asif, 2020).

21.3 Decarbonization

Decarbonization of the energy sector is at the heart of the energy transition as reduction in CO_2 and other GHG emissions is fundamental in the fight against climate change. The energy sector can be decarbonized through a range of technologies and solutions such as renewable energy, electric vehicles (EVs), hydrogen and fuel cells, carbon capture and storage (CCS), and phasing out of fossil fuels. Renewable energy is the single most critical component of the decarbonization drive.

Through the Paris Agreement, the world has adopted the first ever universally legally binding global climate deal to avoid dangerous climate change by limiting global warming to well below 2 °C. Alarm bells are, however, being frequently rung by the concerned circles including the United Nations' Intergovernmental Panel on Climate Change (IPCC), that the world is seriously overshooting this target. In order to be anywhere closer to achieving this target, the world needs to make major changes in four big global systems: energy, land use, cities, and industry. The energy sector is where the greatest challenges and opportunities exist (Gillam & Asplund, 2021). Energy systems have a huge variation in terms of their associated environmental emissions.

21.3.1 *Renewable Energy*

Renewable energy is the primary pathway for the energy sector's low- or zero-carbon transition. Over the last couple of decades, renewable technologies, especially solar PV and wind turbines, have made great progress in terms of technological developments and economic maturity. The global installed capacity of renewables increased from 2581 GW in 2019 to 2838 GW in 2020, exceeding expansion in the previous year by almost 50%. In 2021, renewable energy installed capacity increased globally by 257 GW, reaching 3064 GW. Over the last few years, renewable energy has increased power generation capacity more than fossil fuels and nuclear power put together. For instance, more than 83% of all new power generation capacity installed globally in 2022 came from renewable sources as shown in Fig. 21.1 (IRENA, 2023).

For several years now in a row, renewable energy is adding more power generation capacity compared to the combined addition by fossil fuels and nuclear power. In the year 2020, for example, renewables contributed to more than 80% of all new power generation capacity added worldwide. The growth of the renewable sector is primarily being propelled by solar and wind power, with the two technologies accounting for 91% of the new renewables added during the year. Renewable energy is already supplying 26% of the global electricity needs. According to IEA, to achieve the net-zero emissions by 2050, almost 90% of global electricity generation is to be supplied from renewables. There was over US$ 303 billion invested in

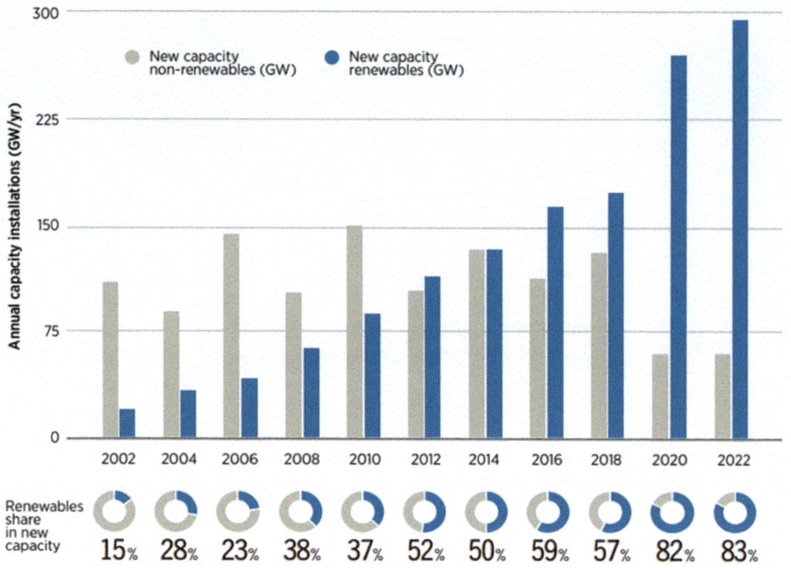

Fig. 21.1 Growth of renewable energy in the power generation capacity

renewable energy projects during the year (REN21, 2020). The upward scale of the renewables' developments can be gauged from the fact that China has started developing the first 100 GW phase of massive solar and wind power initiatives. The initiative is likely to be expanded to several hundreds of GW in capacity as China aims to develop 1200 GW of renewables by 2030 (Scully, 2021). The renewables' growth trends are projected to continue, as the annual capacity addition of solar and wind power is set to grow fourfold between 2020 and 2030 (Gillam & Asplund, 2021). It is also expected that by 2026, solar and wind power will account for around 95% of the total new capacity addition in the power sector.

The success of renewables has been propelled by technological advancements, economy of scale, and supportive policies. Solar and wind power industries are massively benefiting from the scientific and engineering advancements. Solar PV cells, for example, are becoming more efficient and reliable. Concentrated solar PV cells, as shown in Fig. 21.2 (NREL, 2021a), have achieved efficiency figures of over 40%. Figure 21.3 shows the advancements in the PV cell technology (NREL, 2021b). The progress of renewable technologies, especially solar energy systems, is making is significantly being helped by their broadening application domains. The building sector has been a vital area of application for solar PV and solar water heating systems (Dehwah & Asif, 2019; Dehwah et al., 2018; Mahmood et al., 2017; Asif, 2016a, b, 2021b). PV systems are also being installed over agricultural farms, termed as agrivoltaics. For PV and wind turbines, the issue of low-power density is being addressed by offshore applications. Application of PV on water bodies—i.e., lakes, canals, rivers, and oceans—termed as floating PV is becoming popular with added advantage of higher system efficiency. Wind turbines are witnessing

Fig. 21.2 Concentrated solar PV

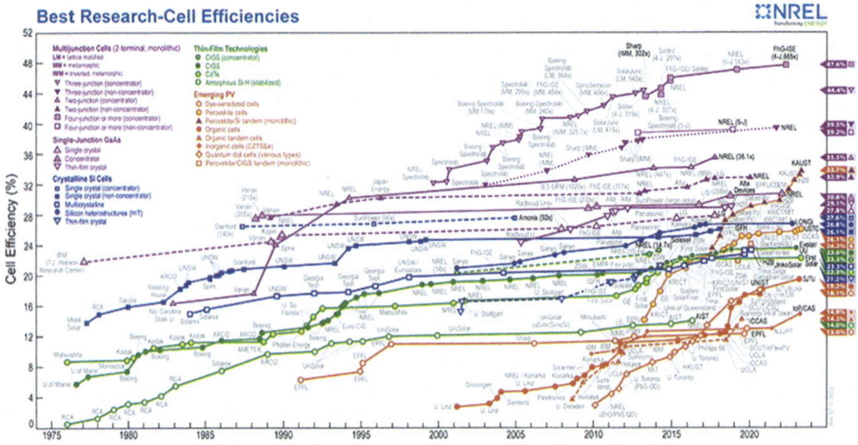

Fig. 21.3 Growth of solar cells in terms of types and efficiency

improvements both at the manufacturing and the installation end. Besides improvements in aerodynamic designs, advanced and sophisticated materials are helping develop larger, lighter, and stronger wind blades. These developments have enabled wind turbines to grow rapidly in size as shown in Fig. 21.4. The offshore application of wind turbines has significantly boosted the capacity factor. Within a couple of decades, larger and more sophisticated wind turbines and better site selection have resulted in the average annual capacity factor increasing from 20–30% to 40–50%. Some offshore wind turbines are now claiming to have a capacity factor of over 60% (Asif, 2022a).

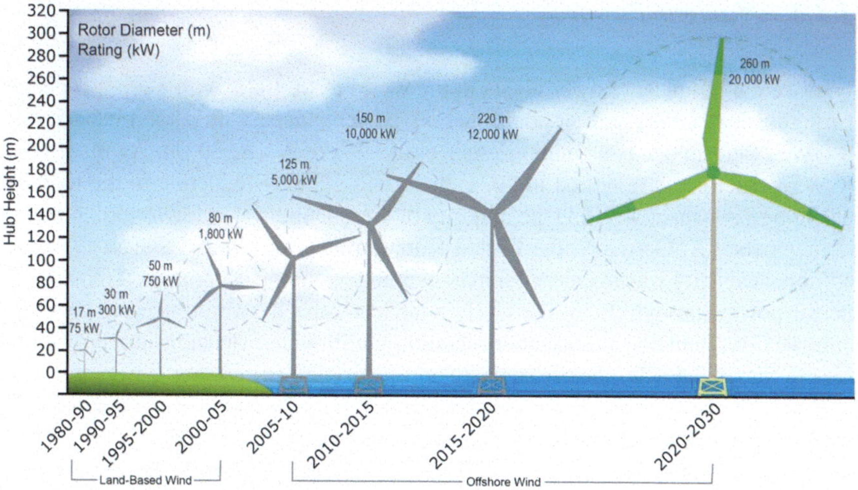

Fig. 21.4 Growth in size of wind turbines

21.3.2 Decarbonization in the Fossil Fuel Sector

Fossil fuels have traditionally led the energy sector, presently contributing to almost 80% of the total supplies. Oil and gas are the main fuels being used in the transport sector. The industrial sector also heavily relies on fossil fuels. Despite the grounds being gained by renewables, in a foreseeable future, fossil fuels are projected to be a major part of the global energy outlook. The emissions associated with fossil fuels however need to be curbed in order to avoid irreparable damage to the planet's ecosystem. Among the three main types of fossil fuels, coal is the most polluting one, while natural gas has the least environmental burdens. Within the fossil fuel sector, there is a gradual shift away from coal especially in power generation, which can also be considered as a decarbonization trend. There is however a need to have more profound decarbonization efforts. There can be two major pathways in this respect, which are carbon capture and storage and the transformation of fossil fuels into hydrogen or hydrogen-rich fuels. Removal of carbon from fossil fuels—which are fundamentally hydrocarbons—allows for the generation of hydrogen as a secondary energy source that does not produce CO_2. Carbon capture and storage (CCS), also termed as carbon capture and sequestration, involves the removal of CO_2 from the direct combustion of fossil fuels for power generation or industrial processes. The idea is to prevent the release of CO_2 into the atmosphere and instead store it in underground geological formations for the long term (Asif, 2022a).

21.3.3 Nuclear Power

There were 442 nuclear power plants in the world as of 2021, with a gross capacity of roughly 394 GWe, and an additional 57 power plants, with a total capacity of 60 GWe, were under construction (Asif et al., 2017; WNA, 2021). As more than 8 GW of reactors were retired in 2021, there was a 3 GW reduction in the world's nuclear power capacity. While the majority of these lengthy shutdowns took place in G7 members, i.e., Germany, the United Kingdom, and the United States, all of the additional capacity was in emerging market and developing economies. To maintain nuclear power through 2030, the world's nuclear capacity would need to grow by about 10 GW annually. G7 members should give lifetime extensions top priority in order to maximize new nuclear capacity and fortify the current low-emission infrastructure. The goals embodied in net-zero targets have fostered the development of novel nuclear power technologies, including as small modular reactors (SMRs), which have a reduced size of under 300 MW per reactor, down to 10 MW. SMRs have the potential to be more affordable, easier, and quicker to build than conventional large reactors. More than 70 designs are currently being worked on. SMRs might be produced in a factory and shipped to the ultimate destination, reducing financial requirements and accelerating project deadlines. With the decarbonization of power systems and the rise in the proportion of solar and wind energy, SMRs could be an important part of solving the increased flexibility required in power generation. They can also be used to produce hydrogen and heat. Expanding nuclear power is necessary to balance off the need for fossil fuels and enhanced renewable energy generation in order to realize the net-zero scenario. Nuclear power has reduced the need for coal, natural gas, and oil since it has been used to produce electricity for more than 50 years, which has led to a reduction of 66 Gt of CO_2 emissions globally (IEA).

21.3.4 Electric Vehicles

Electric vehicles are leading the decarbonization efforts in the transport sector. Electric vehicles are environment-friendly, require low maintenance due to fewer components, are quiet to operate, and offer convenience in urban use. The growth of electric vehicles is being supported by wide-ranging policies. These include standards (such as requiring a certain share of clean vehicles or setting limits for fleet-wide average emission intensity), purchase price subsidies (i.e., tax exemptions or tax credits), incentives encouraging the usage of clean vehicles (financial or nonfinancial incentives including free parking, zero-low road tax, and bus lane usage), pricing of externalities (such as carbon pricing), and scrappage policies targeting emitting vehicles. Active support for research and development (R&D) and infrastructure development are also instrumental in this respect (Asif, 2022b).

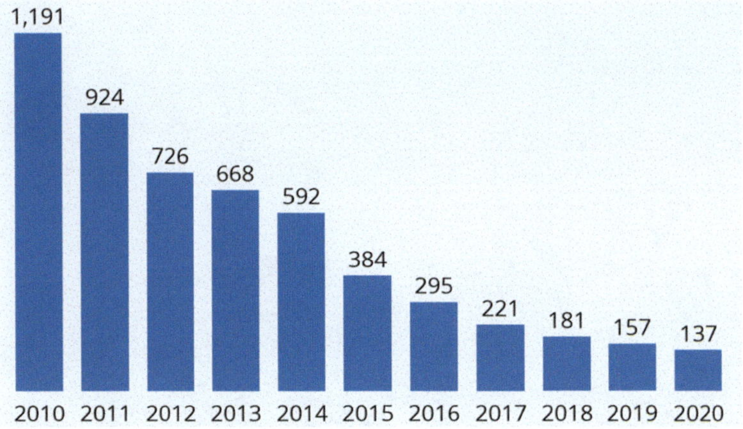

Fig. 21.5 Declining price trend of lithium batteries

Numerous policies are helping to promote the use of electric vehicles. Standards (such as requiring a specific percentage of clean vehicles or setting limits for the fleet's average emissions intensity) are included, as are purchase price subsidies (such as tax exemptions or tax credits), incentives encouraging the use of clean vehicles (such as free parking, zero-low road tax, and bus lane usage), pricing of externalities (such as carbon pricing), and scrappage policies that target emitting vehicles. Additionally, important in this regard are active support for R&D and infrastructure development. Electric vehicles (EVs) are making steep rise in car market share. In 2020, the worldwide sale of EVs, for example, increased by 41% despite the COVID-related economic downturn and a drop of 6% in the overall sale of vehicle. During the same year, Europe recorded the registration of new electric cars increase by 100%, and the number of electric car models available worldwide increased from 260 to 370. The global electric vehicle market saw a 65% YoY growth in 2022 as EV sales climbed over 10.2 million units. EVs accounted for over 14% of the world's passenger vehicle sales in 2022, compared to 9% in 2021 (Counterpoint). Battery price has been an important cost factor in the economics of EVs. Battery prices are experiencing a rapid decline, recording a drop from $1191 in 2010 to $137/kWh in 2020 as shown in Fig. 21.5 (Roper, 2020). While electric mobility is also paving its way in the aviation and ship industry, the sale of electric cars is expected to increase from around 3.5 million in 2020 to over 23 million by 2030 (Adler, 2021).

21.3.5 Energy Storage

Energy storage is an important aspect of the energy value chain. The modern electricity infrastructure, especially in the wake of increasing supplies from

Fig. 21.6 Utility-scale lithium-ion battery

renewable technologies, is finding energy storage critically important for its smooth operations. Energy storage is becoming of particular interest to power sector stakeholders including utilities, end users, grid system operators, and regulators.

Modern renewables such as solar energy and wind power, being dependent on weather conditions, have the intermittency drawback. Solar radiation, for example, is available only during the daytime. However, the daytime availability of solar radiation can be hindered by multiple weather conditions such as rain, snow, fog, and overcast conditions. Issues such as dust storms, smog, haze, and smoke from wildfires also affect the intensity of solar radiation. Similarly, availability of wind is not a constant phenomenon either. Furthermore, even during their spells of availability, solar radiation and wind speed can fluctuate quickly and heavily, accordingly affecting the output from the respective systems. Renewable energy thus needs a backup storage to serve as a reliable source of energy.

Pumped storage hydropower projects have traditionally been used as the optimum large-scale energy storage solution. In recent years, battery technology has seen major techno-economic breakthroughs to become another option for large-scale energy storage. The 100 MW lithium battery storage developed by Tesla in Australia in 2017 has been a turning point in the field of large-scale battery storage. Large- and utility-scale battery storage systems, as shown in Fig. 21.6 (NREL, 2021c), have become a viable option. The United States has planned a 1500 MW/6000 MWh lithium-ion battery project, the first phase of which with a capacity of 300 MW/ 1200 MWh started to operate in December 2020. Australia and the United Kingdom are also developing major battery storage projects. Some of these projects include a 1200 MW project in New South Wales, a 700 MW system by Origin Energy Ltd, a 500 MW system in New South Wales, and a 300 MW facility in Victoria. The United Kingdom has over 1.1 GW of battery storage capacity in operation, while projects of 600 MW of cumulative capacity are under construction.

21.3.6 Hydrogen and Fuel Cells

Hydrogen as a fuel has unique characteristics. Its use does not release any toxic emissions and it has the highest calorific value compared to other commonly used fuels. Hydrogen is the simplest and one of the most plentiful elements in the universe. Despite its abundance, however, hydrogen does not occur independently—it exists bonded with other elements and is available in compound forms such as water, hydrocarbons, and carbohydrates (Asif et al., 2009). Hydrogen can be produced from fossil fuels as well as renewables. In the former case, hydrogen can be produced through various routes such as reformation of natural gas, partial oxidation of heavy fossil fuels, and coal gasification. Production of hydrogen from fossil fuels—making up over 90% of the current hydrogen supplies—however leads to GHG emissions and is typically termed as "blue hydrogen." An environmentally clean option is to produce hydrogen through electrolysis which involves splitting water into hydrogen and oxygen with the help of electricity. Ideally, the electricity used for electrolysis should be from renewables, making it a "clean hydrogen" which is also termed as "green hydrogen." Hydrogen can be stored, transported, and used for energy applications through various technologies. Hydrogen in the capacity of energy vector also has the potential to become the optimum solution for the renewables' intermittency by providing an energy storage solution. The vision of building an energy infrastructure that uses hydrogen as a fuel and energy carrier, a concept called hydrogen economy, is the path toward the full commercial application of hydrogen energy technologies (Muneer et al., 2005). Fuel cell is an important complementary technology. It is a device that converts hydrogen directly into electricity through electrochemical oxidation while generating pure water as its byproduct. Fuel cells come in many types and are classified mainly in terms of the kind of electrolyte they use which in turn influences other features such as the involved electrochemical reaction, required catalysts, the cell's operating temperature range, and the needed fuel (DOE).

21.4 Energy Efficiency

The fast growth in the global demand for energy is pressurizing the entire energy value chain. A one-dimensional approach, of matching the growing energy demands with corresponding capacity addition, is not a sustainable solution, especially when the planet is already overshooting its bio-capacity by almost 70%. A sustainable energy pathway, aiming to satisfy the global energy requirements while protecting the environment, has to begin with decreasing the use of energy through energy efficiency (EE) measures. Energy efficiency is regarded as a better solution to address energy shortages than adding a new capacity. To industrial and commercial entities, energy efficiency delivers economic and environmental gains, besides offering a competitive edge.

Energy efficiency is a broad domain in terms of the nature of facilities, the types of energy losses and wastes, and the range of solutions available to improve efficiency. Given the technological and policy advancements, the field of energy efficiency is continuously evolving. In terms of fundamental approach, energy efficiency solutions can be broadly classified into three types: change of attitude, performance enhancement, and new technologies. An energy efficiency program, typically involving eradication of the unnecessary use of energy and improvement in the efficiency of the required energy, starts with an energy audit exercise. The type of energy audit to be carried out primarily depends upon the scope and objectives of the intended energy efficiency program. The energy audit process is also influenced by factors such as available resources (funding, manpower, and time), type of facility, and provision of data and support. Detailed energy management programs can also include execution of the recommended solutions and post-implementation measurement and verification work to ensure the desired energy-saving goals are achieved.

The use of energy can be reduced across all major sectors including buildings, industry, and transport. Buildings account for over one third of the global energy consumption (Hamida et al., 2021; Ahmed & Asif, 2020; Alawneh et al., 2019; Alazameh & Asif, 2021). Energy use in buildings can be reduced through a range of energy efficiency measures. Energy-efficient solutions for buildings can be broadly classified as active and passive energy saving measures (Ahmed & Asif, 2021). The choice of energy efficiency solutions depends on factors such as the nature of the facility, site condition and local climate, desired levels of comfort and improvement, and financial situation. Through energy efficiency measures, energy demand in existing as well as new buildings can be reduced by 30–80% (Khan & Asif, 2017; Khan et al., 2017b; Mahmood et al., 2017; Ahmed et al., 2019). Energy efficiency in the transport sector can be improved through measures such as incorporating fuel economy standards and eco-driving (REN21, 2022). It is estimated that with the help of proven and commercially viable technologies, energy use in the manufacturing industry can be reduced by 18–26% (Hassan et al., 2017). Energy efficiency is beneficial both at the micro- and macro-levels. While it offers savings on bills to individual consumers, it also helps commercial and industrial entities in terms of economic competitiveness. While energy efficiency helps the utility and energy sector in terms of demand-side management and peak-load shaving, it also helps foster national security. The global economy could increase by $18 trillion by 2035 if energy efficiency is adopted as the "first choice" for new energy supplies, which would also achieve the emission reductions required to limit global warming to 2 °C (Khan et al., 2017b; UNEP). Besides enabling economic growth and improving energy security, energy efficiency can also play a vital role in the fight against climate change as it can deliver more than 40% of the reduction in energy-related greenhouse gas emissions over the next 20 years (IEA, 2020).

21.5 Distributed Generation

Distributed or decentralized generation (DG) is the energy generated close to the point of use. Figure 21.7 highlights how a DG system differs from a centralized generation system. Distributed generation avoids/minimizes transmission and distribution setup, thus saving on cost and losses. It offers better efficiency, flexibility, and economy as compared to large and centralized generation systems. There are several energy technologies that can be used in DG systems depending on the application and type of project. Based on the type of energy resource, DG technologies can be classified into two categories: renewable-based systems and nonrenewable-based systems. Renewable-based DG systems employ technologies such as solar energy, wind power, hydropower, biomass, and geothermal energy. Some of these technologies can be further classified into different types. Solar technologies, for example, can be categorized into solar PV, solar thermal power, and solar water heating. Similarly, biomass can be used to deliver solid fuels, liquid fuels such as biodiesel and bioethanol, and gaseous fuels. Renewable-based DG systems offer several benefits such as reduced greenhouse gas emissions and lower operation and maintenance costs. These systems, however, are typically intermittent and need energy storage to offer reliable solutions. Nonrenewable-based DG technologies are also available in a wide range and may include internal combustion (IC) engine, combined heat and power (CHP), gas turbines, micro-turbines, Stirling engines, and fuel cells. These technologies can use different types of fossil fuels.

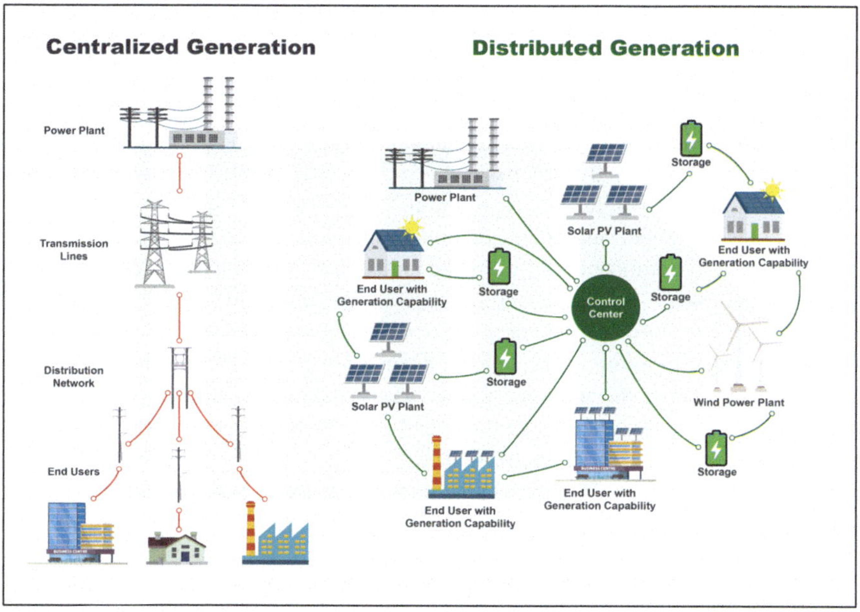

Fig. 21.7 Overview of central and distributed generation systems

Fig. 21.8 Rooftop PV systems

Renewables such as solar and wind power systems are leading the DG landscape. DG is playing an important role in the global electrification efforts and is presenting viable solutions for meeting modern energy needs and enabling the livelihoods of hundreds of millions who still lack access to electricity or clean cooking solutions (Ahmed et al., 2019; Asif, 2021a). Solar PV is one of the most successful DG technologies, especially at small-scale and off-grid levels. The building sector offers a tremendous potential for DG PV systems (Asif, 2016a, b). It is estimated that since 2010, over 180 million off-grid solar systems have been installed including 30 million solar home systems. In 2021, around 7.4 million off-grid solar products were sold. Rooftop PV systems, as shown in Fig. 21.8 (REN21, 2022; NREL, 2021d), make up 40% of the total PV installations worldwide.

21.6 Digitalization

Digitalization is driving the needed fundamental shift in the energy industry which is also disrupting traditional market players (DNV-GL, 2019). Digitalization, also referred as the fourth Industrial Revolution, is a broad term in the context of the energy sector. An important dimension of digitalization is the collection and analysis of energy data to optimize energy demand and supply, to achieve system efficiency and cost-effectiveness. While decarbonization, decentralization, and decreasing use of energy are transforming the energy sector, digitalization—through the proliferation of sensors, computing, communication, and predictive and control techniques—is also set to change the way energy services are realized and delivered. This is accomplished through a range of established and emerging technologies, above all artificial intelligence (AI). Digitalization from the perspective of business opportunities created in the energy sector can be regarded as the use of digital technologies to change a business process and enhance efficiency and revenue; it is the process of moving to a digital business. However it is defined, digitalization is having a profound impact on the global energy scenario. While digitalization is leading to new business models, it is also disrupting existing models of generation, consumption, markets, and businesses and employment, potentially pushing some of the established ones on their way out (IEA, 2017). Digitalization of the energy sector employs technologies such as artificial intelligence, machine learning, big data and data analytics, Internet of Things, cloud computing, blockchain, and robotics and automation. These technologies are at various degrees of techno-economic maturity for their application in the energy sector. Digitalization is revolutionizing the energy sector by improving the productivity, safety, accessibility, and overall sustainability of energy systems. New, smarter ways of modeling, monitoring, analyzing, and forecasting energy production and consumption are helping the sustainable energy transition. With the range of advantages it offers, digitalization is also posing several challenges. Utilities and energy industry are getting exposed to cybersecurity risks as they heavily rely on large datasets in the digital transformation.

21.7 Policy and Investment Trends

Decarbonization is becoming the central part of the national and international energy policy frameworks across the world. As part of greenhouse gas emission targets, which mandate a reduction in overall emissions and can include net-zero and carbon-neutral targets, there have been new emission reduction commitments covering around 47% of total global emissions. Some examples in this respect are of China aiming to become carbon neutral by 2060, Japan by 2050, and the Republic of Korea by 2050. A major development is from the European Union (EU) committing to reduce carbon emissions by 55% from the 1990 levels by 2030 and to become net zero by 2050. The United Kingdom has planned to reduce carbon emissions by 68% by 2030, compared to the 1990 level to go net zero by 2050. A major decarbonization boost with far-reaching impacts at the global level has been US's rejoining of Paris Agreement in 2021. The United States, after having walked out of the Paris Agreement in 2020, under the nationally determined contribution (NDC) has committed to reduce emissions by 50–52% below 2005 levels by 2030, which is equivalent to a reduction of 40–43% below 1990 levels (CAT, 2021). Overall, over 30 nations have incorporated climate neutrality by 2050 (or 2060) in laws, proposed legislation, or a national policy document (Sachs et al., 2021). There have been significant decarbonization efforts on the part of other stakeholders as well including energy, banking, and the corporate sector. The world's leading corporations are becoming increasingly aware of the threats associated with climate change and the business opportunities in taking action. It will help reduce not only their own emissions but also from their business associates.

The decarbonization efforts are being crucially helped by the vital policy advancements in the renewable sector. Conducive policy support has been instrumental in the strong growth of renewable energy in recent years. Governments all over the world have launched a variety of programs and policies in recent decades to encourage the creation and uptake of renewable energy technology. Governments are adopting policies that subtly encourage the use of renewable energy, such as targets for net-zero greenhouse gas emissions, economic and environmental

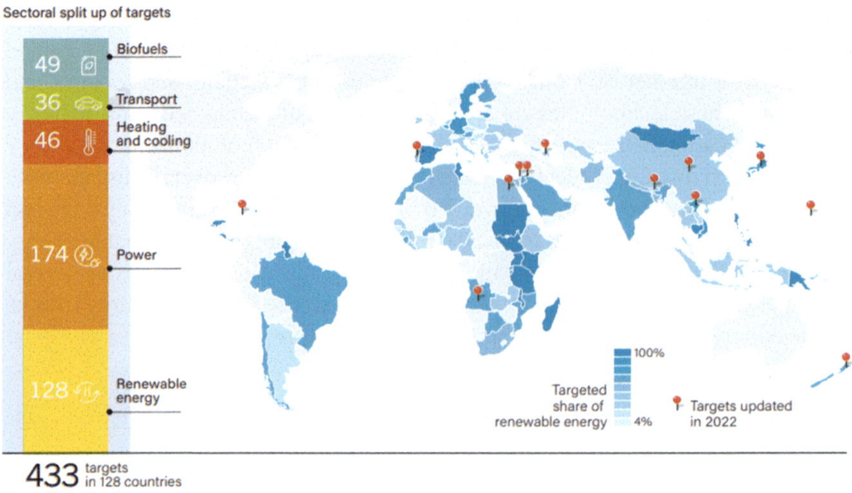

Fig. 21.9 Account of policies targeting various renewable technologies

recovery plans, and regulations addressing climate change. Such measures may have an effect on the economy's supply and demand sides. Government initiatives, financial incentives, and market forces have all been instrumental in driving the cost drop of renewable energy sources. By encouraging investment and innovation in the renewable sector, competitive bidding procedures, feed-in tariffs, and tax incentives have helped to further drive down costs. The level of ambition that nations have to quicken the energy transition is demonstrated by their renewable energy objectives, laws, and committed funds for implementation. Although only 31 nations declared plans for 100% renewable energy, most of them for the year 2050, 128 countries have economy-wide targets for renewable energy by the end of 2022 as shown in Fig. 21.9 (REN21, 2023). Power was the primary goal of sector-specific renewable energy targets, which were made public in 133 nations and 41 subnational authorities.

Targets for reducing greenhouse gas emissions, commitments to climate neutrality, and policies for net-zero emissions are some examples of policies aimed at decarbonization that also indirectly promote the use of renewable energy. One hundred forty-six nations had declared or implemented a net-zero aim as of May 2023 (REN21, 2023). The net-zero goal was mentioned in legislation in 26 nations, including the EU, in 47 policy documents, 52 discussions or proposals, and 15 declarations or pledges. In order to achieve net-zero emissions, renewable energy is essential, and 94 nations have both net-zero and economy-wide renewable energy objectives. It is also noteworthy that energy efficiency policies are instrumental in driving the energy transition in all sectors (including buildings, heating, and cooling) and complement renewable energy policies. Policies for carbon pricing place a cost on carbon emissions to promote the adoption of low-carbon technologies and consequently lower greenhouse gas emissions. The policies, which can be implemented by governments and/or regional groups, can take the shape of carbon taxes or cap-and-trade programs. By offering incentives for businesses to cut their carbon footprints, they can assist in generating income for investments in renewable energy and other low-carbon technologies.

In recent years, renewable energy has been making headlines in the energy world. In 2022, renewable energy received over USD 495.4 billion of investment worldwide. Renewable energy sources, particularly solar and wind power, have taken huge strides in terms of technical sophistication, commercial viability, and popular acceptance. In terms of growth rate, installed capacity, and investment, renewable energy has been driving the global energy landscape for the past few years. The global total final energy consumption (TFEC) increased by 16% between 2011 and 2021.

21.8 Energy Transition: Challenges

Energy transition faces technology challenges on multiple fronts. Renewable energy, the backbone of the energy transition, for example, faces intermittency, energy storage, and grid integration challenges before it could be relied upon for base load requirements. Grid integration issues arise when switching from centralized fossil fuel-based power generation to decentralized renewable sources. The electrical infrastructures in use today are frequently ill-prepared to manage intermittent energy sources such as wind and solar. Infrastructure needs to be upgraded and expanded to handle the production of renewable energy, which will need substantial investments and careful planning. To solve these problems, new energy storage technologies and smart grid development are crucial. Some of the other technologies energy transition is heavily counting on—carbon capture and storage and hydrogen—face their own set of technical and economic barriers and are far from being viable options for large-scale deployment.

Energy transition have enormous economic implications. It, for example, requires massive investments in new technologies and infrastructure, as well as employment displacement in traditional energy industries. Implementing just transition policies that give displaced employees support and retraining opportunities is how policymakers must solve these issues. Financing the changeover is still a significant barrier, particularly in developing nations.

Many countries around the world, both developed and developing, have established ambitious targets regarding GHG emissions, renewable energy, and energy transition. A clarity on the pathway to these targets however remains an issue and, in many cases, pathways are based on smooth projections. In many cases, there is a lack of transparent, long-term policies and regulatory frameworks. There is also a lack of inter-stakeholder coordination on critical aspects of these targets. Businesses experience uncertainty as a result of frequent legislative reversals, which makes it challenging to prepare for sustainable energy initiatives. The vagueness and inconsistency of government policy is thus one of the main obstacles to energy transition programs.

One of the biggest aspects yet relatively less highlighted aspect of energy transition is related to the UN Sustainable Development Goal 7 (SDG 7). While nearly 800 million people lack access to grid and many more faces unreliable grid, and around three billion people rely on crude biomass to meet cooking and heating requirements, the developed and rich countries' switch to cleaner fuels becomes meaningless at the global level. To truly realize a sustainable energy transition, developing countries' major challenge of access to refined energy resources has to be addressed. It needs to be a just transition which implies that all people have fair access to clean energy. Initiatives for the energy transition must consider the possibility of energy inequality, making sure that vulnerable populations are not left behind.

Energy transition also faces challenges on geostrategic fronts. By altering the dynamics of energy production and supply, the shift to renewable energy could change the geopolitical landscape. Countries that export fossil fuels may experience economic difficulties when demand declines. Other major issues are the implementation and practice of the enacted polices and commitments and bringing about the behavioral challenge. A change in behavior and cultural norms is necessary to make the transition to cleaner energy sources. Nations, societies, and individuals need to embrace sustainable lifestyles and energy-efficient practices, which is one of the hardest things to do. The UN General Secretary Antonio Guterres stated: "Government or private sector commitments to net-zero cannot be a mere public relations exercise" and "Using bogus 'net-zero' pledges to cover up massive fossil fuel expansion is reprehensible. It is rank deception. This toxic cover-up could push our world over the climate cliff. The sham must end" (UN, 2022) speak volumes about the actual state of affairs when it comes to current models of energy transition and sustainable development.

21.9 Conclusions

In the twenty-first century, the energy sector is experiencing an unprecedented transition in order to address challenges such as growing demand, depleting fossil fuel reserves, issues with security of supplies, fluctuating prices, and above all climate change. This energy transition is primarily regarded as a pathway of shifting away from fossil fuel-based energy systems by the middle of the century. It is fundamentally a sustainable or low-carbon energy transition, which is having profound impacts across the entire energy value chain. The transition is not just about becoming carbon neutral or zero carbon; it is rather much more vibrant and impactful, thanks to the changes and advancements occurring on the fronts of energy resources and their consumption, technological solutions, socioeconomic adjustments, and political and policy responses.

The twenty-first-century energy transition has four main dimensions: decarbonization, energy efficiency, decentralization, and digitalization. Decarbonization of the energy sector is the most important dimension of the energy transition. Reduction in CO_2 and other GHG emissions is fundamental to the fight against climate change. The energy sector can be decarbonized through a range of technologies and solutions including renewable energy, electric vehicles, hydrogen and fuel cells, carbon capture and storage, and phasing out of fossil fuels. The replacement of fossil fuels with renewable energy is the most critical part of the decarbonization drive. Renewable energy is already supplying over 25% of the global electricity needs. To achieve the net-zero emissions by 2050, almost 90% of global electricity generation is to be supplied from renewables. Making up for over 80% of the global newly added power generation capacity in 2021, renewable energy has already become a leading stakeholder in the energy sector. Energy efficiency is an imperative part of the energy transition with massive scope across various sectors especially in buildings, industry, and transport. The global economy could increase by $18 trillion by 2035 if energy efficiency is prioritized as a solution to addressing energy supply issues. Decentralized or distributed generation offers efficiency, flexibility, and economy and is thus regarded as an integral part of the unfolding energy transition. Digitalization is revolutionizing the energy sector by improving the productivity, safety, accessibility, and overall sustainability of energy systems. New, smarter ways of modeling, monitoring, analyzing, and forecasting energy production and consumption are helping the sustainable energy transition.

While the energy transition is being advanced by major developments on the technology and policy fronts, it faces a range of challenges as well. While the existing challenges are not insurmountable, the energy transition is set to remain a demanding and complex pathway. While the advanced countries such as EU and OECD member states are relatively well positioned to decarbonize their economies, developing countries critically lack the needed resources including finance, technical infrastructure, knowledge and awareness, and policy frameworks. There is thus a need for strong international partnerships and support toward the developing nations to achieve the global decarbonization targets. Given the fact that in the wake of supply disruptions due to the Russia-Ukraine war, many of the countries championing energy transition have reverted to coal, highlighting the need for greater resilience in the energy transition pathways. It also emphasizes a greater and broader global census on energy transition that could be difficult to be derailed by geopolitical conflicts.

References

Adler, K. (2021, May 3). *Global electric vehicle sales grew 41% in 2020, more growth coming through decade: IEA*. HIS Markit. https://ihsmarkit.com/research-analysis/global-electric-vehicle-sales-grew-41-in-2020-more-growth-comi.html

Ahmed, W., & Asif, M. (2020). BIM-based techno-economic assessment of energy retrofitting residential buildings in hot humid climate. *Energy and Buildings, 227*, 110406. https://doi.org/10.1016/j.enbuild.2020.110406

Ahmed, W., & Asif, M. (2021). A critical review of energy retrofitting trends in residential buildings with particular focus on the GCC countries. *Renewable and Sustainable Energy Reviews, 144*, 111000. https://doi.org/10.1016/j.rser.2021.111000

Ahmed, W., Asif, M., & Alrashed, F. (2019). Application of building performance simulation to design energy-efficient homes: Case study from Saudi Arabia. *Sustainability, 11*(21), 6048. https://doi.org/10.3390/su11216048

Alawneh, R., Mohamed Ghazali, F. E., Ali, H., & Asif, M. (2019). A new index for assessing the contribution of energy efficiency in LEED certified green buildings to achieving UN sustainable development goals in Jordan. *International Journal of Green Energy, 6*, 490–499. https://doi.org/10.1080/15435075.2019.1584104

Alazameh, A., & Asif, M. (2021). Commercial building retrofitting: Assessment of improvements in energy performance and indoor air quality. *Case Studies in Thermal Engineering, 26*, 100946. https://doi.org/10.1016/j.csite.2021.100946

Asif, M. (2013). Energy and environmental security. In *Encyclopaedia of environmental management* (Vol. II, pp. 833–842). Taylor & Francis.

Asif, M. (2016a). Urban scale application of solar PV to improve sustainability in the building and the energy sectors of KSA. *Sustainability, 8*, 1127. https://doi.org/10.3390/su8111127

Asif, M. (2016b). Growth and sustainability trends in the GCC countries with particular reference to KSA and UAE. *Renewable and Sustainable Energy Reviews, 55*(2016), 1267–1273.

Asif, M. (2019a). *Energy and environmental security* (Handbook of Environmental Management). Taylor & Francis.

Asif, M. (2019b). An empirical study on life cycle assessment of double-glazed aluminium-clad timber windows. *International Journal of Building Pathology and Adaptation, 37*(5), 547–564. https://doi.org/10.1108/IJBPA-01-2019-0001

Asif, M. (2020). Role of energy conservation and management in the 4D sustainable energy transition. *Sustainability, 12*, 10006. https://doi.org/10.3390/su122310006

Asif, M. (2021a). *Energy and environmental outlook for South Asia*. CRC Press. ISBN: 978-0-367-67343-7.

Asif, M. (2021b). *Energy and environmental security in developing countries*. Springer. ISBN: 978-3-030-63653-1.

Asif, M. (2022a). *Handbook of energy transitions*. CRC Press. ISBN: 978-0-367-68859-2.

Asif, M. (2022b). *The 4Ds of energy transition: Decarbonization, decentralization, decreasing use, and digitalization*. Wiley. ISBN: 978-3-527-34882-4.

Asif, M., & Muneer, T. (2007). Energy supply, its demand and security issues for developed and emerging economies. *Renewable and Sustainable Energy Reviews, 11*(7), 1388–1413.

Asif, M., Currie, J., Muneer, T., & Kubie, J. (2009). Security assessment of importing solar electricity for the EU. *Journal of Energy Institute, 82*, 102–105.

Asif, M., Dehwa, A., Ashraf, F., Shaukat, M., Khan, H., & Hassan, M. (2017). Life cycle assessment of a three-bedroom house in Saudi Arabia. *Environments, 4*(3), 52. https://doi.org/10.3390/environments4030052

CAT. (2021, April 23). *Ambitious US target upgrade reduces the 2030 global emissions gap by 5–10%*. Climate Action Tracker. https://climateactiontracker.org/climate-target-update-tracker/usa/

Counterpoint. *Global passenger electric vehicle market share, Q1 2021 – Q4 2022* (counterpointresearch.com). https://www.counterpointresearch.com/global-electric-vehicle-market-share/#:~:text=The%20global%20electric%20vehicle%20market%20saw%20a%2065%25,models%20in%20Q4%202022%20were%20from%20Chinese%20brands

Dehwah, A., & Asif, M. (2019). Assessment of net energy contribution to buildings by rooftop PV systems in hot-humid climates. *Renewable Energy, 131*, 1288–1299. https://doi.org/10.1016/j.renene.2018.08.031

Dehwah, A., Asif, M., & Tauhidurrahman, M. (2018). Prospects of PV application in unregulated building rooftops in developing countries: A perspective from Saudi Arabia. *Energy and Buildings, 171*, 76–87.

DNV-GL. (2019). *Digitalization and the future of energy.* DNV-GL (smartenergycc.org). https://smartenergycc.org/wp-content/uploads/2019/07/Digitalization_report_pages.pdf

DOE. *Types of fuel cells.* Energy Efficiency and Renewable Energy, Department of Energy. https://www.energy.gov/eere/fuelcells/types-fuel-cells

EIA. (2019, September 24). *EIA projects nearly 50% increase in world energy usage by 2050, led by growth in Asia.* Today in Energy – U.S. Energy Information Administration (EIA). https://www.eia.gov/todayinenergy/detail.php?id=41433

Gillam, E., & Asplund, R. (2021, August). *Will solar take the throne.* Invesco.

Hamida, M., Ahmed, W., Asif, M., & Almaziad, F. (2021). Techno-economic assessment of energy retrofitting educational buildings: A case study in Saudi Arabia. *Sustainability, 13*(1), 179. https://doi.org/10.3390/su13010179

Hassan, M. T., Burek, S., & Asif, M. (2017). Barriers to industrial energy efficiency improvement – Manufacturing SMEs of Pakistan. *Energy Procedia, 113*, 135–142.

IEA. (2017, November). *Digitalization and energy – Analysis* (Technical Report). International Energy Agency. https://www.iea.org/reports/digitalisation-and-energy

IEA. (2020, December). *Energy efficiency 2020* (Final Report). International Energy Agency.

IEA. (2021, May). *Net Zero by 2050: A roadmap for the global energy sector* (Flagship Report). International Energy Agency.

IEA. *Nuclear power.* International Energy Agency. https://www.iea.org/reports/nuclear-electricity

IRENA. (2021). *World adds record new renewable capacity in 2020.* International Renewable Energy Agency. https://www.irena.org/newsroom/pressreleases/2021/Apr/World-Adds-Record-New-Renewable-Energy-Capacity-in-2020

IRENA. (2023). *World energy transitions outlook 2023: 1.5°C pathway.* International Renewable Energy Agency.

IRENA. *Energy transition.* International Renewable Energy Agency (irena.org). https://www.irena.org/energytransition

Khan, H., & Asif, M. (2017). Impact of green roof and orientation on the energy performance of buildings: A case study from Saudi Arabia. *Sustainability, 9*(4), 640. https://doi.org/10.3390/su9040640

Khan, M., Asif, M., & Stach, E. (2017a). Rooftop PV potential in the residential sector of the Kingdom of Saudi Arabia. *Buildings, 7*(2), 46. https://doi.org/10.3390/buildings7020046

Khan, H., Asif, M., & Mohammed, M. (2017b). Case study of a nearly zero energy building in Italian climatic conditions. *Infrastructures, 2*(4), 9. https://doi.org/10.3390/infrastructures2040019

Mahmood, A., Asif, M., Hassanain, M., & Babsail, M. (2017). Energy and economic evaluation of green roof for residential buildings in hot-humid climates. *Buildings, 7*, 30. https://doi.org/10.3390/buildings7020030

Muneer, T., Asif, M., & Munawwar, S. (2005). Sustainable production of solar electricity with particular reference to the Indian economy. *Renewable and Sustainable Energy Reviews, 9*(5), 444–473.

Nahiduzaman, K., Al-Dosary, A., Abdallah, A., Asif, M., Kua, H., & Alqadhib, A. (2018). Change-agents driven interventions for energy conservation at the Saudi households: Lessons learnt. *Journal of Cleaner Production, 185*, 998–1014.

NREL. (2021a). *Concentrated solar PV.* Photo by Dennis Schroeder, National Renewable Energy Laboratory.

NREL. (2021b). *Best research-cell efficiency.* National Renewable Energy Laboratory.

NREL. (2021c). *Utility scale lithium ion battery.* Photo: 56318 by Dennis Schroeder, National Renewable Energy Laboratory.

NREL. (2021d). *Rooftop PV systems.* Photos: 45231 and 45180 by Dennis Schroeder, National Renewable Energy Laboratory.

Qudratullah, H., & Asif, M. (2020). *Dynamics of energy, environment and economy: A sustainability perspective*. Springer. ISBN: 978-3-030-43578-3.
REN21. (2020). *Renewables 2020 global status report*. Renewable Energy Network.
REN21. (2022). *Renewables 2022 global status report*. World Renewable Energy Network.
REN21. (2023). *Renewables 2023 global status report: Global overview*. IRENA.
Roper, W. (2020). *Lithium battery prices plunge*. Statista. https://www.statista.com/chart/23807/lithium-ion-battery-prices/
Sachs, J. D., Kroll, C., Lafortune, G., Fuller, G., & Woelm, F. (2021). *Sustainable Development Report 2021: The decade of action for the SDGs*. Cambridge University Press. https://doi.org/10.1017/9781009106559, https://s3.amazonaws.com/sustainabledevelopment.report/2021/2021-sustainable-development-report.pdf
Scully, J. (2021, October 12). *China signals construction start of 100GW, first phase of desert renewables rollout*. PV-Tech (pv-tech.org). https://www.pv-tech.org/china-signals-construction-start-of-100gw-first-phase-of-desert-renewables-rollout/
UN. (2022, November 8). *Bogus net-zero pledges "rank deception", sham must end, secretary-general stresses at launch of report by high-level expert group on non-state actors' commitments* (SG/SM/21576). Meeting Coverages and Press Releases. https://press.un.org/en/2022/sgsm21576.doc.htm
UN News. (2023, September 20). *Climate and environment*. https://news.un.org/en/story/2023/09/1141082
UNEP. *Energy efficiency: The game changer*. United Nations Environment. https://www.unep.org/explore-topics/energy/what-we-do/energy-efficiency
WNA. (2021). *The nuclear fuel report: Global scenarios for demand and supply availability 2021–2040*. World Nuclear Association.

Chapter 22
Solar and Wind Energy Will Supply More than 50% of World Electricity by 2030

Ali Sayigh

22.1 Global Warming and Climate Change

It is now indisputable that climate change and global warming are happening on a vast scale that we cannot ignore. In many parts of the world, the situation is causing blackouts and more than 3000 self-ignited fires, glaciers are melting faster than predicted, and flooding is happening everywhere. In the Antarctic, sea ice has been melting at an alarming rate and shrank by 1463 km^2 between 2010 and 2016 (Fig. 22.1).

Many UN meetings in the last three decades have warned of the dangers of global warming and have urgently called for carbon emissions to be significantly reduced. Targets have been set, targets have been agreed, targets have been cut back, and targets have been ignored. Yet the economic and social costs and environmental dangers have increased. Global warming has accelerated while it has been disputed or ignored. Those who pay the price are the elderly and the young and overwhelmingly the poor.

It is estimated that 10,000 old people in France died in 2014 a year when Europe experienced a record breaking heat wave. During August 2017, the daytime temperature reached 54 °C in Kuwait, while in January 2019, the coastal nondesert regions of Australia experienced prolonged temperatures of 40 °C where formerly it rarely exceeded 27 °C. Global warming is happening faster than expected (Fig. 22.2).

I remember in July 2012 Prof Neil Ferguson of Harvard gave a BBC radio Reith lecture about the rule of law. He was unable to come up with a solution to the complex problem of how to apply the rule of law across different cultures and varying economic systems and concluded that perhaps the answer lay in asking

A. Sayigh (✉)
WREC, WREN Chairman, Brighton, UK
e-mail: asayigh@wrenuk.co.uk; https://www.wrenuk.co.uk/

A. Sayigh (ed.), *Transition Towards a Carbon Free Future*, Innovative Renewable Energy, https://doi.org/10.1007/978-3-031-61660-0_22

Australia (1) Forest fire (2) Flooding in the USA (3) South Pole melting (4)

Fig. 22.1 Climate change. (1) Institute of Engineering & Technology Journal, March (2019a); (2) http://www.borealforest.org/world/innova/forest_fire.htm; (3) Union of Concerned Science publication (2018); (4) The Guardian Newspaper, June (2018)

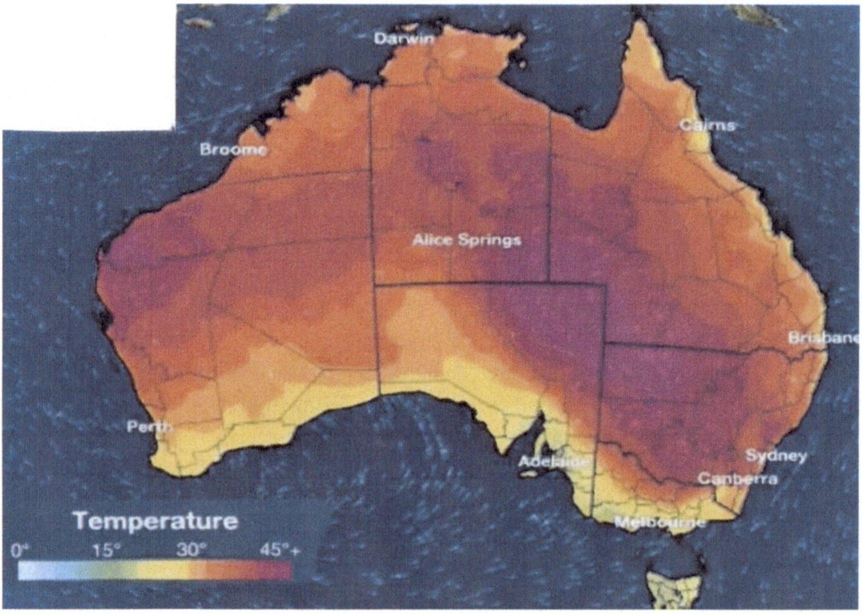

Fig. 22.2 Picture of Global Warming in Australia, January 2019

our children what they would do, perhaps, he surmised, they would have the answers.

Now it has taken 16-year-old Greta Thunberg to wake up the world about climate change and global warming. She told UK Members of Parliament: "We children have no future because of global warming," and on another occasion in Sweden, she said: "We cannot wait to 2030 to do something about it, I shall be 28 years old?" Out of the mouth of babes ....

See her picture in Fig. 22.3.

Another major problem of which we have only recently become aware of its scale is that of the melting of the permafrost in the north and south polar regions. The exposed ground will release millions of tons of methane and CO_2 (Schaefer).

Fig. 22.3 Greta Thunberg

Globally, there is much less methane in the atmosphere than carbon dioxide, but it traps heat at about 20 times as efficiently as carbon dioxide. What are the methods of methane in the polar regions? "The first source of methane is called methyl clathrate. Methyl clathrates are molecules of methane that are frozen into ice crystals. They can form deep in the Earth or underwater, but it takes very special conditions, with high pressure and low temperature, to make them. If the temperature or pressure changes, the ice that imprisons the methane will break apart, and the methane will escape. We are not sure how much methane is trapped in methyl clathrates, or how much is in danger of escaping." The other major source of methane in the Arctic is due to decayed and buried organic matter, such as dead plants and animals, frozen in permafrost for thousands of years. As long as they remain frozen, they will stay in the permafrost. However, if they thaw, they will further decay, releasing carbon dioxide or methane into the atmosphere (Schaefer).

Furthermore, it is important to realize that even if we stop the use of fossil fuels, it will only result in 50% cut in CO_2 emissions because the other 50% is due to our methods of producing food on an industrial scale and current farming policy and methods. It is due to soil carbon sequestration. It is imperative that we all think of not only how to reduce GHG emissions but also how to rationalize our electricity consumption. It is a myth beloved of twentieth-century political economists that growth can and should occur in each and every year. It is commonsense that this cannot be sustainable or true ad infinitum. It is this myth that drives climate change denial. Each one of us must think positively of how we can reduce GHG emissions and how we can reduce our personal electricity consumption; we can demand and use public transport, walk or cycle to our workplace, and cut back our meat and dairy consumption.

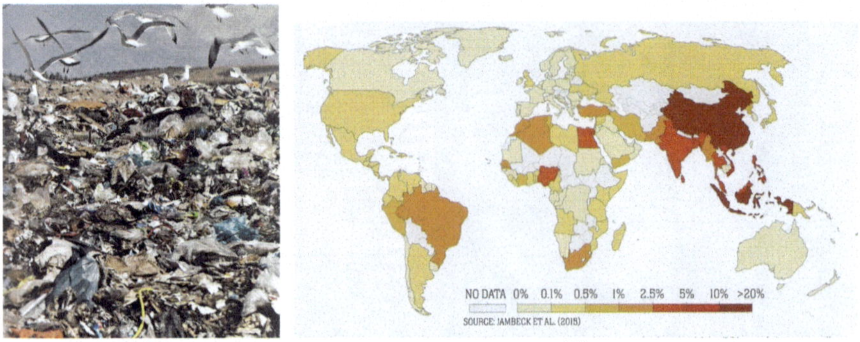

(a) (b)

Fig. 22.4 (**a**) Plastics in everyday life in the UK. (**b**) Projection of mismanaged plastics in 2025

It was a wonderful achievement when plastics were developed in the petrochemical industry in the 1960s, but their indestructible nature was not foreseen, a truly tragic case of unforeseen consequences. Most plastics in common use commercially and domestically are nonbiodegradable. Our profligate use of the "friendly" product has resulted in pollution of land and sea on a monstrous scale which impacts on the health and viability of humans, plants, and sea and land animals, in other words our natural environment (see Fig. 22.4). In the UK alone, 2.28 million tons of plastics were produced in 2018 (Engineering & Technology, 2019b).

Another example (PHOTON Newsletter, 2019a) is that of 2015; the global polymer and resin production was 381 million tons (see Fig. 22.5).

22.2 The Reality of Using Renewable Energy to Reduce the Effect of Climate Change

All countries are waking up and taking action and moving toward actively using renewable energy, mostly in the form of electricity. Some of them are seriously replacing fossil fuels with clean energy sources. For example, if we take the Gulf region, Saudi Arabia has pledged to have 40 GW capacity from renewables by 2030. Presently, its power generation is 55 GW, 8 GW from renewable energy. The Dubai electricity power supply in 2017 was about 10 GW mostly from gas (PHOTON Newsletter, 2019a). It is planning to have a solar park of 5 GW by 220 from renewable energy. The Emir recently announced that Dubai will have 75% of its power generated from renewables by 2030 (see Fig. 22.6).

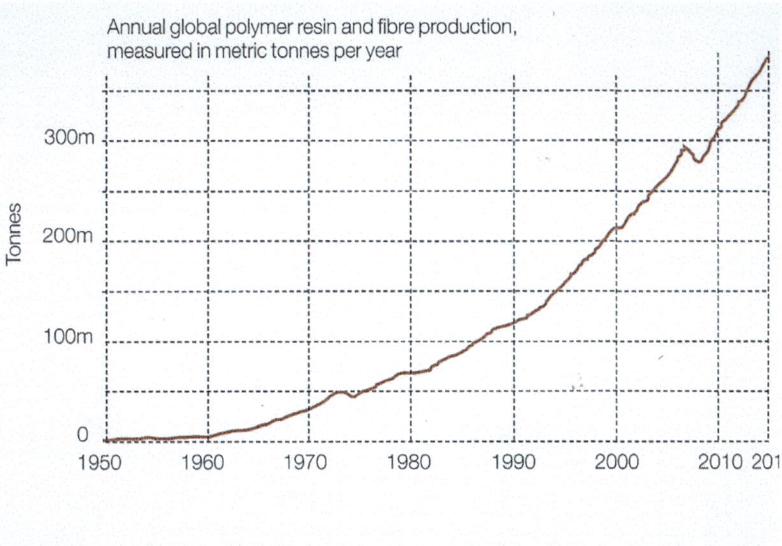

Fig. 22.5 Polymer resin and fiber annual production globally in 2017 was 381 million tons (Journal of Institute of Engineering & Technology, April 2019b)

Fig. 22.6 Dubai renewable energy drive

22.3 UK Substantially Changing to Renewable Energy

The UK power capacity in 2019 is 37.44 GW, 28.1% from renewable energy, while about 51.1% is from fossil fuels. This is planned to change with the new drive by the UK government to make offshore within renewable energy wind will supply more than 30% of electricity by 2022. It is expected by 2030 that offshore wind will supply 30 GW of power and 70% of UK's electricity will be generated from low carbon sources (Energy World, 2019a). The government has pledged that by 2050, no fossil electricity will be available. Table 22.1 shows the power percentage generated in the UK during 2019.

Table 22.1 Power percentage generated in the UK by 2029

Energy source	%
Gas-fired power station	40.7
Renewables	28.1
Nuclear plants	22.5
Coal-fired power station	1.3
Electricity imported	7.4

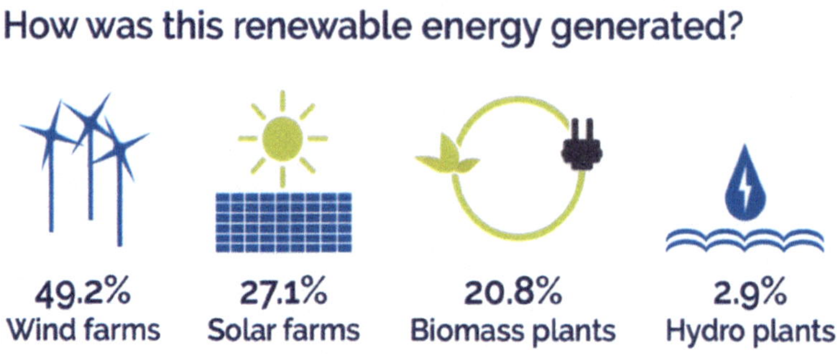

Fig. 22.7 Renewable energy types supplying 28.1% of UK's electricity in 2019 (https://smarterbusiness.co.uk/uk-renewable-energy-percentage-2018/)

Fig. 22.8 Blackfriars Bridge 5 MW, London and solar farm near Reading

Presently, the UK has the largest offshore wind farm capacity in the world of 8 GW. Wind farms now generate nearly14.0 GW of electric power (https://smarterbusiness.co.uk/uk-renewable-energy-percentage-2018/) (Fig. 22.7).

For the first time in 2018, no coal was used to generate electricity for 1000 h. Figure 22.8 shows some of the PV applications in the UK.

22.4 India's Renewable Energy Program

India has started to actively generate electricity by solar and wind energy. The government plans to have 227 GW by 2022. Both onshore wind and solar will form the backbone of this plan. However, offshore wind is expected to supply 30 GW of power by 2022.

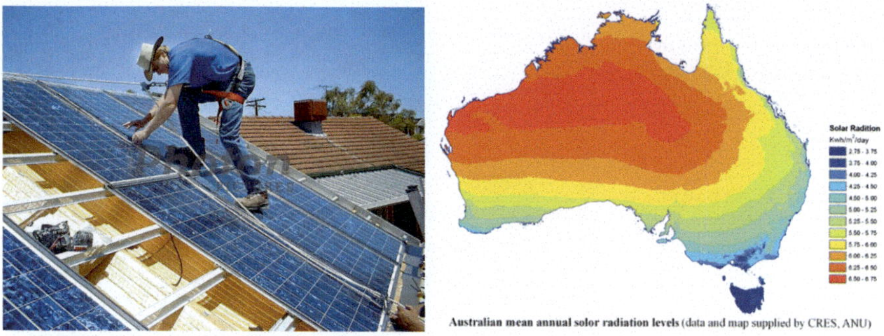

Fig. 22.9 Rooftop installation in Australia and solar radiation map (4.00–6.75) kWh/m^2/day (Jones, 2018)

The electricity sector in India has one main national grid with an installed capacity of 356.100 GW as on 31 March 2019. Renewable power plants, which also includes large hydro, constituted 34.5% of the total installed capacity. It is clear that by 2030, India will have more than 50% of its electricity from renewable energy.

22.5 Australia's Renewable Energy

One of the most effective programs in renewable energy is the Australian one. It plans to install 2 GW PV on rooftops in 2019 (see Fig. 22.9) (PHOTON Newsletter, 2019b). From Fig. 22.9, it can be seen that Australia has a vast potential to turn away from fossil fuel usage since it has one of the highest levels of solar energy intensity.

At the end of 2018, 14.8 GW of new generation had been built, and the finance of $24.5 billion (Australian dollar) in investment created more than 13,000 jobs (Clean Energy Council, 2024). In 2018, the power generation was 29.8 GW. It is clear that renewable energy is already generating more than 50% of its electrical power.

22.6 Brazil's Renewable Energy Program

Brazil was the third largest installer of PV after China and India during 2017 and 2018. By the end of 2018 alone, Brazil had installed more than 2.0 GW PV. Figure 22.10 shows one of the largest installations in Brazil (SolarPower Europe, 2018). Brazil has 137 GW of electric power so far in 2019. This compares with the expected growth of global PV installation capacity of 505.2 GW by the end of 2019, and by end of 2022, it could reach 1026.2 GW (SolarPower Europe, 2018).

According to the report in "The Brazil Business," hydropower can supply 65% of Brazil's energy needs but only 25% is utilized. PV supplies 2 GW, wind energy

Fig. 22.10 A 300 MW PV farm installed in Brazil in 2019 at a cost of US$ 300

supplies only 2.2 GW, while the potential supply is 140 GW. Biomass can provide 7.5% of Brazil's power. Thus, renewable energy could easily meet more than 50% of its electricity demand in 2030 (The Brazil Business, 2015).

22.7 China Trying to Reduce Emission by Using Renewables

In 2017, China declared that it would invest $360 billion in renewable energy by 2020. It will scrap the plan to build 85 coal-fired power plants. In March 2017, Chinese authorities stated that the country was already exceeding official targets for implementing energy efficiency, carbon emission reduction, and its increasing share of clean energy sources. China is experiencing an energy transformation, which is being driven by technological change and the falling cost of renewables. China is not just investing in renewables to phase out coal, but aims to be a major exporter of renewable energy. In China, various other factors have been introduced, including increased energy efficiency in residential, industrial, and commercial buildings and lowering energy demand for transportation, owing to the increase of autonomous vehicles and ride sharing, similar to that of Cuba (World Economic Forum, 2017).

China has been investing yearly $100 million in its energy sources since 2017 which is twice of that of the USA, of which $32 million was on renewable energy. China expects to invest more than $360 billion in renewable energy by 2020. China has more than 1000 companies manufacturing renewable energy equipment. Figure 22.11 shows the export record of one of the leading PV manufacturers.

China's renewable energy capacity reached 728 GW by the end of 2018 (Reuters, 2019). It is estimated that 65% of power generation in China comes from hydro. It is worth noting that China is leading the rest of the world in using electric cars (see Fig. 22.12).

Fig. 22.11 JinKO PV major Chinese PV company

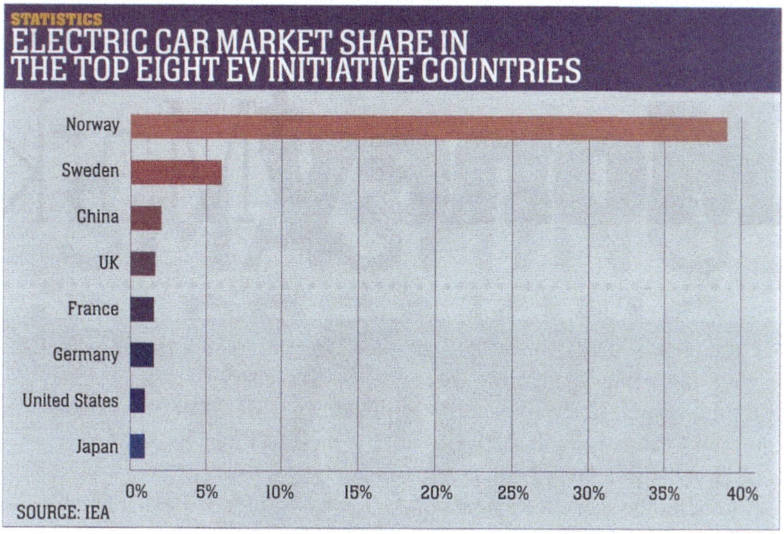

Fig. 22.12 Electric car percentage used in the top eight countries (IEA Electric Vehicles, 2017)

Fig. 22.13 Large PV installation in China, Photon Newsletter 23 April, 2019 and January 31, 2022

China is number one in using PV; in 2018, it installed more than 44.1 GW, (PV Magazine, 2019), while the global figure was 174.63 GW. During the first 3 months of 2019, China installed 5.2 GW (PHOTON Newsletter, 2019c). Thus by 2030, more than 50% of its electricity will be generated by renewable energy.

If one takes EV number of cars, then (Fig. 22.13):

China = 1,200,000
Europe = 1,000,000
USA = 850,000
Rest of the world = 150,000
The total exceeds three million cars

22.8 USA

Despite the present president of the USA not agreeing to sign the Climate Change UN Treaty and being against the use of renewable energy, 45 main cities out of 57 are doubling their renewable energy applications (PHOTON Newsletter, 2019c).

Figure 22.14 illustrates this drive to double their facilities in 2019.

As of April of 2019, the US has more than 60 GW of PV.

California set the scene for the rest of the USA by announcing that all their power will come from renewable energy by 2045. This is a great challenge for them to achieve (Ref. S&P Intelligence Report). Presently, renewable energy supplies 35% of California power. Coal has already been phased out and nuclear power is going to be phased out by 2024. The only remaining fossil fuel to provide power is gas.

Fig. 22.14 Los Angeles Department of Water and Power (LADWP) plant

Wind and solar, however, do require storage in order to be a viable continuous supplier of energy and this is being tackled actively. Also hydro and geothermal resources are sufficient to supply a base load in California. Therefore, the prediction of 50% power supply from renewables in the USA could be easily achieved when the current policy under President Trump is replaced by green technology policy. Figure 22.15 shows the power generation in the USA and in California. The US power production in 2017 is about 1300 GW. In 2019, 260 GW will be generated from renewable energy (Energy World, 2019b).

Wood Mackenzie predicts the USA to have three million solar installations by 2021 and four million by 2023. Currently, in May 2019, the number of solar installations in the USA has reached more than two million, according to the Wood Mackenzie Power & Renewables and the Solar Energy Industries Association (SEIA). This represents more than 70 GW of capacity (PHOTON Newsletter, 2019h).

Many regions are promising to reach 100% renewable power to cover their needs by 2050. These include Nigeria, Puerto Rico, Hawaii, and California.

Recently, Iraq announced a tender to install 755 MW of PV to be built by 2020 across five provinces of Babel, Wasit, Karbala, Al-Muthanna, and Diwaniyah. This will meet approximately 50% of their present electricity generation. The Minister of Energy has said that Iraq aims to be a global center of renewables (PHOTON Newsletter, 2019d).

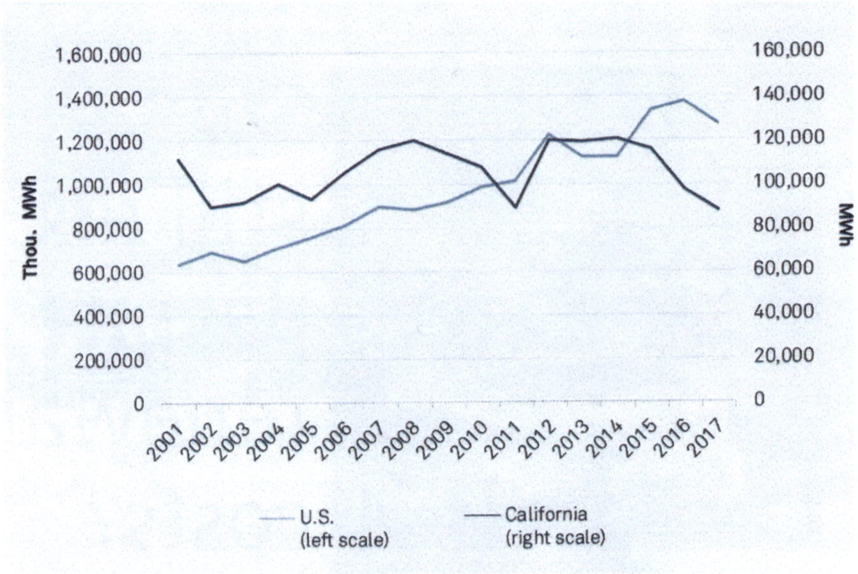

Fig. 22.15 Power generation in the USA and in California in 2017

22.9 Chile's Renewable Energy Program

Most South American countries have very active programs in renewable energy and Chile is no exception. One scheme is to use floating PV in a small mining community (see Fig. 22.16) (PHOTON Newsletter, 2019e). Chile installed 715 MW of renewables into their electricity system in 2018, which is 92% of all electricity installation in that year, which was divided between solar (59%) and wind (32%). Presently, only 20.8% of total electricity in Chile comes from renewables (Renewables Now, 2019).

22.10 Egypt's Renewable Energy Program

Egypt has a substantial hydropower generation, notably the Aswan Dam, but during the last 5 years, Egypt has invested in wind energy and is now pressing ahead with PV applications (Fig. 22.17) (PHOTON Newsletter, 2019d).

Egypt has managed to produce a national wind atlas, which shows that Egypt enjoys strong wind at an average speed on 10.5 m/s. This has led the Egyptian government to aim to provide 12% of the generated electricity (the equivalent of 6.8 GW) through wind energy by 2022; this goal is part of Egypt's 2030 strategic plan to have no coal-based electricity. Siemens will also deliver up to 12 wind farms

Fig. 22.16 Chile floating 84 kW, PV farm on a lake north of Santiago supplying 153 MWh a year to the mining community at the beginning of 2019

Fig. 22.17 400 MW PV as part of 1.6 GW solar park in south of Egypt

in the Gulf of Suez and West Nile areas, comprising around 600 wind turbines and an installed capacity of 2 GW. Currently, 29 solar power projects have been financed at a total investment of $1.8 billion, producing almost 1.8 GW of solar power, on a 14.3-square-mile plot of land. Egypt is buying the electricity at a rate of 7.8 ¢/kWh, for 25 years (Fig. 22.18).

Fig. 22.18 Wind farm in Egypt

Fig. 22.19 Netherlands PV and wind energy and Zimbabwe gathering

The hydropower currently generates 2.64 GW. While Benban Solar Park generates 1.6 GW. It is expected that by 2030, Egypt will have more than 18 GW from renewables (Egypt Today, 2018). At the moment, the total power electricity in Egypt is about 32 GW.

22.11 Examples from Other Countries

Figure 22.19 shows examples from various countries. For example, at the end of 2018, the Netherlands had 4.4 GW PV and 20.7 GW from wind power (PV Magazine, 2019). Oman has signed an agreement with the Saudi Arabian authority, ACWA, to build 500 MW PV jointly with the Gulf Investment

Corporation, west of Muscat (PHOTON Newsletter, 2019f). Poland PV installation has reached a capacity of 603 MW at the end of March 2019 (PHOTON Newsletter, 2019c). Zimbabwe is planning with a US Labacorp Power to build a solar project of 900 MW (PHOTON Newsletter, 2019g).

22.12 Conclusions

According to the IRENA report (IRENA, 2019), at the end of 2018, renewable energy power reached 2351 GW: hydropower accounted for nearly 50% of this (1293 GW), wind power 564 GW, solar power 486 GW, biomass 115 GW, geothermal 13 GW, and marine power 0.5 GW. The total represents 33% of the global generating power.

If the aim is to have 50% of power generated from renewable sources by 2030, then we have just over a decade left to achieve this. This will mean significant changes in energy and agricultural policies and consumption patterns and the removal of fossil fuels from the power equation. Should we fail to achieve this, then we are faced with a very bleak future and a terrible legacy to hand on to our children and grandchildren. As far as humans are concerned, vast tracts of the planet on which we currently depend will have become virtually uninhabitable and the oceans and air polluted.

References

Clean Energy Council. (2024). *Clean Energy Australia report.* https://www.cleanenergycouncil.org.au/resources/resources-hub/clean-energy-australia-report

Egypt Today. (2018). http://www.egypttoday.com/Article/2/62730/2018-The-year-of-new-renewable-energy

Energy World, April 2019a.

Energy World No 478, February 2019b.

Engineering & Technology, Volume 14, March 2019.

Engineering & Technology, April 2019b.

http://www.borealforest.org/world/innova/forest_fire.htm

https://smarterbusiness.co.uk/uk-renewable-energy-percentage-2018/

IEA Electric Vehicles. (2017). *Number of cars globally exceeds 3 million.*

IRENA Report, March 2019.

Jones, A. (2018). City of Sydney decentralize renewable energy master plan 20 July 2018. In *Proceedings of World Renewable Energy Congress – 18*, University of Kingston, London, UK, 30 July–3 August 2018.

PHOTON Newsletter, March 25, 2019a.

PHOTON Newsletter, April 16, 2019b.

PHOTON Newsletter, April 23, 2019c.

PHOTON Newsletter, May 8, 2019d.

PHOTON Newsletter, April 2, 2019e.

PHOTON Newsletter, May 7, 2019f.

PHOTON Newsletter, April 24, 2019g.
PHOTON Newsletter, May 10, 2019h.
PV Magazine – Global Edition. (2019). https://www.pv-magazine.com/2019/01/21/china-installed-44-1-gw-of-solar-in-2018-according-to-official-nea-data/
Renewables Now. (2019). https://renewablesnow.com/news/renewables-grow-to-208-share-in-chiles-power-mix-638961/
Reuters. (2019). https://uk.reuters.com/article/us-china-renewables/chinas-2018-renewable-power-capacity-up-12-percent-on-year-idUKKCN1PM0HM
Schaefer, K. National Snow & Ice Data Center. https://nsidc.org/cryosphere/frozenground/methane.html
SolarPower Europe. (2018). *PV Global Market Outlook 2018–2022*. http://www.solarpowereurope.org/wp-content/uploads/2018/09/Global-Market-Outlook-2018-2022.pdf
The Brazil Business. (2015). https://thebrazilbusiness.com/article/energy-sources-in-brazil
The Guardian News Paper. (2018, April 2).
Union of Concerned Science. (2018). https://www.ucsusa.org/sites/default/files/attach/2018/07/gw-fact-sheet-epif.pdf
World Economic Forum. (2017). https://www.weforum.org/agenda/2017/08/how-china-is-leading-the-renewable-energy-revolution

Chapter 23
Renewable Energy Sustainability in the Top EU's Strong Economy Countries

Shoeib Faraji Abdolmaleki, Danial Esfandiary Abdolmaleki, and Pastora M. Bello Bugallo

23.1 Introduction

The transition to renewable energy is a critical component of global efforts to address climate change and achieve sustainable development. Renewable energy sources, such as solar, wind, and hydropower, offer significant advantages over fossil fuels in terms of their environmental impact, energy security, and economic benefits. The EU has been at the forefront of this transition, setting ambitious renewable energy targets and implementing policies to promote renewable energy production.

The EU has set a target to achieve a 32% share of renewable energy in its final energy consumption by 2030. The EU's member countries have also set their renewable energy targets, which vary in ambition and scope. Some countries, such as Germany and France, have set ambitious targets to achieve a significant share of renewable energy in their energy mix. While the EU has made significant progress in promoting renewable energy production, achieving sustainable renewable energy production is not an easy task. Achieving renewable energy sustainability requires a holistic approach that considers economic, social, and environmental aspects and involves careful planning, evaluation, and monitoring.

Despite the widespread adoption of renewable energy sources in the EU, their sustainability assessment remains a complex issue. The sustainable development of renewable energy requires a comprehensive evaluation of various factors, such as their environmental impacts, social and economic benefits, and the potential trade-offs between them. However, there is a lack of consensus on how to assess the

S. F. Abdolmaleki · D. E. Abdolmaleki · P. M. Bello Bugallo (✉)
TECH-NASE Research Group, Department of Chemical Engineering, Universidade de Santiago de Compostela, Santiago de Compostela, Spain
e-mail: pastora.bello.bugallo@usc.es

A. Sayigh (ed.), *Transition Towards a Carbon Free Future*, Innovative Renewable Energy, https://doi.org/10.1007/978-3-031-61660-0_23

sustainability of renewable energy in the EU's strong economy countries (e.g., Marques & Fuinhas, 2012; Anna, 2016; Wang & Yang, 2020).

The existing literature on the sustainability assessment of renewable energy in the EU's strong economy countries is limited and fragmented (e.g., Fouquet, 2013; Rickels et al., 2016). While some studies have focused on the environmental impacts of renewable energy, others have examined the social and economic benefits (Ifaei et al., 2017). There is a need for an integrated approach that considers all three pillars of sustainability. To address this gap, this study proposes an integrated sustainability assessment framework for renewable energy in the EU's strong economy countries. The framework considers the environmental, social, and economic aspects of sustainability and aims to provide a comprehensive and holistic evaluation of renewable energy sources.

The motivation behind this study is to contribute to the ongoing discussion on the sustainable development of renewable energy in the EU's strong economy countries. By providing an integrated sustainability assessment framework, this study can inform policymakers and stakeholders about the potential benefits and trade-offs associated with renewable energy sources (Santika et al., 2019). Sustainable indicators are a useful tool for evaluating renewable energy sustainability performance, as they provide a quantitative measure of performance across multiple dimensions. Sustainable indicators are based on the United Nations' Sustainable Development Goals (SDGs), which provide a framework for measuring sustainable development across economic, social, and environmental dimensions.

The purpose of this study is to provide a comprehensive and objective assessment of renewable energy sustainability performance in the top EU countries with strong economies. This study builds on previous research that has assessed renewable energy sustainability using different methods and indicators. By using sustainable indicators and the fuzzy analytical hierarchy process, this study aims to provide a robust and objective assessment of renewable energy sustainability performance in these countries.

23.2 Materials and Methods

23.2.1 Framework and Policy

Efforts have been made to shape the concept of sustainable development and to evaluate the sustainability of a system. One of the critical components in this regard is the framework on which assessments are based. For countries to achieve sustainable development goals such as supplying 20% of energy using renewable energy in the EU by 2020, they must formulate specific policies for their short-term and long-term development plans. Success in such long-term projects requires a stable political framework and a well-tailored support system, which includes financial, technical, and administrative support, to overcome obstacles in distorted energy markets (Fouquet, 2013). The joint efforts of scientists and politicians from various fields are necessary for this debate, as the development of sustainability indicators is a process that involves both scientific knowledge production and political norm creation (Rametsteiner et al., 2011). The concept of renewable energy is crucial in the EU policy, and the EU is strongly committed to promoting renewable energy. It is reasonable to claim that for the union, renewable energy is not just a means to combat climate change but also an end in itself. This commitment is evident from the setting of renewable target policies, which overlap and may hinder the effectiveness of the EU Emissions Trading System (Braathen, 2011; IETA, 2015), the cornerstone of the EU's policy to combat climate change (European Commission, 2020). Additionally, public policies such as incentives and subsidies can play a decisive role in the development of renewable energy (Marques & Fuinhas, 2012). They have examined various drivers on a large scale for EU countries, which support this claim. Theories on sustainable development and the proliferation of sustainable development indicator processes result in the development and application of various conceptual frameworks (IISD, 2005).

The successful deployment of renewable energy has a strong synergy with some of the Sustainable Development Goals (SDGs), which highlights the potential impact of renewable energy on achieving these goals. However, implementing the SDGs into local and national development planning requires significant effort and energy, particularly for the energy sector, which is a key enabler of all development activities (Simsek et al., 2020). Therefore, it is essential to analyze the specific context of each country and the required effort and energy to implement these goals. Achieving the SDGs will undoubtedly affect the energy sector, and some countries may require much more effort and energy to attain these goals (Santika et al., 2019).

Frameworks provide the logical structures for developing indicators, serving as a high-level reference for the basic concepts of sustainable development. Renewable energy is a dominant concept in energy policy, and legal frameworks are specifically designed to classify energy sources as renewable or nonrenewable, with the aim of promoting sustainability and reducing climate change (Harjanne & Korhonen, 2019). While policies are the foundation for producing indicators, it is essential to link indicators to policy processes to ensure their relevance in public debates. Failure to do so could render the indicators politically problematic and lead to their discontinuation. Frameworks play a crucial role in making indicators accessible to nonexperts while staying true to technical requirements and sustainable development goals (Innes & Booher, 2000). The differences among frameworks lie in how they conceptualize the dimensions of sustainable development and their interlinkages, grouping of issues, and selection and aggregation of indicators.

23.2.2 *Data Gathering and Development*

23.2.2.1 Indicators

Indicators aim to be a flexible tool for decision-makers at all levels to gain a better understanding of their situation, policy impacts, and potential policy changes. General sustainability indicators must provide numerical values for sustainability results and dimensions, enabling users to compare systems and choose the most sustainable option at the lowest cost. These numerical values help simplify decision-making and provide an overall picture of progress toward Sustainable Development Goals (Liu, 2014; Anna, 2016). Wang and Yang (2020) proposed three objectives to address practical and theoretical challenges in SD indicators and decision-making:

- Provide a framework for measuring SD based on the SDGs and systems thinking theory.
- Analyze latent correlations among variables from different SDGs.
- Demonstrate an empirical application of the Systemic Indicator of Sustainable Development (SISD) framework for measuring sustainability.

In order to assess the sustainability of renewable energy, the criteria group selected must be clear, comprehensive, and nonredundant, with each criterion being counted only once (Bouyssou, 1990). While a small set of indicators is more relevant for decision-makers, it is unrealistic to expect decision-makers to remember a large number of indicators (Charfeddine & Kahia, 2019). Composite indicators can complement single indicators and allow for an explicit assessment of trade-offs between policies. While individual indicators serve the detailed assessment of specific policy measures, composite indicators enable an assessment of the overall impact of policies. The current outline of the SDG indicator framework may be interpreted as dedicated to a concept of strong sustainability, where sustainable development requires maintaining all indicators at least at their current level (Rickels et al., 2016). Ideally, indicators are selected using well-defined and broadly accepted methods based on empirical studies of their historical influence on the desired objective, the influence of policy measures on the indicator, and correlations between indicators (Alfsen & Greaker, 2007; Schultz et al., 2008).

This study aims to identify appropriate indicators for the assessment process based on several criteria. Firstly, it considers criteria that have been established by institutions, organizations, governments, and literature (Faraji Abdolmaleki & Bello Bugallo, 2021). Secondly, the indicators must reflect the circumstances at a country level and be relevant in both local and global contexts. Thirdly, the direction of the effect of the criteria on renewable energy (RE) and sustainable development (SD) should be unambiguous, whether positive or negative. Fourthly, the indicators should be readily available with sufficient existing data. Finally, the selected indicators should be easily understandable, and their relationship to RE and SD should be clear.

23.2.2.2 Means of Data Collection, Questionnaire Design, and Information Acquisition

The study has chosen the top ten economies countries of the EU, which were identified through the International Monetary Fund. These countries include Germany, France, Italy, Spain, the Netherlands, Poland, Sweden, Belgium, and Ireland, and they were considered as viable alternatives. The majority of the data was obtained from various sources, including Eurostat, the World Bank, BP Statistical Review, and the International Energy Agency's latest report in 2022, the World Population Review, Trading Economics, and the United Nations Development Program. After reviewing the literature and the selection criteria, 13 indicators were selected from various databases.

As a result, a total of 13 criteria were gathered and their orientation toward sustainability was determined (Table 23.1).

To participate in our study, these experts were selected based on their expertise and experience in the areas of renewable energy, sustainability, and energy policy. Their input was critical in determining the weights of each indicator using the Saaty scale and in making comparisons between the alternative countries based on the selected criteria (Saaty, 1984). The insights and feedback provided by the experts were essential in ensuring the validity and reliability of our evaluation process.

The Saaty scale, also known as the analytic hierarchy process (AHP) scale, is often presented in a table format. The table typically consists of a square matrix with rows and columns representing the same set of criteria or alternatives being evaluated. The main diagonal of the matrix contains 1 s, indicating that a criterion or alternative is equally important to itself.

The upper triangle of the matrix contains the pairwise comparisons between criteria or alternatives, represented by values ranging from 1 to 9, indicating the degree of importance of one criterion or alternative compared to another. These

Table 23.1 Selected indicators toward sustainability assessment of renewable energy

Indicator	Direction	Unit
Employment rate (job creation-labor impact)	+	%
Human Development Index (HDI)	+	%
Gross Domestic Product (GDP)	+	Euro per capita
Greenhouse Gas (GHG) emission	–	Grams per capita
Economy growth rate	+	%
Crime rate	–	–
Corruption	–	%
Dependency on fossil fuels	–	Terawatt-hours
Inflation rate	–	–
Energy intensity	+	Kilograms of oil equivalent
Energy consumption per capita	–	Gigajoules per capita
General Energy Production/Population	+	–
Share of renewable energy	+	–

Table 23.2 Fuzzy judgment using the Saaty scale

Scale	Expression of the comparative components	Triangular fuzzy scale
1	Preferred equally	(1,1,1)
2	In between	(1,2,3)
3	Preferred moderately	(2,3,4)
4	In between	(3,4,5)
5	Preferred strongly	(4,5,6)
6	In between	(5,6,7)
7	Very strongly preferred	(6,7,8)
8	In between	(7,8,9)
9	Extremely preferred	(9,9,9)

values are determined through a series of judgments or comparisons made by experts or stakeholders in the decision-making process using the Saaty scale (Table 23.2).

The lower triangle of the matrix contains the reciprocals of the values in the upper triangle, ensuring that the matrix is consistent and satisfies the mathematical properties required by the AHP method. Once the matrix is completed, it is used to calculate the weights or priorities of the criteria or alternatives, which can then be used to make informed decisions.

A fuzzy questionnaire using the Saaty scale is a type of questionnaire that is used in decision-making. The Saaty scale is a measurement that allows for pairwise comparison of items based on their importance or priority. In a fuzzy questionnaire, the Saaty scale is used to assess the degree of importance or priority of certain criteria or factors in a decision-making process. The fuzzy logic approach allows for imprecision or uncertainty in the data, which is common in decision-making scenarios. It allows for more nuanced and subjective assessments of importance or priority, rather than a simple binary or numerical ranking. The scale is a useful tool in decision-making processes where there is a high degree of uncertainty or complexity. It allows decision-makers to consider a range of factors and prioritize them in a nuanced and subjective manner, which can lead to more informed and effective decisions.

23.2.2.3 Fuzzy Analytic Hierarchy Process (FAHP)

FAHP is a decision-making method that can handle complex and uncertain information by using linguistic terms and fuzzy logic. FAHP allows decision-makers to evaluate multiple criteria and alternatives and identify the best course of action based on their relative importance (Haddad et al., 2017; Shaaban et al., 2018; Mastrocinque et al., 2020). The FAHP method consists of four main steps:

1. Establishing the hierarchy: In this step, the decision problem is structured hierarchically, with the objective at the top, criteria in the middle, and alternatives at the bottom. The hierarchy is used to break down the problem into smaller and more manageable subproblems.

2. Pairwise comparison: In this step, decision-makers evaluate the importance of each criterion with respect to the objective and the importance of each alternative with respect to each criterion. This is done by pairwise comparison, where the decision-maker assigns a linguistic term to the relative importance of each criterion or alternative compared to others.
3. Fuzzy judgment aggregation: In this step, the fuzzy judgments obtained in the pairwise comparison are aggregated into a group decision. This is done by using Chang's method. This method is widely used in decision-making applications where the input data is uncertain or vague. The method allows decision-makers to integrate their subjective judgments and preferences with quantitative data to obtain a comprehensive evaluation of the alternatives (Chang, 1996). The method is also flexible and can be used with different types of fuzzy sets and membership functions. It consists of three steps:

   (a) Fuzzy geometric mean: The fuzzy geometric mean is used to calculate the aggregated weights of each criterion or alternative. The fuzzy geometric mean is defined as:

$$w_i = \left[(w_{i1})^{m1*} (w_{i2})^{m2*} \ldots * (w_{ik})^{mk} \right]^{(1/M)}, \qquad (23.1)$$

   where w_i is the aggregated weight of criterion or alternative i, w_{ij} is the linguistic term assigned by the decision-maker for the pairwise comparison between i and j, m_j is the weight assigned to j, k is the number of alternatives or criteria, and M is the sum of all weights.

   (b) Consistency measure: The consistency measure is used to check the consistency of the pairwise comparisons. The consistency measure is defined as:

$$CR = (\lambda_{\max} - n)/(n - 1), \qquad (20.2)$$

   where $\lambda_{\max}$ is the maximum eigenvalue of the pairwise comparison matrix and n is the number of criteria or alternatives. A consistency ratio (CR) of less than 0.1 is considered acceptable.

   (c) Normalization: The weights obtained from the fuzzy geometric mean are normalized to sum up to 1.

4. Synthesizing the results: In this step, the weights obtained from the pairwise comparison are used to rank the alternatives based on their overall performance.

The FAHP method is a useful tool for decision-makers to evaluate complex and uncertain information and make informed decisions based on a comprehensive and objective assessment of the available alternatives. The use of Chang's method for fuzzy judgment aggregation helps ensure the consistency and reliability of the decision-making process.

Figure 23.1 demonstrates a diagram stating the methodology of this study. After reviewing the EU's policies and establishing a sustainable framework, the indicator

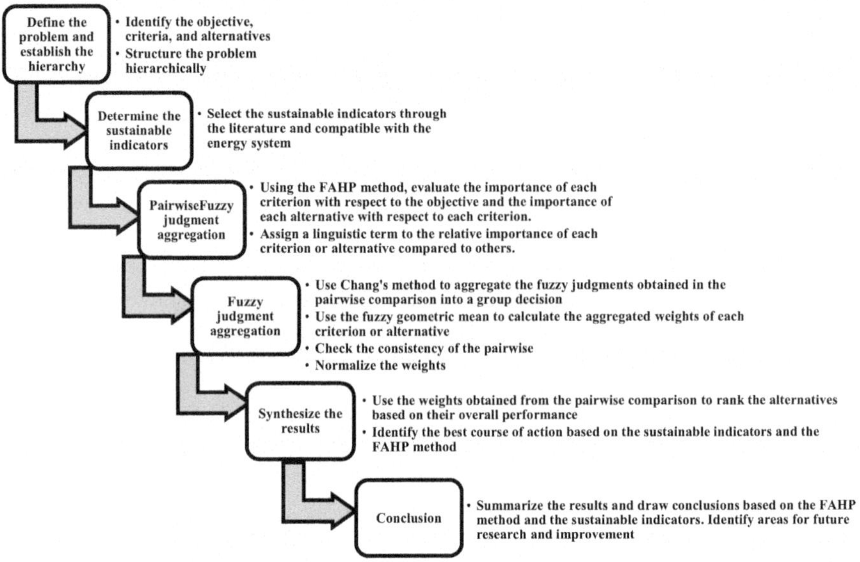

Fig. 23.1 Methodology flowchart

set is used for sustainability measurement. The sustainable indicators used in this study are carefully chosen to ensure a comprehensive assessment of sustainability in the context of the EU's policies. The FAHP is employed to determine the weights of each indicator and to calculate the final sustainability score. The fuzzy steps in the methodology allow for the inclusion of uncertain or imprecise data, which is a common feature of sustainability assessments. MATLAB is used to facilitate the calculations and to automate the fuzzy logic operations. The outputs obtained from the analysis are then interpreted to provide insights into the sustainability performance of the EU's policies. Finally, policy adjustments may be made based on the results, with the aim of improving the sustainability of the EU's policies over time.

23.3 Results and Discussions

The EU has set ambitious targets for renewable energy generation and consumption and renewable energy sustainability. Evaluation is an essential tool to track and ensure progress toward these targets. This study provides an overview of the sustainability aspects related to renewable energy sustainability in the EU countries with high level of economy. It analyzes the current status of renewable energy consumption and production, as well as the policies and measures in place to promote sustainable renewable energy development and address any potential negative impacts. It also assesses the socioeconomic and environmental impacts of renewable energy, including job creation, economic growth, and greenhouse gas

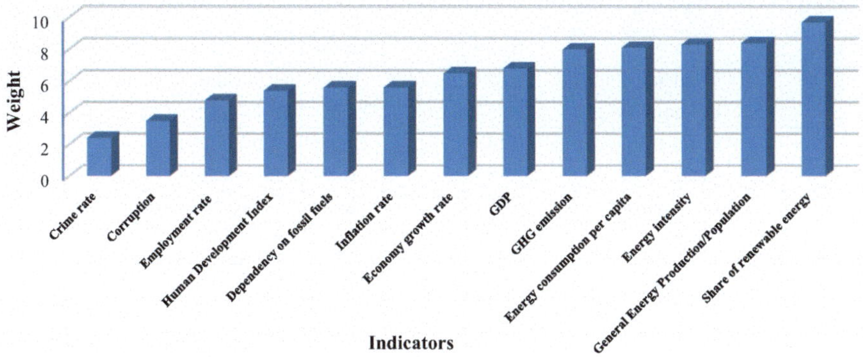

Fig. 23.2 Indicators' analysis and expert viewpoints

emission reduction. Furthermore, it identifies key challenges and opportunities for sustainable renewable energy development in the EU and provides recommendations for policymakers, industry stakeholders, and civil society to enhance the sustainability of renewable energy.

In order to properly define renewables and incorporate the concept of sustainability, it is important to consider the primary drivers. Various literature on the sustainability of renewable energy sources have explored different examples, but this study has identified main drivers that will enable the analysis of sustainability in selected nations based on renewables while also fulfilling indicator selection criteria. There are several factors that can impede a successful assessment, including enough selected indicators, issues with the evaluation steps, and dealing with viewpoints.

The given information outlines the indicators and criteria used to evaluate the sustainability of renewable energy in a selected sample of EU countries. A total of 13 indicators were used, which were divided into five drivers, namely, energy, environment, economy, social, and institutional.

The energy driver included five indicators, namely, dependency on fossil fuels, energy intensity, energy consumption per capita, general energy production per population, and share of renewable energy. The environment driver included only one indicator, i.e., greenhouse gas (GHG) emissions. The economy driver included three indicators, i.e., gross domestic product (GDP), economy growth rate, and inflation rate. The social driver included three indicators, i.e., employment rate, Human Development Index (HDI), and crime rate. Finally, the study considered corruption as an institutional (governmental) indicator.

Moreover, the study experts provided the weight of each indicator, indicating their relative importance in the evaluation process (Fig. 23.2). The weight ranged from 2.4 for crime rate to 9.7 for the share of renewable energy. The highest weights were assigned to indicators related to renewable energy, general energy production per population, and energy intensity, indicating their significant role in the evaluation of sustainability.

Table 23.3 Ranking the strongest EU economy in terms of sustainability of renewable energy

Rank	Country	Score	Sustainability condition	Group
1	Sweden	0.2249	Higher sustainability	I
2	Belgium	0.1556	Higher sustainability	I
3	Ireland	0.1135	Higher sustainability	I
4	France	0.0884	Medium sustainability	II
5	Germany	0.0878	Medium sustainability	II
6	Spain	0.0871	Medium sustainability	II
7	Netherlands	0.0762	Lower sustainability	III
8	Poland	0.0596	Lower sustainability	III
9	Italy	0.0530	Lower sustainability	III

The information provided suggests that a comprehensive approach was taken to evaluate the sustainability of renewable energy in the selected EU countries. The use of multiple indicators from various drivers ensures a well-rounded evaluation of the sustainability of the renewable energy systems. The weight assigned to each indicator also provides an indication of their relative importance in the evaluation process, which can be useful for policymakers and stakeholders to prioritize their efforts toward achieving sustainable energy systems.

Sweden, Belgium, Ireland, France, Germany, Spain, the Netherlands, Poland, and Italy were evaluated using the FAHP method in MATLAB to determine their sustainability in the renewable energy sector (Table 23.3). The results showed that Sweden had the highest score, while Italy had the lowest. It revealed that the countries with higher scores had a greater focus on renewable energy, with lower dependency on fossil fuels, lower energy intensity, and higher general energy production per population.

Dependency on fossil fuels was one of the critical indicators used in the energy driver (Table 23.4). Countries with a lower dependency on fossil fuels tend to have a more sustainable energy system (Vakulchuk et al., 2020; Bekhrad et al., 2020). Renewable energy sources can provide a more sustainable solution for the future, as they are inexhaustible and have a lower impact on the environment. The share of renewable energy was assigned the highest weight, indicating its significant role in evaluating the sustainability of renewable energy systems.

Energy intensity is another critical indicator in the energy driver, as it represents the amount of energy required to produce a unit of GDP. Countries with a lower energy intensity tend to be more sustainable in the renewable energy sector, as they require less energy to produce the same amount of goods, leading to lower emissions and greater energy efficiency (Wang et al., 2020). This indicator highlights the importance of implementing energy-efficient practices and technologies in industries and households to reduce the energy intensity of production and consumption, ultimately contributing to the overall sustainability of renewable energy systems. Additionally, general energy production per population was found to be an essential indicator, indicating that countries with a higher general energy production per population tend to have more sustainable renewable energy systems, as they have

Table 23.4 Energy driver indicators

Indicator/country	Germany	France	Italy	Spain	Netherlands	Poland	Sweden	Belgium	Ireland
Dependency on fossil fuels	31,647	19,916	23,873	22,769	48,688	30,111	17,525	48,104	28,473
Energy intensity	2.76	3.29	2.54	2.64	3.05	3.42	3.80	3.87	1.32
Energy consumption per capita	152.00	144.00	107.20	117.40	198.20	117.20	218.90	235.80	125.00
General energy production/population	48.37	73.96	24.53	30.85	64.91	59.75	136.48	49.12	28.92
Share of renewable energy	0.054	0.013	0.048	0.066	0.035	0.012	0.030	0.040	0.074

the capacity to produce more energy from renewable sources (Strantzali & Aravossis, 2016; Cuesta et al., 2020).

Countries with higher scores had a greater focus on renewable energy, with a higher share of renewable energy and higher general energy production per population. This suggests that the promotion of renewable energy can play a significant role in achieving sustainable energy systems. Renewable energy sources, such as solar, wind, hydro, and geothermal, have the potential to provide a reliable, affordable, and sustainable source of energy that can contribute to reducing GHG emissions and mitigating climate change (Kourkoumpas et al., 2018; Wang et al., 2020, 2021).

GHG emissions is a critical indicator in evaluating the sustainability of renewable energy, as the reduction of GHG emissions is a crucial goal of sustainable development (Al Garni et al., 2016; Mastrocinque et al., 2020). Countries with lower GHG emissions tend to have a more sustainable energy system, indicating that policies and measures aimed at reducing GHG emissions can contribute to achieving sustainable energy systems (Table 23.5).

The economy driver (Table 23.6) included three indicators, i.e., GDP, economy growth rate, and inflation rate. While these indicators are not directly related to the sustainability of renewable energy, they can provide insights into the economic implications of promoting renewable energy (Lee & Zhong, 2015; Ligus, 2017). The study showed that countries with a higher GDP tend to have a higher score, indicating that promoting renewable energy can contribute to economic growth. Furthermore, countries with a lower inflation rate tend to have a more stable economic environment, which can be conducive to promoting sustainable energy systems.

The employment rate is a critical indicator in evaluating the social implications of promoting renewable energy (Boran, 2018; Armeanu et al., 2017). The study showed that countries with a higher employment rate tend to have a higher score, indicating that promoting renewable energy can create jobs and contribute to improving the social welfare of the population. The HDI and crime rate are also critical indicators in evaluating the social implications of promoting renewable energy (Cuesta et al., 2020). Countries with a higher HDI tend to have a more educated and healthy population, which can be conducive to promoting sustainable energy systems (Table 23.7).

The results of the study showed that countries with a greater focus on renewable energy, lower dependency on fossil fuels, lower energy intensity, and higher general energy production per population tend to have a more sustainable energy system. Policymakers and stakeholders can use the weight assigned to each indicator to prioritize their efforts toward achieving sustainable energy systems. Furthermore, the study showed that promoting renewable energy can contribute to economic growth, job creation, and improving the social welfare of the population.

Table 23.5 Environmental (GHG emission) and institutional (corruption) indicators

Indicator/country	Germany	France	Italy	Spain	Netherlands	Poland	Sweden	Belgium	Ireland
GHG emission	9.56	6.01	6.62	5.80	10.29	10.33	4.80	9.56	12.90
Corruption	80	71	56	61	82	56	85	73	74

Table 23.6 Economy driver indicators

Indicator/country	Germany	France	Italy	Spain	Netherlands	Poland	Sweden	Belgium	Ireland
GDP	0.355	0.325	0.267	0.235	0.419	0.138	0.448	0.360	0.705
Economy growth rate	2.60	6.80	6.70	5.50	4.90	6.80	5.10	7.60	13.60
Inflation rate	3.2	2.1	1.9	3.0	2.8	5.2	2.7	3.2	2.4

Table 23.7 Social driver indicators

Indicator/country	Germany	France	Italy	Spain	Netherlands	Poland	Sweden	Belgium	Ireland
Employment rate	0.111	0.097	0.086	0.074	0.117	0.102	0.099	0.095	0.105
Human Development Index	0.102	0.098	0.097	0.098	0.102	0.095	0.102	0.101	0.102
Crime rate (C_6)	35.79	51.99	44.85	33.32	27.16	30.50	48.00	44.58	45.51

23.4 Conclusion

To evaluate the sustainability of renewable energy in a country level or as a part of energy system, it is crucial to assess the stability of energy systems. This study created a sustainable framework with a hierarchy model to evaluate the sustainability of renewable energy. The evaluation has been guided by five drivers and 13 criteria and the study's alternatives. The rank of sustainable countries regarding renewable energy has been determined by running the FAHP code in MATLAB, with Sweden, Belgium, Ireland, France, Germany, Spain, the Netherlands, Poland, and Italy ranking at higher and lower levels of sustainability, respectively.

The ranking of these countries reflects various factors contributing to their sustainable energy systems, such as Sweden's ambitious renewable energy targets, Belgium's heavy investment in offshore wind energy, and France's investment in renewable energy despite its long history with nuclear energy. The investigation concludes that the third group III, which includes the Netherlands, Poland, and Italy, is the least stable group in the field of renewable energy, and countries of groups I and II should help improve the level of sustainability in group III through cooperation.

To achieve greater sustainability of the EU member states in renewable energy, the study recommends accelerated new mechanisms to determine technological possibilities for achieving carbon neutrality, financial management aid, guideline frameworks, collective responsibility analysis, research and development, and criteria and supervision to establish a continuous feedback channel from metrics to policy. Therefore, energy security problems must be managed at a European level through a functioning internal market and more regional and European collaboration. The EU must take more cogent external action to achieve effective energy security.

References

Al Garni, H., Kassem, A., Awasthi, A., Komljenovic, D., & Al-Haddad, K. (2016). A multicriteria decision making approach for evaluating renewable power generation sources in Saudi Arabia. *Sustainable Energy Technologies and Assessments, 16*, 137–150. https://doi.org/10.1016/j.seta.2016.05.006

Alfsen, K. H., & Greaker, M. (2007). From natural resources and environmental accounting to construction of indicators for sustainable development. *Ecological Economics, 61*(4), 600–610. https://doi.org/10.1016/j.ecolecon.2006.06.017

Anna, B. (2016). Classification of the European Union member states according to the relative level of sustainable development. *Quality and Quantity, 50*(6), 2591–2605. https://doi.org/10.1007/s11135-015-0278-x

Armeanu, D. Ş., Vintilă, G., & Gherghina, Ş. C. (2017). Does renewable energy drive sustainable economic growth? Multivariate panel data evidence for EU-28 countries. *Energies, 10*(3). https://doi.org/10.3390/en10030381

Bekhrad, K., Aslani, A., & Mazzuca-Sobczuk, T. (2020). Energy security in Andalusia: The role of renewable energy sources. *Case Studies in Chemical and Environmental Engineering, 1*, 100001. https://doi.org/10.1016/j.cscee.2019.100001

Boran, F. E. (2018). A new approach for evaluation of renewable energy resources: A case of Turkey. *Energy Sources, Part B: Economics, Planning and Policy, 13*(3), 196–204. https://doi.org/10.1080/15567249.2017.1423414

Bouyssou, D. (1990). Building criteria: A prerequisite for MCDA. In *Readings in multiple criteria decision aid* (pp. 58–80). Springer. https://doi.org/10.1007/978-3-642-75935-2_4

Braathen, N. A. (2011). Interactions between emission trading systems and other overlapping policy instruments. OECD Green Growth Papers, 2011-02, OECD Publishing, Paris. https://doi.org/10.1787/5k97gk44c6vf-en

Chang, D.-Y. (1996). Applications of the extent analysis method on fuzzy AHP. *European Journal of Operational Research, 95*(3), 649–655. https://doi.org/10.1016/0377-2217(95)00300-2

Charfeddine, L., & Kahia, M. (2019). Impact of renewable energy consumption and financial development on CO_2 emissions and economic growth in the MENA region: A panel vector autoregressive (PVAR) analysis. *Renewable Energy, 139*, 198–213. https://doi.org/10.1016/j.renene.2019.01.010

Cuesta, M. A., Castillo-Calzadilla, T., & Borges, C. E. (2020). A critical analysis on hybrid renewable energy modeling tools: An emerging opportunity to include social indicators to optimise systems in small communities. *Renewable and Sustainable Energy Reviews, 122*. https://doi.org/10.1016/j.rser.2019.109691

European Commission. (2020). *EU SDG Indicator set, 2020*, 1–28.

Faraji Abdolmaleki, S., & Bello Bugallo, P. M. (2021). Evaluation of renewable energy system for sustainable development. *Renewable Energy and Environmental Sustainability, 6*, 44. https://doi.org/10.1051/rees/2021045

Fouquet, D. (2013). Policy instruments for renewable energy – From a European perspective. *Renewable Energy, 49*, 15–18. https://doi.org/10.1016/j.renene.2012.01.075

Haddad, B., Liazid, A., & Ferreira, P. (2017). A multi-criteria approach to rank renewables for the Algerian electricity system. *Renewable Energy, 107*, 462–472. https://doi.org/10.1016/j.renene.2017.01.035

Harjanne, A., & Korhonen, J. M. (2019). Abandoning the concept of renewable energy. *Energy Policy, 127*, 330–340. https://doi.org/10.1016/j.enpol.2018.12.029

IETA. (2015, July 13). *Overlapping policies with the EU ETS*. https://www.ieta.org/resources/EU/IETA_overlapping_policies_with_the_EU_ETA.pdf

Ifaei, P., Karbassi, A., Lee, S., & Yoo, C. K. (2017). A renewable energies-assisted sustainable development plan for Iran using techno-econo-socio-environmental multivariate analysis and big data. *Energy Conversion and Management, 153*, 257–277. https://doi.org/10.1016/j.enconman.2017.10.014

IISD. (2005, December). Sustainable development indicators – Proposals for a way forward. International Institute for Sustainable Development. Google Search. https://www.iisd.org/publications/report/sustainable-development-indicators-proposals-way-forward

Innes, J. E., & Booher, D. E. (2000). Indicators for sustainable communities: A strategy building on complexity theory and distributed intelligence. *Planning Theory and Practice, 1*(2), 173–186. https://doi.org/10.1080/14649350020008378

Kourkoumpas, D. S., Benekos, G., Nikolopoulos, N., Karellas, S., Grammelis, P., & Kakaras, E. (2018). A review of key environmental and energy performance indicators for the case of renewable energy systems when integrated with storage solutions. In *Applied energy* (Vol. 231, pp. 380–398). Elsevier. https://doi.org/10.1016/j.apenergy.2018.09.043

Lee, C. W., & Zhong, J. (2015). Construction of a responsible investment composite index for renewable energy industry. In *Renewable and sustainable energy reviews* (Vol. 51, pp. 288–303). Elsevier. https://doi.org/10.1016/j.rser.2015.05.071

Ligus, M. (2017). Evaluation of economic, social and environmental effects of low-emission energy technologies development in Poland: A multi-criteria analysis with application of a fuzzy analytic hierarchy process (FAHP). *Energies, 10*(10), 1550. https://doi.org/10.3390/en10101550

Liu, G. (2014). Development of a general sustainability indicator for renewable energy systems: A review. In *Renewable and sustainable energy reviews* (Vol. 31, pp. 611–621). Elsevier. https://doi.org/10.1016/j.rser.2013.12.038

Marques, A. C., & Fuinhas, J. A. (2012). Are public policies towards renewables successful? Evidence from European countries. *Renewable Energy, 44*, 109–118. https://doi.org/10.1016/j.renene.2012.01.007

Mastrocinque, E., Ramírez, F. J., Honrubia-Escribano, A., & Pham, D. T. (2020). An AHP-based multi-criteria model for sustainable supply chain development in the renewable energy sector. *Expert Systems with Applications, 150*. https://doi.org/10.1016/j.eswa.2020.113321

Rametsteiner, E., Pülzl, H., Alkan-Olsson, J., & Frederiksen, P. (2011). Sustainability indicator development-science or political negotiation? *Ecological Indicators, 11*(1), 61–70. https://doi.org/10.1016/j.ecolind.2009.06.009

Rickels, W., Dovern, J., Hoffmann, J., Quaas, M. F., Schmidt, J. O., & Visbeck, M. (2016). Indicators for monitoring sustainable development goals: An application to oceanic development in the European Union. *Earth's Future, 4*(5), 252–267. https://doi.org/10.1002/2016EF000353

Saaty, T. L. (1984). The analytic hierarchy process: Decision making in complex environments. In *Quantitative assessment in arms control* (pp. 285–308). Springer. https://doi.org/10.1007/978-1-4613-2805-6_12

Santika, W. G., Anisuzzaman, M., Bahri, P. A., Shafiullah, G. M., Rupf, G. V., & Urmee, T. (2019). From goals to joules: A quantitative approach of interlinkages between energy and the Sustainable Development Goals. *Energy Research and Social Science, 50*, 201–214. https://doi.org/10.1016/j.erss.2018.11.016

Schultz, J., Brand, F., Kopfmüller, J., & Ott, K. (2008). Building a "theory of sustainable development": Two salient conceptions within the German discourse. *International Journal of Environment and Sustainable Development, 7*(4), 465–482. Inderscience Publishers. https://doi.org/10.1504/IJESD.2008.022390

Shaaban, M., Scheffran, J., Böhner, J., & Elsobki, M. (2018). Sustainability assessment of electricity generation technologies in Egypt using multi-criteria decision analysis. *Energies, 11*(5), 1117. https://doi.org/10.3390/en11051117

Simsek, Y., Santika, W. G., Anisuzzaman, M., Urmee, T., Bahri, P. A., & Escobar, R. (2020). An analysis of additional energy requirement to meet the sustainable development goals. *Journal of Cleaner Production, 272*. https://doi.org/10.1016/j.jclepro.2020.122646

Strantzali, E., & Aravossis, K. (2016). Decision making in renewable energy investments: A review. In *Renewable and sustainable energy reviews* (Vol. 55, pp. 885–898). Elsevier. https://doi.org/10.1016/j.rser.2015.11.021

Vakulchuk, R., Overland, I., & Scholten, D. (2020). Renewable energy and geopolitics: A review. In *Renewable and sustainable energy reviews* (Vol. 122). Elsevier. https://doi.org/10.1016/j.rser.2019.109547

Wang, Q., & Yang, X. (2020). Investigating the sustainability of renewable energy – An empirical analysis of European Union countries using a hybrid of projection pursuit fuzzy clustering model and accelerated genetic algorithm based on real coding. *Journal of Cleaner Production, 268*, 121940. https://doi.org/10.1016/j.jclepro.2020.121940

Wang, Y., Zhang, D., Ji, Q., & Shi, X. (2020). Regional renewable energy development in China: A multidimensional assessment. *Renewable and Sustainable Energy Reviews, 124*. https://doi.org/10.1016/j.rser.2020.109797

Wang, Z., Bui, Q., Zhang, B., Nawarathna, C. L. K., & Mombeuil, C. (2021). The nexus between renewable energy consumption and human development in BRICS countries: The moderating role of public debt. *Renewable Energy, 165*, 381–390. https://doi.org/10.1016/j.renene.2020.10.144

Chapter 24
Assessing the Sustainability of Spain's Renewable Energy System

Danial Esfandiary Abdolmaleki, Shoeib Faraji Abdolmaleki, and Pastora M. Bello Bugallo

24.1 Introduction

In recent years, renewable energy systems have gained increasing importance due to growing concerns over climate change and energy security (Fang et al., 2018; Baskutis et al., 2021; Muneer & Dowell, 2022). As one of the sunniest countries in Europe with significant wind potential, Spain is particularly well-positioned to harness renewable energy sources such as solar and wind power. However, evaluating the performance of renewable energy systems in Spain is a complex task due to the diverse characteristics of its autonomous communities. Each region has unique geographical and climatic conditions that can affect the efficiency and effectiveness of renewable energy systems. Therefore, a comprehensive and systematic evaluation of the performance of renewable energy systems in Spain is needed to identify the strengths and weaknesses of the existing infrastructure and to inform decision-making for future investments in renewable energy.

The widespread adoption of renewable energy sources has encountered significant challenges that hinder their seamless integration into existing energy systems. One of the most prominent issues is the inadequate infrastructure and technology to support the large-scale deployment of renewable energy systems (Di Felice et al., 2018; Bardhan et al., 2019). Although some countries have made considerable progress in developing renewable energy systems, many others still depend on fossil fuels because of a lack of investment in renewable energy infrastructure. Furthermore, the intermittent nature of some renewable energy sources, such as solar and wind power, poses significant challenges to maintaining a stable and reliable energy grid (Razmjoo et al., 2021; Zahedi et al., 2022). Addressing these challenges will

D. E. Abdolmaleki · S. F. Abdolmaleki · P. M. Bello Bugallo (✉)
TECH-NASE Research Group, Department of Chemical Engineering, Universidade de Santiago de Compostela, Santiago de Compostela, Spain
e-mail: pastora.bello.bugallo@usc.es

A. Sayigh (ed.), *Transition Towards a Carbon Free Future*, Innovative Renewable Energy, https://doi.org/10.1007/978-3-031-61660-0_24

require collaborative efforts from stakeholders such as governments, businesses, and communities to invest in and support the development of renewable energy systems that can cater to the energy needs of the twenty-first century.

The aim of this study is to evaluate the current state of renewable energy systems in Spain using the TOPSIS method by analyzing and ranking the different autonomous communities in Spain. This study identifies the strengths and weaknesses of each region's renewable energy infrastructure. This research is significant because renewable energy systems have the potential to play a critical role in reducing greenhouse gas emissions and addressing climate change. However, the widespread adoption of renewable energy systems has encountered significant challenges that hinder their seamless integration into existing energy systems, such as the lack of adequate infrastructure and the intermittency of certain renewable energy sources. By evaluating the renewable energy systems in Spain and identifying the gaps and challenges in the sector, this study will provide valuable insights for policymakers and stakeholders to develop targeted plans for improving renewable energy systems in Spain and addressing the challenges that hinder their widespread adoption. The findings of this study can also serve as a model for other countries and regions seeking to expand their renewable energy infrastructure.

24.2 Methodology

This research aims to identify the most appropriate renewable energy options for Spain utilizing the TOPSIS approach in MATLAB. The methodology encompasses four main steps: (1) determining project parameters, (2) collecting information from national and local databases, (3) adjusting the questionnaire weighting by ten experts, and (4) implementing the TOPSIS method in MATLAB (Fig. 24.1).

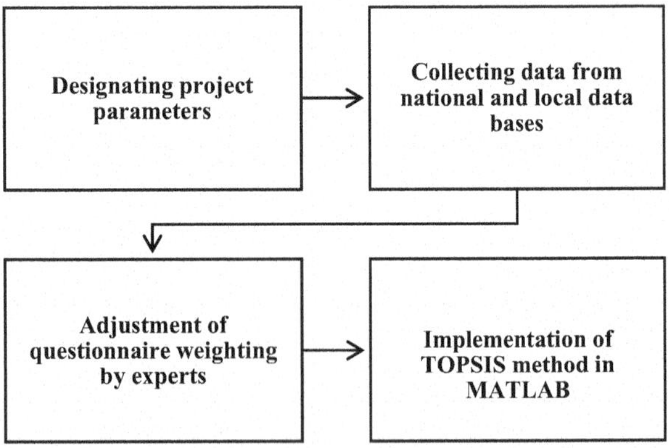

Fig. 24.1 Main steps of research methodology

24.2.1 Determining Project Parameters

In this phase of the research methodology, specific parameters were established to guide the study. The parameters included the selection of criteria for the renewable energy system. To ensure a comprehensive approach, a total of 30 indicators were selected, based on an extensive review of relevant literature and prior studies (e.g., Onat & Bayar, 2010; Stougie et al., 2018; Shaaban et al., 2018; Mastrocinque et al., 2020; Faraji Abdolmaleki & Bello Bugallo, 2021). These indicators were chosen to align with the four dimensions including energy criteria and three key pillars of sustainability, namely, social, economic, and environmental. Furthermore, specific criteria were established for evaluating renewable energy system. To develop these criteria, experts were consulted to select the most relevant and impactful factors. After a rigorous process, ten final criteria were chosen to guide the analysis and evaluation of the research (Fig. 24.2).

By incorporating a range of criteria, this approach ensured that the research was grounded in a thorough and systematic methodology, leading to robust and reliable findings. After establishing the final parameters, experts determined the directions for them (Table 24.1). The directions for the analysis were developed based on the specific research questions and goals of the study, with the intention of generating insightful and useful findings.

Energy	**Environmental**	**Economic**	**Social**
• Electricity consumption • Total generation • Share of renewable energy	• CO_2 emission	• GDP • Unemployment rate • Budget deficit	• HDI • Life expectancy

Fig. 24.2 Studied dimensions and criteria for sustainability assessment of renewable energy

Table 24.1 Criteria and their direction

Criteria	Direction
Share of renewable energy	+
Installed capacity	+
GDP	+
HDI	+
Life expectancy	+
Unemployment rate	–
CO_2 emission	–
Total generation	+
Electricity consumption	–
Budget deficit	–

24.2.2 Collecting Information from National and Local Databases

The process of collecting information from national and local databases in Spain necessitated a comprehensive comprehension of the available data sources, meticulous planning, and a thorough verification process to ensure that the data gathered was accurate, relevant, and appropriate for the research inquiry at hand. Additionally, all data was obtained from websites and organizations in Spain that provided valuable insights into specific topics of interest. It should be noted that all data was obtained from the year 2022, if available; where data was not available for 2022, the most recent data available from 2021 was used. This ensures that the findings and conclusions drawn from the data are based on the most current and accurate information available at the time of the study.

24.2.3 Adjusting the Questionnaire Weighting by Ten Experts

Ten experts were chosen to provide their views on the importance of each criterion. The experts' inputs were then utilized to adjust the weighting of the questionnaire accordingly. It consists of a list of the ten indicators, and the experts were asked to rate each indicator based on its significance on a scale of 1 to 10 (with 10 being the most important). The results were also compared with previous studies and validated by experts in the field.

24.2.4 Implementing the TOPSIS Method in MATLAB

The TOPSIS methodology is a multi-criteria decision-making method used to determine the best alternative based on multiple criteria. It is commonly used in various fields, including engineering, management, and social sciences. Recently, Hosseini Dehshiri and Amiri (2023) have conducted a research using TOPSIS to address the challenge of energy and hydrogen generation due to population growth and industrial expansion. This method facilitated realizing the most efficient strategy between energy investing companies under uncertainty. Karamountzou and Vagiona (2023) used this methodology to examine the suitability and sustainability of onshore wind farms in Greece. A number of seven exclusion criteria studied for the term "suitability" and nine environmental, technical-economic, and social criteria in five different scenarios were selected for the sustainability assessment. To create a long-term renewable energy control system with enhanced system, Alghassab (2022) employed fuzzy analytical hierarchy process—the TOPSIS method—to analyze the factors affecting sustainable renewable energy resources. As a result, wind energy was the most sustainable and preferable renewable energy source. Therefore, TOPSIS is considered as an appropriate method considering the sustainability of renewable energy in different scopes.

In this study, the TOPSIS method is implemented in eight steps in MATLAB.

Step 1: Construct the decision matrix

$$X = \left(x_{ij}\right) \tag{24.1}$$

where $i = 1, 2, \ldots, m$ and $j = 1, 2, \ldots, n$.

This matrix contains the performance ratings of each option for each criterion. The variable x_{ij} represents the performance rating of the i-th option on the j-th criterion.
Step 2: Normalize the decision matrix

$$R = \left(r_{ij}\right) \tag{24.2}$$

where $i = 1, 2, \ldots, m$ and $j = 1, 2, \ldots, n$. The normalized decision matrix is calculated by dividing each element in the decision matrix by the square root of the sum of squares of the corresponding column:

$$r_{ij} = x_{ij}/\mathrm{sqrt}\left(\mathrm{sum}\left({x_{ij}}^2\right)\right) \quad \text{for } j = 1, 2, \ldots, n \tag{24.3}$$

Step 3: Determine the weight vector $W = (w_j)$ where $j = 1, 2, \ldots, n$. The weight vector represents the relative importance of each criterion based on ten expert view points and the normalized weight vector is calculated by dividing each element in the weight vector by the sum of all elements:

$$w_j = w_j/\mathrm{sum}\left(w_j\right) \quad \text{for } j = 1, 2, \ldots, n \tag{24.4}$$

Step 4: Calculate the weighted normalized decision matrix $V = (v_{ij})$ where $i = 1, 2, \ldots, m$ and $j = 1, 2, \ldots, n$. The weighted normalized decision matrix is calculated by multiplying each element in the normalized decision matrix by the corresponding weight:

$$v_{ij} = r_{ij} \times w_j \quad \text{for } i = 1, 2, \ldots, m \text{ and } j = 1, 2, \ldots, n \tag{24.5}$$

Step 5: Determine the positive ideal solution (PIS) A^+ and negative ideal solution (NIS) A^-.
For maximization problems:

$$A_j^+ = \max(v_{ij}) \quad \text{for } i = 1, 2, \ldots, m \text{ and } j = 1, 2, \ldots, n \tag{24.6}$$

$$A_j^- = \min(v_{ij}) \quad \text{for } i = 1, 2, \ldots, m \text{ and } j = 1, 2, \ldots, n \tag{24.7}$$

For minimization problems:

$$A_j^+ = \min(v_{ij}) \quad \text{for } i = 1, 2, \ldots, m \text{ and } j = 1, 2, \ldots, n \tag{24.8}$$

$$A_j^- = \max(v_{ij}) \quad \text{for } i = 1, 2, \ldots, m \text{ and } j = 1, 2, \ldots, n \tag{24.9}$$

Step 6: Calculate the distance of each option to the PIS and NIS:

$$D_i^+ = \text{sqrt}\left(\text{sum}\left(\left(v_{ij} - A_j^+\right)^2\right)\right) \quad \text{for } i = 1, 2, \ldots, m \tag{24.10}$$

$$D_i^- = \text{sqrt}\left(\text{sum}\left(\left(v_{ij} - A_j^-\right)^2\right)\right) \quad \text{for } i = 1, 2, \ldots, m \tag{24.11}$$

Step 7: Calculate the relative closeness to the ideal solution (C_i):

$$C_i = D_i^- / (D_i^+ + D_i^-) \quad \text{for } i = 1, 2, \ldots, m \tag{24.12}$$

Step 8: Rank the alternatives in the order of preference based on their relative closeness to the ideal solution. The normalized scores are used to rank the alternatives in the order of preference:

$$(C_i / \text{sum}(C_i)) \quad \text{for } i = 1, 2, \ldots, m \tag{24.13}$$

where the alternative with the highest normalized score is considered the best alternative.

24.3 Results

Sustainable development is crucial for the future of our planet, and the renewable energy sector plays a significant role in achieving it. Spain has set ambitious goals for the development of renewable energy, and the government has been actively working toward achieving them. To assess the performance of different Spanish communities, this study used the TOPSIS method and also considered ten parameters related to social, economic, environmental, and renewable energy aspects by assigning weights to each parameter based on their relative importance. The results (Fig. 24.3) show that Castile and León and La Rioja are the top performers in the

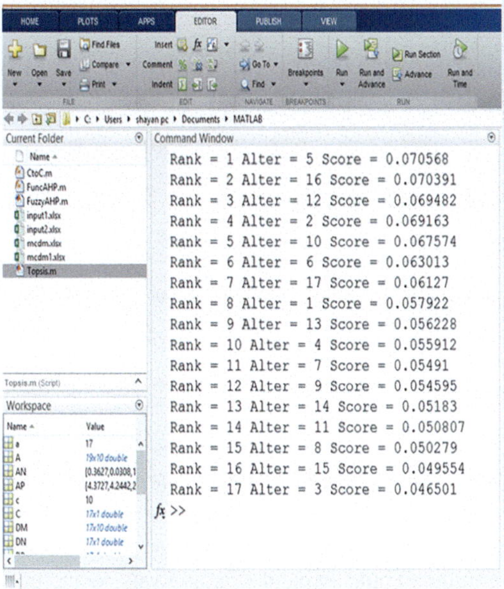

Rank	Community	Score
1	Castile and León	0.070568
2	La Rioja	0.070391
3	Murcia	0.069482
4	Aragon	0.069163
5	Galicia	0.067574
6	Castilla La Mancha	0.063013
7	Comunidad Valenciana	0.06127
8	Andalusia	0.057922
9	Madrid	0.056228
10	Cantabria	0.055912
11	Canary Islands	0.05491
12	Estremadura	0.054595
13	Navarre	0.05183
14	Balearic Islands	0.050807
15	Catalonia	0.050279
16	Basque Country	0.049554
17	Asturias	0.046501

Fig. 24.3 Ranking Spain's communities in renewable energy system using TOPSIS

renewable energy field, followed by Murcia and Aragon. Asturias is the lowest-ranked community (Asturias, with its low share of renewable energy, low installed capacity, and high CO_2 emissions, has been identified as a community with weak performance in the renewable energy sector). Based on the rankings obtained from the TOPSIS analysis, Castile and León, La Rioja, and Murcia are the top-performing communities in the renewable energy field in Spain. These communities have demonstrated their strong commitment to renewable energy by achieving high scores in the parameters evaluated in this study. Castile and León, in particular, has shown remarkable performance in all dimensions, indicating that it has great potential for further development in the renewable energy sector.

The findings also suggest that economic and social factors play a vital role in the success of renewable energy initiatives (Fig. 24.4). Communities with high GDP, low unemployment rates, and high HDI, such as Castile and León, tend to perform better in the renewable energy field. On the other hand, communities with weak economic and social indicators (like Asturias) face challenges in implementing renewable energy projects.

The outcomes provide valuable insights that can be used by policymakers and stakeholders to make informed decisions regarding renewable energy in Spain. By identifying the strengths and weaknesses of different communities, this research facilitates the development of targeted strategies aimed at improving the renewable energy sector in underperforming regions. Additionally, investors and companies can utilize the rankings to determine potential areas for investment and development. Therefore, policymakers and stakeholders should prioritize improving economic and

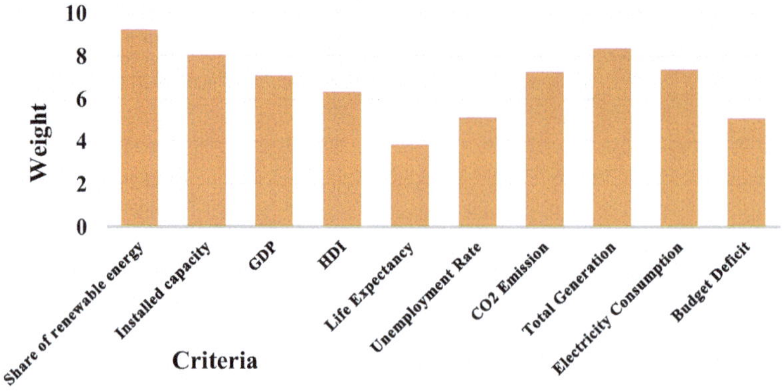

Fig. 24.4 Weight of criteria using expert viewpoint

social conditions in weaker performing communities to facilitate the growth of the renewable energy sector; the final results provide a useful benchmark for assessing the performance of different communities, which can serve as a reference point for future research and analysis (future studies could expand this analysis to include other countries and regions).

This study suggests that other countries or regions can use the TOPSIS method to evaluate their renewable energy sectors and consider the parameters and dimensions that were used in this study. However, it is essential to note that the parameters and weights used in this study were based on the specific context of Spain, and the results may not be directly applicable to other countries or regions.

24.4 Conclusion

Energy systems are complex and interconnected, with each component playing a critical role in ensuring the system's overall sustainability. A weakly or poorly coordinated part of the system could jeopardize its effectiveness. To assess the strengths and weaknesses of the Spanish energy system, this research focused on the individual parts that make it up—the autonomous communities. The study evaluated the sustainability of different renewable energy options and ranked the autonomous communities based on ten sustainability indicators, using the TOPSIS method for multi-criteria decision-making.

The findings revealed significant differences in the sustainability of renewable energy among the autonomous communities. While some regions demonstrated strong sustainability performance and were deemed sustainable, others had lower sustainability rankings and require further improvement in their renewable energy systems. It is worth noting that these less sustainable regions may have untapped potential for renewable energy development, and targeted investment and policy interventions could help improve their sustainability rankings. Factors such as government policies, investment in renewable energy infrastructure, and public attitudes toward renewable energy contribute to the variation in sustainability rankings.

This study emphasizes the significance of sustainability evaluation in renewable energy systems and offers insightful information for stakeholders and policymakers to establish practical plans for a more robust and sustainable energy system in Spain. The interrelationships between the sustainability indicators can be explored in more detail, and the driving forces for the sustainability of renewable energy in autonomous communities can be found.

References

Alghassab, M. (2022). Quantitative assessment of sustainable renewable energy through soft computing: Fuzzy AHP-TOPSIS method. *Energy Reports, 8*, 12139–12152. https://doi.org/10.1016/j.egyr.2022.09.049

Bardhan, R., Debnath, R., & Jana, A. (2019). Evolution of sustainable energy policies in India since 1947: A review. *WIREs Energy and Environment, 8*(5). https://doi.org/10.1002/wene.340

Baskutis, S., Baskutiene, J., Navickas, V., Bilan, Y., & Cieśliński, W. (2021). Perspectives and problems of using renewable energy sources and implementation of local "green" initiatives: A regional assessment. *Energies, 14*(18), 5888. https://doi.org/10.3390/en14185888

Di Felice, L., Ripa, M., & Giampietro, M. (2018). Deep Decarbonisation from a biophysical perspective: GHG emissions of a renewable electricity transformation in the EU. *Sustainability, 10*(10), 3685. https://doi.org/10.3390/su10103685

Fang, K., Zhou, Y., Wang, S., Ye, R., & Guo, S. (2018). Assessing national renewable energy competitiveness of the G20: A revised Porter's diamond model. *Renewable and Sustainable Energy Reviews, 93*, 719–731. https://doi.org/10.1016/j.rser.2018.05.011

Faraji Abdolmaleki, S., & Bello Bugallo, P. M. (2021). Evaluation of renewable energy system for sustainable development. *Renewable Energy and Environmental Sustainability, 6*, 44. https://doi.org/10.1051/rees/2021045

Hosseini Dehshiri, S. J., & Amiri, M. (2023). An integrated multi-criteria decision-making framework under uncertainty for evaluating sustainable hydrogen production strategies based on renewable energies in Iran. *Environmental Science and Pollution Research*. https://doi.org/10.1007/s11356-023-25489-5

Karamountzou, S., & Vagiona, D. G. (2023). Suitability and sustainability assessment of existing onshore wind farms in Greece. *Sustainability, 15*(3), 2095. https://doi.org/10.3390/su15032095

Mastrocinque, E., Ramírez, F. J., Honrubia-Escribano, A., & Pham, D. T. (2020). An AHP-based multi-criteria model for sustainable supply chain development in the renewable energy sector. *Expert Systems with Applications, 150*. https://doi.org/10.1016/j.eswa.2020.113321

Muneer, T., & Dowell, R. (2022). Potential for renewable energy–assisted harvesting of potatoes in Scotland. *International Journal of Low-Carbon Technologies, 17*, 469–481. https://doi.org/10.1093/ijlct/ctac012

Onat, N., & Bayar, H. (2010). The sustainability indicators of power production systems. *Renewable and Sustainable Energy Reviews, 14*(9), 3108–3115. https://doi.org/10.1016/j.rser.2010.07.022

Razmjoo, A., Gakenia Kaigutha, L., Vaziri Rad, M. A., Marzband, M., Davarpanah, A., & Denai, M. (2021). A technical analysis investigating energy sustainability utilizing reliable renewable energy sources to reduce CO_2 emissions in a high potential area. *Renewable Energy, 164*, 46–57. https://doi.org/10.1016/j.renene.2020.09.042

Shaaban, M., Scheffran, J., Böhner, J., & Elsobki, M. (2018). Sustainability assessment of electricity generation technologies in Egypt using multi-criteria decision analysis. *Energies, 11*(5), 1117. https://doi.org/10.3390/en11051117

Stougie, L., Giustozzi, N., van der Kooi, H., & Stoppato, A. (2018). Environmental, economic and exergetic sustainability assessment of power generation from fossil and renewable energy sources. *International Journal of Energy Research, 42*(9), 2916–2926. https://doi.org/10.1002/er.4037

Zahedi, R., Zahedi, A., & Ahmadi, A. (2022). Strategic study for renewable energy policy, optimizations and Sustainability in Iran. *Sustainability, 14*(4), 2418. https://doi.org/10.3390/su14042418

Index

A. Sayigh (ed.), *Transition Towards a Carbon Free Future*, Innovative Renewable Energy, https://doi.org/10.1007/978-3-031-61660-0

The manufacturer's authorised representative in the EU is Springer Nature Customer Service Centre GmbH, Europaplatz 3, 69115 Heidelberg, Germany. If you have any concerns regarding our products, please contact ProductSafety@springernature.com

Printed and bound by CPI Group (UK) Ltd, Croydon, CR0 4YY
01/10/2026
02225969-0014